# Conference Proceedings of the Society for Experimental Mechanics Series

**Series Editor**

Kristin B. Zimmerman
Society for Experimental Mechanics, Inc.,
Bethel, USA

The Conference Proceedings of the Society for Experimental Mechanics Series presents early findings and case studies from a wide range of fundamental and applied work across the broad range of fields that comprise Experimental Mechanics. Series volumes follow the principle tracks or focus topics featured in each of the Society's two annual conferences: IMAC, A Conference and Exposition on Structural Dynamics, and the Society's Annual Conference & Exposition and will address critical areas of interest to researchers and design engineers working in all areas of Structural Dynamics, Solid Mechanics and Materials Research.

Walter D'Ambrogio · Dan Roettgen · Maarten van der Seijs
Editors

# Dynamic Substructuring & Transfer Path Analysis, Vol. 4

Proceedings of the 43rd IMAC, A Conference and Exposition on Structural Dynamics 2025

River Publishers

*Published, sold and distributed by:*

River Publishers
Broagervej 10
9260 Gistrup
Denmark

www.riverpublishers.com

ISBN 97887-438-0149-8 (Hardback)

ISBN 97887-438-0161-0 (eBook)

https://doi.org/10.13052/9788743801498

Conference Proceedings of the Society for Experimental Mechanics

An imprint of River Publishers

# Preface

*Dynamic Substructuring & Transfer Path Analysis* represents one of twelve volumes of technical papers presented at the $43^{rd}$ IMAC, A Conference and Exposition on Structural Dynamics, organized by the Society for Experimental Mechanics, and held February 10-13, 2025. The full proceedings also include volumes on Nonlinear Structures & Systems; Dynamics of Civil Structures; Model Validation and Uncertainty Quantification; Special Topics in Structural Dynamics & Experimental Techniques; Computer Vision & Laser Vibrometry; Dynamic Environments Testing; Sensors & Instrumentation and Aircraft/Aerospace Testing Techniques; Topics in Modal Analysis & Parameter Identification I&II; Data Science in Engineering; and Structural Health Monitoring & Machine Learning.

Each collection presents early findings from experimental and computational investigations on an important area within Structural Dynamics. Coupled structures or, substructuring, is one of these areas.

Substructuring is a general paradigm in engineering dynamics where a complicated system is analyzed by considering the dynamic interactions between subcomponents. In numerical simulations, substructuring allows one to reduce the complexity of parts of the system in order to construct a computationally efficient model of the assembled system. A subcomponent model can also be derived experimentally, allowing one to predict the dynamic behavior of an assembly by combining experimentally and/or analytically derived models. This can be advantageous for subcomponents that are expensive or difficult to model analytically. Substructuring can also be used to couple numerical simulation with real-time testing of components. Such approaches are known as hardware-in-the-loop or hybrid testing.

Whether experimental or numerical, all substructuring approaches have a common basis, namely the equilibrium of the substructures under the action of the applied and interface forces and the compatibility of displacements at the interfaces of the subcomponents. Experimental substructuring requires special care in the way the measurements are obtained and processed in order to assure that measurement inaccuracies and noise do not invalidate the results. In numerical approaches, the fundamental quest is the efficient computation of reduced order models describing the substructure's dynamic motion. For hardware-in-the-loop applications difficulties include the fast computation of the numerical components and the proper sensing and actuation of the hardware component. Recent advances in experimental techniques, sensor/actuator technologies, novel numerical methods, and parallel computing have rekindled interest in substructuring in recent years leading to new insights and improved experimental and analytical techniques.

The organizers would like to thank the authors, presenters, session organizers, and session chairs for their participation in this track.

Editors:
Walter D'Ambrogio–*University of L'Aquila*, L'Aquila, Italy; Dan Roettgen–*Sandia National Laboratories*, NM, USA; Maarten van der Seijs–*VIBES technology*, Netherlands

# Contents

# Chapter 1
# Modal Vibration Testing of a Frame and Wing Structure

**Joshua E. Blackham and Matthew S. Allen**

**Abstract** Large, multi-component bolted structures can exhibit complicated vibration responses due to the complexity of frictional interactions at the bolt-plate interface. The effect of nonlinearity due to the joints can be characterized by performing a test to quantify how the effective natural frequency and damping ratio of each mode change with vibration amplitude. In this paper, modal hammer impact testing was performed on the Round Robin Frame and Wing Structure. Various amplitude impacts were used to excite any nonlinearities in the structure. Experimental modal analysis was performed to separate the modal responses of the system. Mode shapes and frequencies are examined to understand any nonlinearities in the system.

**Keywords** Nonlinear Vibration · Experimental Modal Analysis · Frame and Wing Structure

## Introduction

The goal of modeling and testing complicated bolted structures is typically to create a model that can accurately replicate the effective natural frequency and damping of a test structure at the vibration amplitudes of interest. In order to make further progress in experimental-analytical substructuring, the Frame and Wing Structure (FWS) was introduced as a testbed structure that is relatively easy to model and test, with the ultimate goal of comparing and learning about different substructuring methods that are commonly used among different researchers and engineers [1].

Since the introduction of the Frame and Wing Structure, groups at Sandia have performed tests and modeled the structure. Moldenhauer and Roettgen [2] performed substructuring on the original FWS and were able to model the structure's mode frequencies and damping successfully. They also explored certain modes for large nonlinearities in the damping and found that most of the modes only exhibited weak nonlinearities. Linderholt et al. [3] also used substructuring to model the FWS, however, they modeled a new "swept" wing that represents a standard plane wing more accurately than the rectangular wing used in earlier models. Linderholt et al. [4] again modeled the FWS using substructuring methods. This time more complicated elements were added to the structure, such as vibration damping studs placed between the frame and the wing, or small payloads added on the wings. These elements were included in an attempt to introduce more nonlinearity into the system and to be able to model these changes through the developed substructuring processes.

While the original purpose of the frame wing project was to develop and compare different substructuring methods, this paper focuses on testing and characterizing the FWS. A roving impact test was performed to excite vibration modes from multiple locations and at various amplitudes. The frequency response functions (FRFs) obtained from both low-amplitude (i.e. assumed to be linear) responses and higher amplitude (i.e. nonlinear) responses were analyzed using the Algorithm of Mode Isolation to quantify the frequency and damping of each vibration mode. Mode shapes were also found and used to help explain any nonlinearities that might occur in each mode. The Hilbert transform [5, 6] was then applied to some of the high-amplitude impact data to quantify nonlinearities in the frequency and damping of certain modes.

Joshua E. Blackham
Department of Mechanical Engineering, Brigham Young University, Provo, UT 84601
e-mail: joshblackham1925@gmail.com

Matthew S. Allen
Department of Mechanical Engineering, Brigham Young University, Provo, UT 84601
e-mail: matt.allen@byu.edu



Walter D'Ambrogio, et al. (eds.), *Dynamic Substructuring & Transfer Path Analysis, Vol. 4*, Conference Proceedings of the Society for Experimental Mechanics Series, https://doi.org/10.13052/9788743801498-1

## Testing Procedure

The Frame and Wing Structure that was tested included both the frame component SN013 and the thin wing component Wing013A. For reference, the weights of the frame and wing components are 629.7 g and 527.7 g, respectively [7]. In this setup, the frame and wing components were attached using 4 10-32 bolts and #10 washers on each side of the frame as seen in Figure 1. The washers were placed only between the head of the bolt and the contact surface of the wing part. No nuts were used since the frame part was manufactured with threaded inserts placed in all bolt holes. All bolts were pretensioned to 3.4 Nm and the structure was suspended with bungees to approximate free-free boundary conditions. All rigid body modes of the suspended structure were found to be below 3 Hz, which was well below the first elastic mode of the structure.

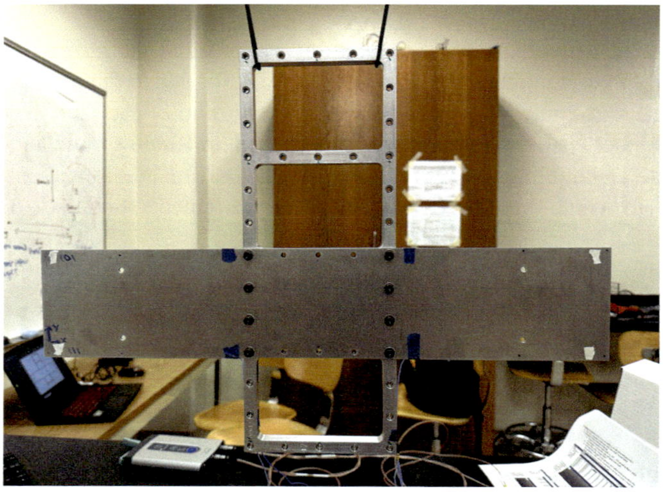

**Fig. 1** Frame and Wing structure suspended by bungees in preparation for modal impact testing

The goal of this testing was to characterize the mode frequencies and shapes of the frame and wing structure and to find modes that exhibit large nonlinearities due to the bolted joints. An initial round of testing focused on both low- and high-amplitude impact responses that were measured up to 1280 Hz. Low-amplitude impacts were used to find a baseline of linear mode frequencies and to find the mode shapes of the structure. High-amplitude impacts exercised the joint nonlinearity and hence could change the frequency and damping observed in a linearized FRF. The locations of the impact and accelerometers are shown in Figure 2, with 4 accelerometers positioned at the corners of the wing and one accelerometer on the frame. Note

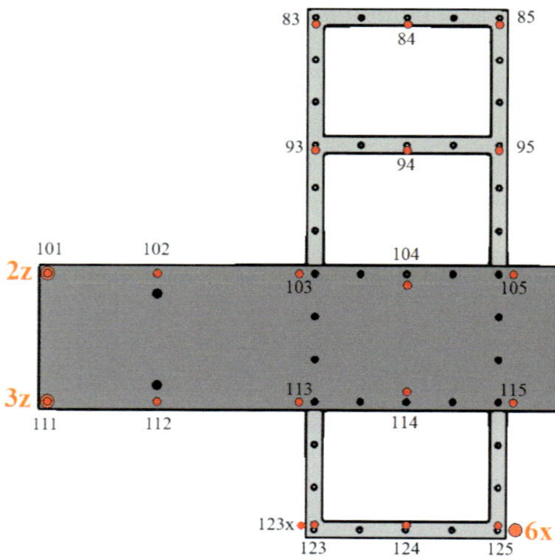


**Fig. 2** Locations of hammer impact points during testing. Impact locations are marked with red dots with a 2 or 3 digit numbering system. Unless specified, all impacts are in the Z direction. Accelerometers are located at points marked with orange circles and numbered 2-6 with directions specified.

that Accelerometer 6 was located on the side of the frame and oriented in the $x$ direction in order to capture the response when impacts were performed in the X direction at Point 123. The impact locations extensively cover the length of the frame and wing parts and were chosen to allow the mode shapes to be found on both the frame and the wing components. After reviewing the initial testing, additional high-amplitude impacts were performed, which excited the structure up to about 700 Hz, in order to investigate nonlinear effects further.

## Frequency Response

The $H_1$ FRF was computed from the impacts at low and high amplitudes and is shown in Figure 3. Figure 4 shows an expanded view for specific modes of interest. While investigating the corresponding mode shapes, the peak at 230 Hz exhibited unusual responses (see Figure 4c) and so two modes were suspected to exist at this frequency. The Complex Mode Indicator Function (CMIF) [8] was computed for the $24 \times 5$ matrix of FRFs from the 5 accelerometers and 24 impact points, and the result is seen in Figure 5. The CMIF confirms that there are two modes that occur at 230 Hz as well as another pair near 650 Hz.

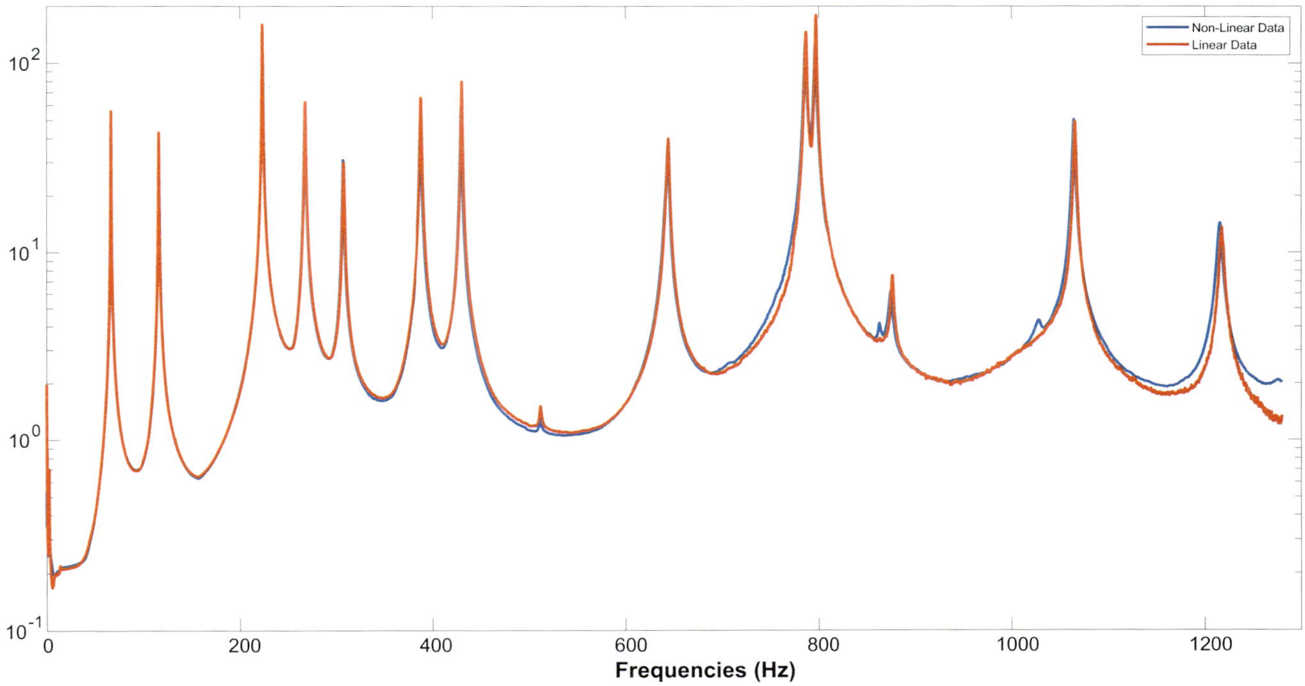


**Fig. 3** Comparison of average FRFs from low and high amplitude impact tests.

## Mode Frequencies and Shapes

After verifying the number of modes located at each frequency, the Algorithm of Mode Isolation (AMI) [9] was used to identify the modal parameters. In the cases where two modes occurred at about the same frequency, AMI used a two-mode fit to extract both modes from the measurements [10]. The mode frequencies and damping are summarized in Table 1, and the first eight mode shapes are seen in Figure 7. Modes 1, 2 and 7 (see Figs. 7a, 7b, and 7g) are predominantly wing bending modes. Modes 4 and 5 (see Figs. 7d and 7e) are frame bending modes with the wings undergoing torsion with Mode 4 in phase and Mode 5 out of phase. Modes 3 and 6 (see Figs. 7c and 7f) involve more complicated wing torsion modes and certain interactions between the wing and the frame. Mode 8 in Fig. 7h is a higher order wing bending mode that interacts with a frame torsion mode.

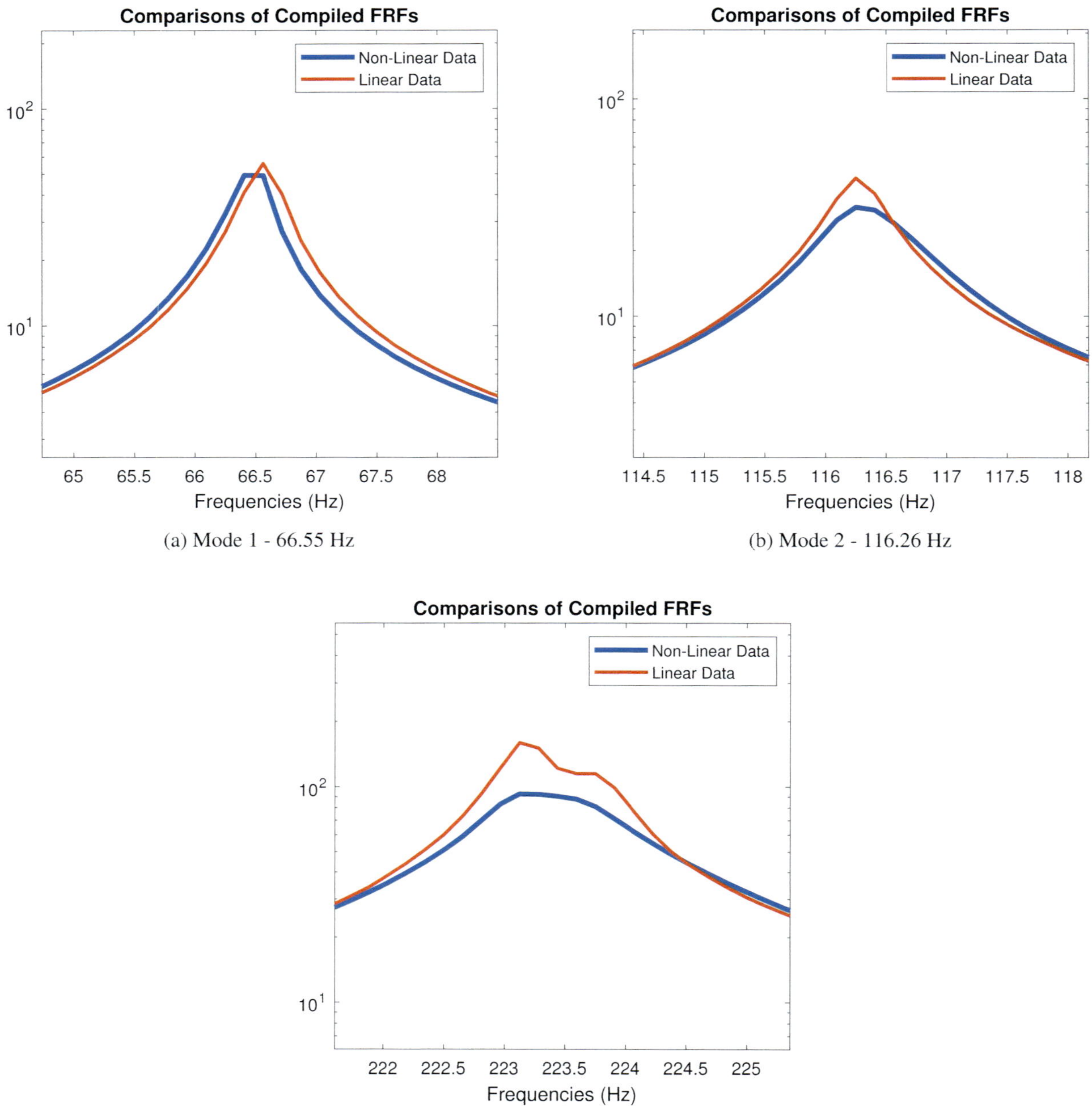


(a) Mode 1 - 66.55 Hz



(b) Mode 2 - 116.26 Hz



(c) Mode 3 - 223.14 Hz and Mode 4 - 223.79 Hz

**Fig. 4** Zoom in on Frequency Response Function for the first 4 modes of the frame and wing structure

A few of the modes included in-plane motion. To identify these modes, the average of the FRFs for each of the accelerometers (i.e. the drive points) was plotted and is shown in Figure 6. Most of the modes are dominant in the accelerometers located on the wing and in the Z-direction (i.e. Acc. 2 - 5). The two exceptions are modes 9 and 14 at about 511 and 876 Hz, which are far more visible in the X-direction accelerometer 6 than in the Z-direction.

The plots of the mode shapes in Figure 7 can be used to ascertain which might be most sensitive to nonlinearity due to the bolted joints. For example, Mode 1 would exert a load that would want to open and close the bolted joints. This type of motion typically does not change the effective frequency or damping much [11], and the FRFs in Fig. 8 seem to support this. Mode 2 also shows little change in frequency or damping, and this seems to be corroborated by the fact that the mode shape

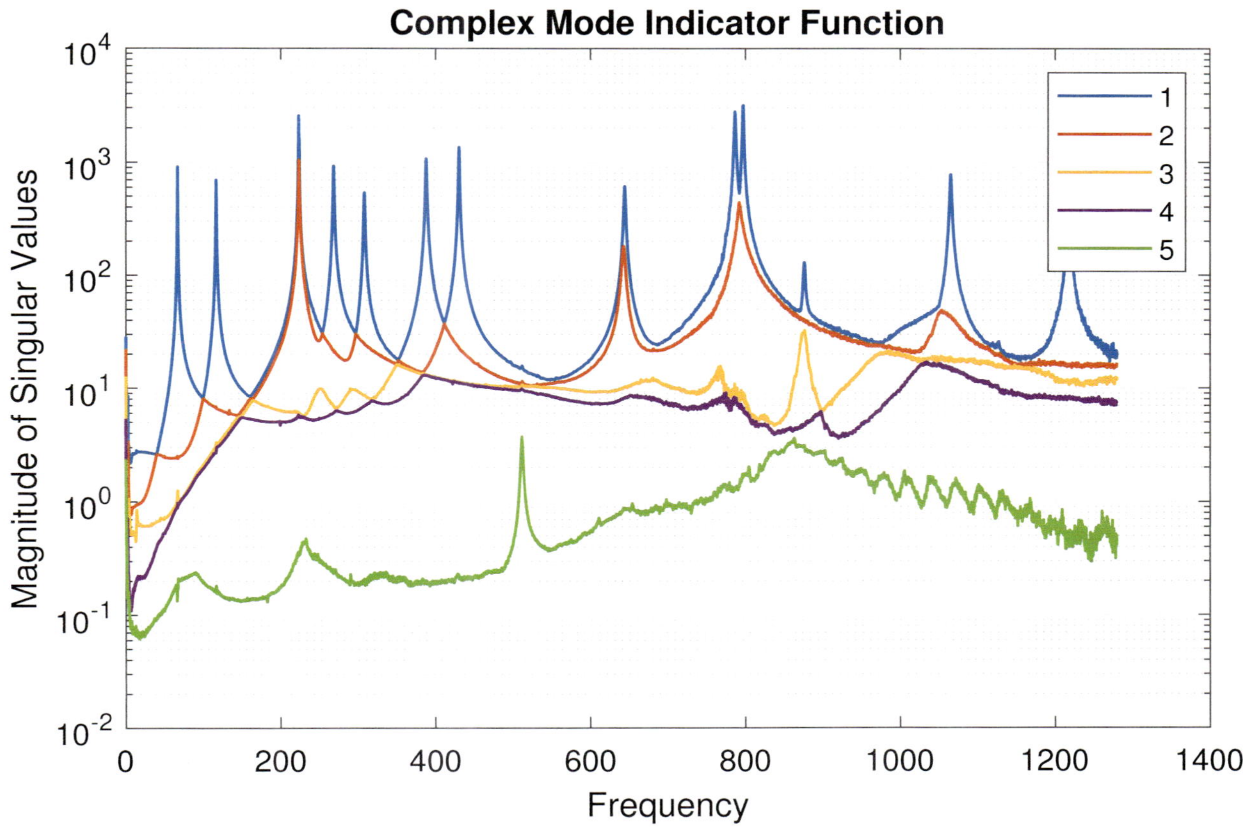


**Fig. 5** Complex Mode Indicator Function for frame and wing low-amplitude impact test.

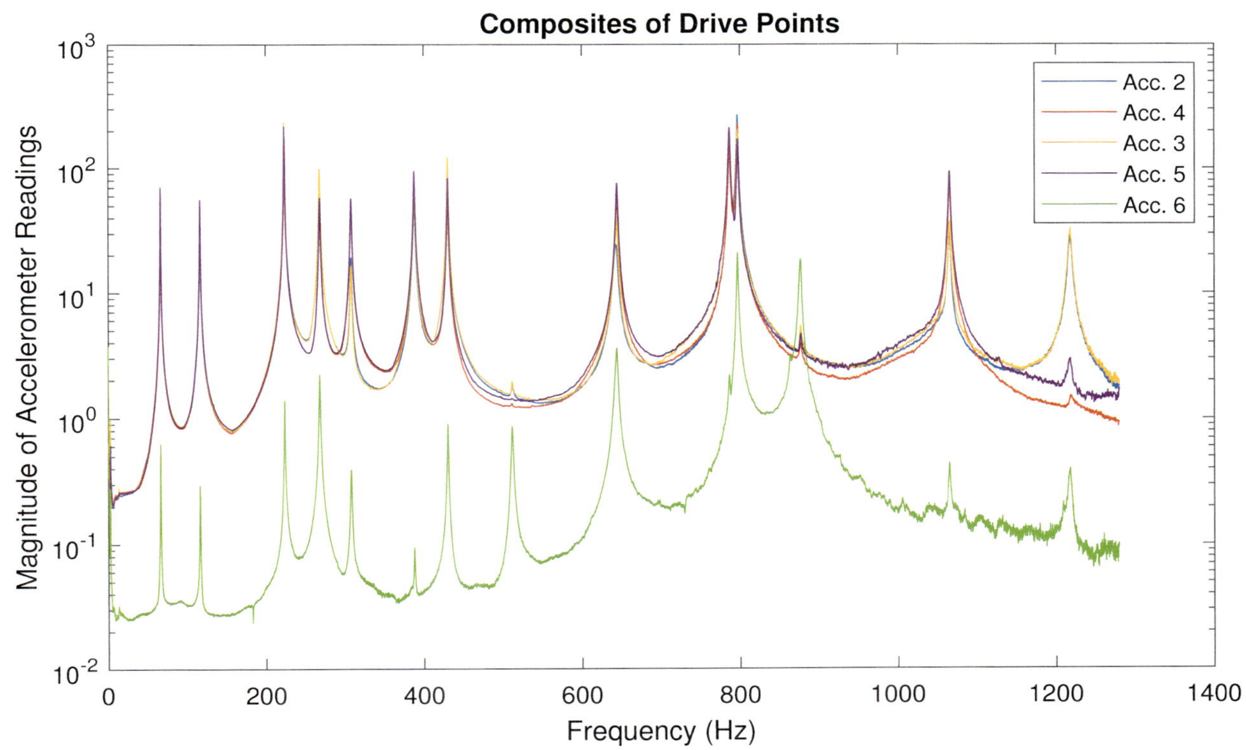


**Fig. 6** Composite of drive points found using AMI.

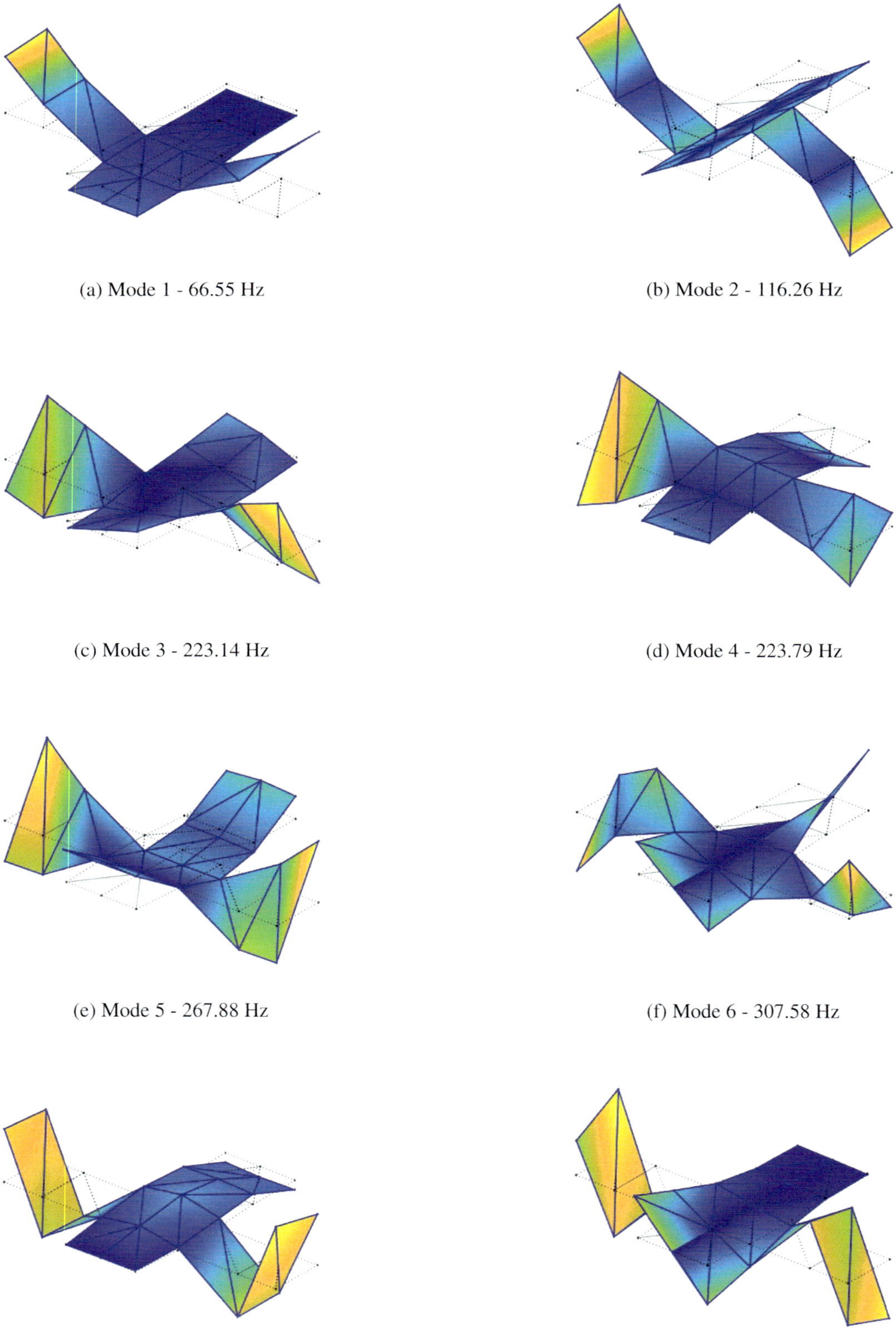

(a) Mode 1 - 66.55 Hz

(b) Mode 2 - 116.26 Hz

(c) Mode 3 - 223.14 Hz

(d) Mode 4 - 223.79 Hz

(e) Mode 5 - 267.88 Hz

(f) Mode 6 - 307.58 Hz

(g) Mode 7 - 387.83 Hz

(h) Mode 8 - 430.17 Hz

**Fig. 7** First eight Mode Shapes of Frame and Wing Structure

**Table 1** Mode frequencies and damping ratios of the frame and wing structure for low and high amplitude testing.

| | | Low Amp. | | High Amp. | | Difference | |
|---|---|---|---|---|---|---|---|
| Mode | Description | Freq. (Hz) | Damp. (%) | Freq. (Hz) | Damp. (%) | Freq. (%) | Damp. (%) |
| 1 | 1st Wing Bending | 66.55 | 0.24 | 66.46 | 0.22 | -0.14 | -8.3 |
| 2 | 2nd Wing Bending | 116.26 | 0.22 | 116.33 | 0.33 | 0.06 | 50 |
| 3 | 1st Wing Torsion | 223.14 | 0.10 | 223.12 | 0.23 | -0.009 | 130 |
| 4 | 1st Frame Bending | 223.79 | 0.15 | 223.67 | 0.23 | -0.054 | 53 |
| 5 | 1st Frame Bending | 267.88 | 0.22 | 267.66 | 0.23 | -0.082 | 4.5 |
| 6 | Torsion of Frame and Wing | 307.58 | 0.24 | 307.16 | 0.22 | -0.14 | -8.3 |
| 7 | 3rd Wing Bending | 387.83 | 0.19 | 387.39 | 0.21 | -0.11 | 10 |
| 8 | 1st Frame Torsion | 430.17 | 0.13 | 429.53 | 0.16 | -0.15 | 23 |
| 9 | In-Plane Mode 1 | 511.28 | 0.20 | 510.77 | 0.12 | -0.10 | -40 |
| 10 | - | 642.63 | 0.32 | 641.30 | 0.38 | -0.21 | 19 |
| 11 | - | 644.18 | 0.16 | 643.73 | 0.17 | -0.070 | 6.2 |
| 12 | - | 786.68 | 0.12 | 786.23 | 0.15 | -0.057 | 25 |
| 13 | - | 797.01 | 0.09 | 796.45 | 0.14 | -0.070 | 55 |
| 14 | In-Plane Mode 2 | 876.25 | 0.14 | 874.50 | 0.21 | -0.20 | 50 |
| 15 | - | 1065.4 | 0.10 | 1064.4 | 0.12 | -0.094 | 20 |
| 16 | - | 1217.4 | 0.18 | 1215.6 | 0.20 | -0.15 | 11 |

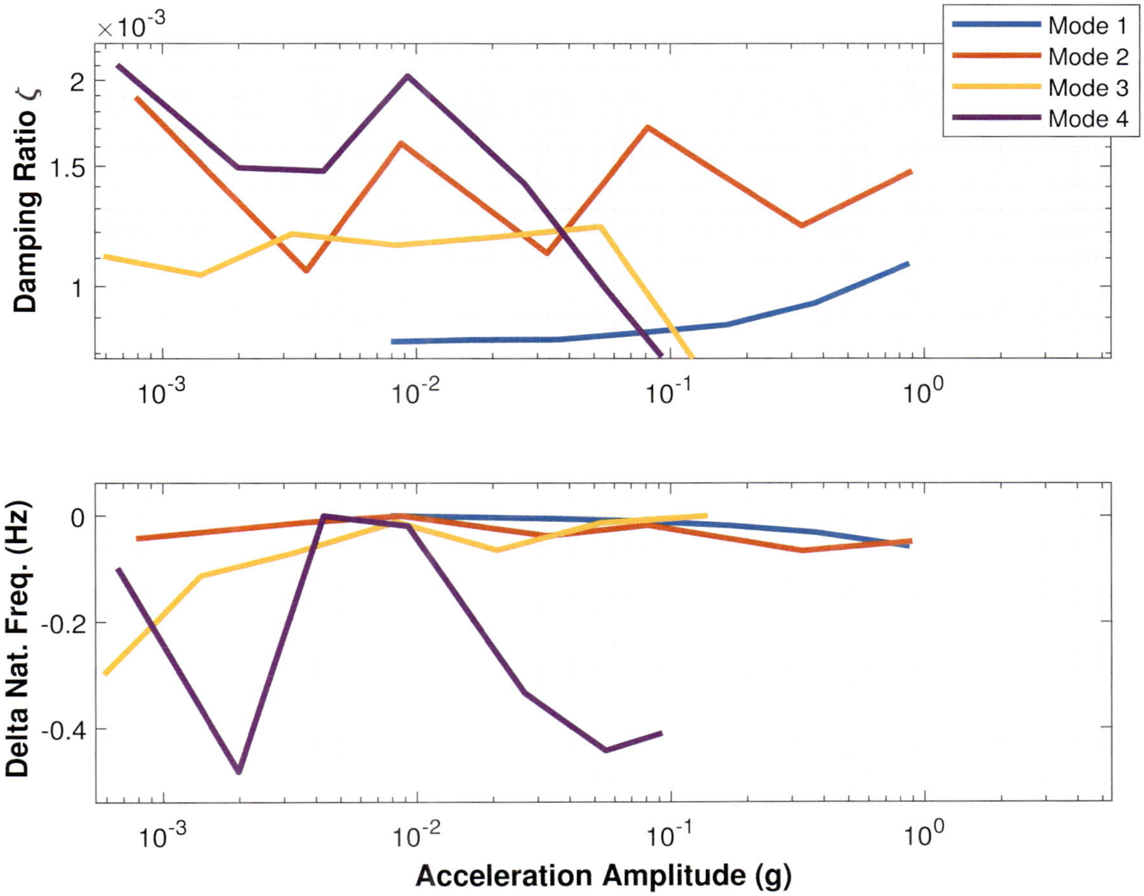


**Fig. 8** Amplitude dependent damping and frequency of the first 4 modes.

would not put much stress on the joints. This trend is similar for many of the bending and torsion modes of the structure. On the other hand, Modes 8 and 14 involve shearing of the joint, suggesting significant nonlinearity, and the results show that they exhibit somewhat larger than normal changes in frequency and damping. It should be noted, however, that these inferences based on the mode shapes are not infallible. For example, Modes 3 and 4 exhibit a relatively large increase in damping, even though their shapes mostly involve opening and closing of the joints as with most of the other modes. On the other hand, these modes are close in frequency so that may introduce error in estimating their damping. To ascertain which modes have nonlinear frequencies and damping, the Hilbert transform was used and will be discussed later.

## Hilbert Transform and Nonlinear Analysis

After mode frequencies and damping were found using AMI, the next step was to investigate any nonlinear effects that might occur due to the bolted joints and friction interfaces in the structure. First, a modal filter was used to separate the responses of the closely spaced pairs of modes. A band-pass filter was also used to obtain only the frequencies for specific modes of interest. After both filters were applied to the high amplitude impact data, the Hilbert transform was then used as described in [6] to find the amplitude dependent frequencies and damping for the first four modes and for Mode 9 of the frame and wing structure The results are shown in Figs. 8 and 9.

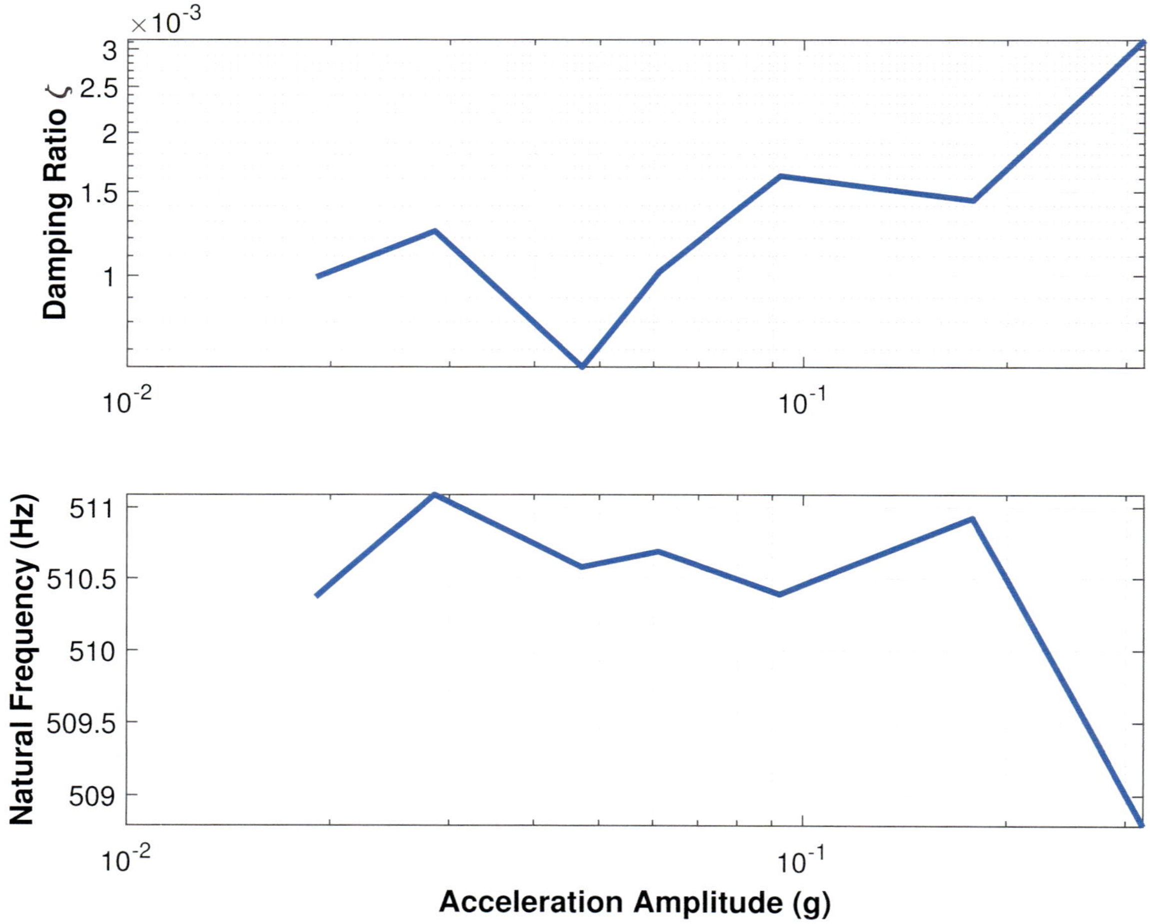


**Fig. 9** Amplitude dependent damping and frequency of Mode 9.

Mode 1 has the cleanest data in both damping and natural frequency and shows a small change across the amplitude range of interest. Such an increase in damping and decrease in stiffness is characteristic of structures with bolted-joint nonlinearities [12]. Mode 2 shows almost no change in natural frequency and a much noisier damping curve, however, the general trend is that the damping is relatively unchanged as well. Modes 3 and 4 are the two modes that occur at the same frequency and so the assumptions that modal coupling does not occur may not have been met. This could be a possible explanation for the large amounts of noise in the Mode 4 curves. Overall, it seems that these four modes are predominantly linear across range of amplitudes that were accessible in these tests. Noisy results such as those shown in Fig. 8 are typical when applying this technique to data when the response is linear.

Figure 9 shows the nonlinear damping and natural frequency for Mode 9. Although slightly noisy, the damping increases by a factor of 3 as amplitude increases. The natural frequency has a less clear trend although the frequency does seem to drop by almost 2 Hz as amplitude increases. This type of response is typical for a nonlinear mode.

## Conclusion

In this paper, the Sandia Frame and Wing structure was tested to classify its vibration modes in order to understand the effect of bolted joints on the structure. Low amplitude hammer impact testing was performed to find the linear frequencies, damping and mode shapes. A second set of FRFs was obtained at higher force levels, to quickly screen for nonlinearity. Mode shapes were also presented to visualize the interaction between the wing and frame through the bolted joints. High amplitude hammer impacts yielded a greater range of amplitude dependent data. By using the Hilbert transform, the amplitude dependent frequency and damping of several modes was investigated. For this structure, nonlinearity was found have only a minor effect on the frequency and damping of the majority of modes in the structure. The fact that many of these modes are linear is notable since the purpose of this structure is to develop and compare different linear substructuring methods, and hence the frame seems well suited for this purpose.

Though the modes in this structure tend to predominantly linear, two pairs of modes were discovered that occur at almost the same frequency. Hence, care must be taken when treating those modes.

## References


1. Roettgen, D., Lopp, G., Jaramillo, A., and Moldenhauer, B. "Experimental Substructuring of the Dynamic Substructures Round-Robin Testbed". In Allen, M., D'Ambrogio, W., and Roettgen, D., editors, *Dynamic Substructures, Volume 4*, pages 119–123, Cham (2023) Springer International Publishing.
2. Moldenhauer, B. and Roettgen, D. "Nonlinear Substructuring of the Dynamic Substructures Technical Division Structure". In *41st International Modal Analysis Conference (IMAC XLI)*, Texas, USA (2023) Society for Experimental Mechanics.
3. Linderholt, A., Roettgen, D., and Moldenhauer, B. "Combining steel and aluminum components of the Benchmark Structure for the Technical Division on Dynamic Substructuring". In *41st International Modal Analysis Conference (IMAC XLI)*, Texas, USA (2023) Society for Experimental Mechanics.
4. Linderholt, A., Roettgen, D., and Moldenhauer, B. "Nonlinear Subcomponent Attachments for the Technical Division on Dynamic Substructuring Benchmark Structure". In *42nd International Modal Analysis Conference (IMAC XLII)*, Florida, USA (2024) Society for Experimental Mechanics.
5. Feldman, M. "Non-linear system vibration analysis using Hilbert transform–I. Free vibration analysis method 'Freevib"'. *Mechanical Systems and Signal Processing*, 8(2):119–127 (1994)
6. Roettgen, D.R. and Allen, M.S. "Nonlinear characterization of a bolted, industrial structure using a modal framework". *Mechanical Systems and Signal Processing*, 84:152–170 (2017)
7. Dynamic Substructuring Focus Group. "SEM 4UF Measured Properties". SEM Wikis.Accessed: 2024-06-10. [Online.] Available: https://wiki.sem.org/wiki/SEM_4UF_Measured_Properties.
8. Shih, C.Y., Tsuei, Y.G., Allemang, R.J., and Brown, D.L. "Complex mode indication function and its applications to spatial domain parameter estimation". *Mechanical Systems and Signal Processing*, 2(4):367–377 (1988)
9. Allen, M.S. and Ginsberg, J.H. "A Global, Single-Input-Multi-Output (SIMO) Implementation of The Algorithm of Mode Isolation and Applications to Analytical and Experimental Data". *Mechanical Systems and Signal Processing*, 20:1090–1111 (2006)
10. Allen, M.S. and Ginsberg, J.H. "Global, Hybrid, MIMO Implementation of the Algorithm of Mode Isolation". In *23rd International Modal Analysis Conference (IMAC XXIII)*, Florida, USA (2005) Society for Experimental Mechanics.
11. Clark, Brennen, Allen, Matthew S., and Pacini, Benjamin. "Nonlinear Normal Modes and Response to Random Inputs of Systems with Bilinear Stiffness". *Journal of Sound and Vibration*, (Submitted April 2024) (2024)
12. Segalman, D.J., Gregory, D.L., Starr, M.J., Resor, B.R., Jew, M.D., Lauffer, J.P., and Ames, N.M. "Handbook on Dynamics of Jointed Structures". Technical report, Sandia National Laboratories, Albuquerque, NM 87185 (2009)

# Chapter 2
# Toward 3D Experimental Impulse-Based Substructuring Using the Virtual Point Transformation

**Oliver M. Zobel, Francesco Trainotti, and Daniel J. Rixen**

**Abstract** The application of Impulse-Based Substructuring (IBS) might be advantageous compared to the classical Frequency-Based Substructuring (FBS) approach because it is a time domain method, e.g., allowing for the addition of non-linear elements. In order to be viable for experimental applications, the method must be able to handle real-world systems and interfaces as well as all the experimental issues commonly encountered with substructuring. For this, the Virtual Point Transformation (VPT), commonly used for FBS to model a rigid interface connection, is adapted for use with IBS. With the shown approach, it is possible to correctly reconstruct some of the initial acceleration response peaks of an experimental test-case. The results show that combining a time domain deconvolution, downsampling with low-pass filter, and the VPT fundamentally enables experimental substructuring of three-dimensional structures in the time domain using the IBS method.

**Keywords** Impulse-based substructuring · Experimental substructuring · Virtual point transformation, · Time domain substructuring · Experimental techniques

## Introduction

Impulse-based substructuring (IBS) is the time-domain counterpart of the well-established frequency-based substructuring method (FBS). The IBS method is especially suited for determining shock responses, where the main goal might be to correctly predict the maximum amplitudes, e.g., of the accelerations. Its advantages originate from it being a time-domain method, i.e., high-frequency content, potentially consisting of many modes, can be represented in a short time series. Additionally, since there is no transformation into the frequency domain, there is no forced prioritization, i.e., leakage does not occur.

While this method has been successfully applied for numerical models in the past, see for instance [1–3], experimental applications are just now starting to be conducted. In the authors' previous work [4], it was shown that the IBS method can be used experimentally to determine the initial response peaks of aluminum and polyoxymethylene (POM) rods considered one-dimensional. This was enabled by using a time domain deconvolution procedure to experimentally estimate the impulse response functions (IRF), and additionally, a downsampling approach was proposed for test cases with limited excitation bandwidth, like the POM rods. Whereas this showed that an experimental application of IBS is generally possible, the practical use-cases are limited when only one-dimensional systems or interfaces can be used. Therefore, in this paper, the fundamentals are laid out to apply the IBS method experimentally to three-dimensional systems.

For this, the Virtual Point Transformation (VPT), as proposed in [5], commonly used within FBS to realize experimental six degree of freedom interface coupling, is adapted for the IBS method. The adapted IBS scheme using VPT is then tested experimentally on a substructuring benchmark structure and compared to reference measurements.

Oliver M. Zobel · Francesco Trainotti · Daniel J. Rixen
Chair of Applied Mechanics, TUM School of Engineering and Design, Technical University of Munich, Boltzmannstr. 15, 85748 Garching, Germany
e-mail: oliver.zobel@tum.de; francesco.trainotti@tum.de; rixen@tum.de



Walter D'Ambrogio, et al. (eds.), *Dynamic Substructuring & Transfer Path Analysis, Vol. 4*, Conference Proceedings of the Society for Experimental Mechanics Series, https://doi.org/10.13052/9788743801498-2

## Theory of Impulse-Based Substructuring

Here, only a short summary of the dually assembled Impulse-Based Substructuring (IBS) theory proposed by RIXEN in 2010 in [2] is given. Further details can be found there or in the authors' previous work, 'Enabling Experimental Impulse-Based Substructuring through Time Domain Deconvolution and Downsampling' [4]. The time-continuous IBS method is given by:

**Definition Impulse-Based Substructuring Method** [2]

$$\begin{cases} \boldsymbol{y}^{(s)}(t) = \displaystyle\int_0^t \boldsymbol{H}^{(s)}(t-\tau)\left(\boldsymbol{f}^{(s)}(\tau) + \boldsymbol{B}_f^{(s)\,\mathrm{T}} \boldsymbol{\lambda}_f(\tau)\right) \mathrm{d}\tau & \text{Equations of Motion} \\ \displaystyle\sum_{s=1}^{N_S} \boldsymbol{B}_y^{(s)} \boldsymbol{y}^{(s)}(t) = \boldsymbol{0} & \text{Compatibility} \end{cases} \quad (1)$$

where:

| | | | | | |
|---|---|---|---|---|---|
| $\boldsymbol{y}(t)$ | System response | $\boldsymbol{H}(t)$ | Matrix of impulse response functions | $(s)$ | Substructure index |
| $\boldsymbol{f}(t)$ | Externally applied forces | $\boldsymbol{B}$ | Signed Boolean constraint matrix | $*_f$ | Force mapping |
| $N_S$ | Number of substructures | $\boldsymbol{\lambda}(t)$ | Global vector of Lagrange multipliers | $*_y$ | Response mapping |

These equations must be discretized in time, and then a time-marching scheme over the discrete time steps $k$ has to be evaluated. The aforementioned scheme first calculates a prediction of the system responses using a convolution product between the impulse response functions (IRFs) and the applied forces while neglecting the compatibility condition for the current time step $k$. Then, based on the existing interface gap, Lagrange multipliers $\boldsymbol{\lambda}$ are calculated that, for a linear system, exactly close the gap when applied to the system in the corrector step [2, 4]. This is summarized in fig. 1.

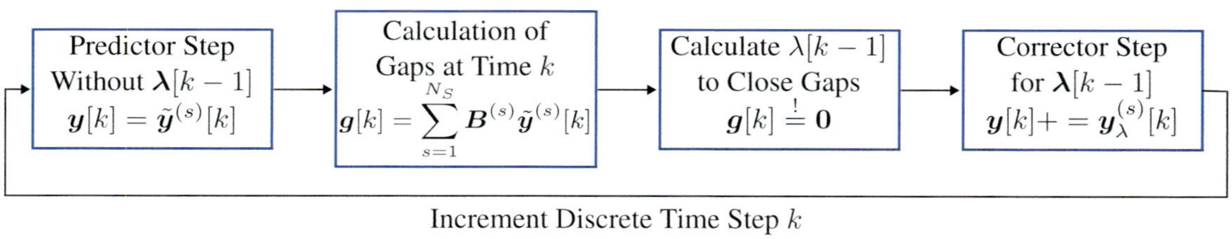


**Fig. 1** Flowchart of IBS scheme calculations within each time step

The IRFs are determined from impact measurements using the time domain convolution procedure proposed by [6]. This procedure allows averaging of multiple impacts and essentially is the time domain counterpart of the $H_1$ FRF estimator.

Depending on the excitation bandwidth compared to the full measurement bandwidth, the system might not be properly excited at all frequencies, see fig. 2. Contrary to FBS, where the higher frequencies can even be omitted after the substructuring procedure, for IBS, the improperly excited frequency content must be removed before the IRF calculation. The authors proposed in [4] to do this with downsampling after applying an appropriate low-pass filter, limiting the considered bandwidth.

## Virtual Point Transformation for IBS

For three-dimensional structures, coupling two triaxial sensors on the interface, one on each substructure, is not sufficient as this does not account for the rotational dofs and only couples the three measured translations, while in reality, the interface connection is a line or surface contact, rather than a point contact. Measuring additional rotational dofs is very difficult because the required torque excitation and rotational sensors are not standard measurement equipment [7]. Instead, it was proposed by DE KLERK ET AL. in [7] to determine the rotational dofs by combining multiple triaxial sensors. Nevertheless, this approach was found to be non-ideal, as the interface problem could be overdetermined, and measurement errors can lead to unwanted stiffening and spurious peaks in the coupled FRFs.

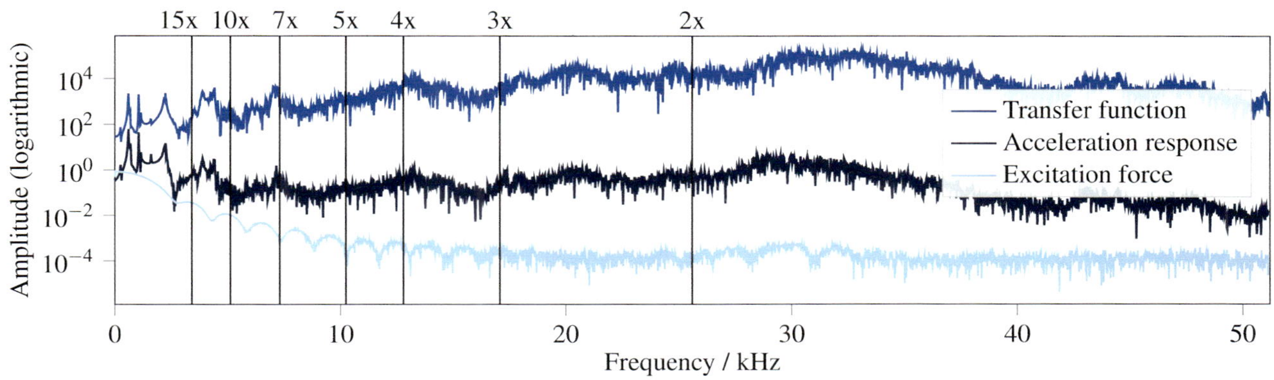


**Fig. 2** Example of transfer function calculated from measured acceleration response and excitation signal of the experimental test case; all quantities transformed into the frequency domain. Vertical lines indicate the cut-off point for a downsampling factor as indicated at the top of the line

To solve this problem, the coupling using a virtual point was introduced in [5] by Van der Seijs et al.. The interface is assumed to be rigid, i.e., only six rigid body modes, three translations and three rotations, are considered, and the remainder or residual $\boldsymbol{\mu}$ is neglected. The measured interface translations (at least nine) are then projected onto the virtual point with six dofs. This projection weakens the interface coupling and is solved for in a least-square sense, averaging out measurement errors [5].

**Projection of Responses** For the virtual point transformation, the triaxial sensors are assumed to be rigidly connected to the interface, i.e., the virtual point. The distance from the virtual point to the sensor $i$ is given by the position vector $\boldsymbol{r}_i$, as shown on the left in fig. 3.

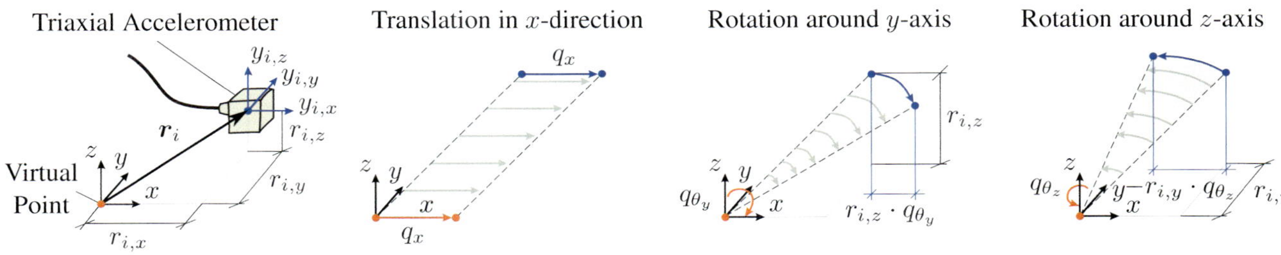


**Fig. 3** Projection of triaxial sensor dofs on virtual point exemplarily shown for the measured $x$-direction $y_{i,x}$ of sensor $i$, figure adapted from *pyFBS* documentation [8]

Using this vector, the responses $\boldsymbol{y}_i$ measured by the sensor $i$ can be projected onto the virtual point dofs $\boldsymbol{q}$, by deriving the required relations, as exemplary shown for the measured response in the $x$-direction in fig. 3:

$$\begin{bmatrix} y_{i,x} \\ y_{i,y} \\ y_{i,z} \end{bmatrix} = \begin{bmatrix} 1 & 0 & 0 & 0 & r_{i,z} & -r_{i,y} \\ 0 & 1 & 0 & -r_{i,z} & 0 & r_{i,x} \\ 0 & 0 & 1 & r_{i,y} & -r_{i,x} & 0 \end{bmatrix} \begin{bmatrix} q_x \\ q_y \\ q_z \\ q_{\theta_x} \\ q_{\theta_y} \\ q_{\theta_z} \end{bmatrix} \tag{2}$$

Equation (2), $\boldsymbol{y}_i = \boldsymbol{R}_{y,i}\boldsymbol{q}$ in short notation, describes how the three sensor dofs $\boldsymbol{y}_i$ can be described using the matrix $\boldsymbol{R}_{y,i}$ containing the rigid body modes of sensor $i$ and the reduced virtual point dofs $\boldsymbol{q}$. In case the orientation of the sensor and the virtual point coordinate system does not line up, a rotation matrix has to be added to the equation. Writing this for all $n_S$ sensors and adding a residual $\boldsymbol{\mu}$ to account for the neglected flexible interface motion, the response $\boldsymbol{y}_i$ of all interface sensors is given by [5]:

$$\begin{bmatrix} \boldsymbol{y}_1 \\ \vdots \\ \boldsymbol{y}_{n_S} \end{bmatrix} = \begin{bmatrix} \boldsymbol{R}_{y,1} \\ \vdots \\ \boldsymbol{R}_{y,n_S} \end{bmatrix} \boldsymbol{q} + \boldsymbol{\mu} \tag{3}$$

An optimal solution for the desired virtual point dofs $\boldsymbol{q}$ can then be found by minimizing the residual $\boldsymbol{\mu}$, which yields, without weighting matrices, the desired transformation matrix $\boldsymbol{T}_y$, projecting the measured responses $\boldsymbol{y}$ on the virtual point dofs $\boldsymbol{q}$ [9]:

$$\boldsymbol{T}_y = \left(\boldsymbol{R}_y^{\mathrm{T}} \boldsymbol{R}_y\right)^{-1} \boldsymbol{R}_y^{\mathrm{T}} \qquad \text{with} \qquad \boldsymbol{q} = \boldsymbol{T}_y \boldsymbol{y} \tag{4}$$

**Projection of Forces** Similar to the projection of the measured responses $\boldsymbol{y}_i$ of sensor $i$, the forces or impacts $\boldsymbol{f}_i$ applied to the system can be projected onto the virtual point [5]:

$$\begin{bmatrix} m_x \\ m_y \\ m_z \\ m_{\theta_x} \\ m_{\theta_y} \\ m_{\theta_z} \end{bmatrix} = \begin{bmatrix} 1 & 0 & 0 \\ 0 & 1 & 0 \\ 0 & 0 & 1 \\ 0 & -r_{i,z} & r_{i,y} \\ r_{i,z} & 0 & -r_{i,x} \\ -r_{i,y} & r_{i,x} & 0 \end{bmatrix} \begin{bmatrix} e_{i,x} \\ e_{i,y} \\ e_{i,z} \end{bmatrix} f_i \tag{5}$$

where the vector $\boldsymbol{r}_i$ describes the distance between the virtual point and the point where the force is acting and the vector $\boldsymbol{e}_i$ the direction of the scalar force value $f_i$. Again, all $n_I$ impacts can be combined in one equation $\boldsymbol{m} = \boldsymbol{R}_f^{\mathrm{T}} \boldsymbol{f}$. An optimal solution for the transformation of the virtual point forces $\boldsymbol{m}$ into the physical space of the measured forces $\boldsymbol{f}$ is given by [9]:

$$\boldsymbol{T}_f^{\mathrm{T}} = \boldsymbol{R}_f \left(\boldsymbol{R}_f^{\mathrm{T}} \boldsymbol{R}_f\right)^{-1} \qquad \text{with} \qquad \boldsymbol{f} = \boldsymbol{T}_f^{\mathrm{T}} \boldsymbol{m} \tag{6}$$

**Adaption of IBS Equations - Interface Compatibility** First, the interface compatibility is no longer written for the physical responses $\boldsymbol{y}^{(s)}$, but for the virtual point dofs $\boldsymbol{q}^{(s)}$ for each virtual point of each substructure $s$:

$$\sum_{s=1}^{N_S} \boldsymbol{B}_q^{(s)} \boldsymbol{q}^{(s)} = \boldsymbol{0} \qquad \text{with} \qquad \boldsymbol{q}^{(s)} = \begin{bmatrix} q_x^{(s)} & q_y^{(s)} & q_z^{(s)} & q_{\theta_x}^{(s)} & q_{\theta_y}^{(s)} & q_{\theta_z}^{(s)} \end{bmatrix}^{\mathrm{T}} \tag{7}$$

For this, the boolean constrain matrix $\boldsymbol{B}_f^{(s)}$ has to be rewritten for the virtual point dofs $\boldsymbol{q}$ as $\boldsymbol{B}_q^{(s)}$. Inserting the transformation matrix from eq. (4), i.e., $\boldsymbol{q} = \boldsymbol{T}_y \boldsymbol{y}$, gives the interface compatibility condition on the virtual point w.r.t. $\boldsymbol{y}^{(s)}$:

**Interface Compatibility on Virtual Point**

$$\sum_{s=1}^{N_S} \underset{\left((n_{\mathrm{V}}\cdot 6)\times\left(n_{\mathrm{aS}}^{(s)}+n_{\mathrm{V}}\cdot 6\right)\right)}{\boldsymbol{B}_q^{(s)}} \quad \underset{\left(\left(n_{\mathrm{aS}}^{(s)}+n_{\mathrm{V}}\cdot 6\right)\times n_{\mathrm{S}}^{(s)}\right)}{\boldsymbol{T}_y^{(s)}} \quad \underset{\left(n_{\mathrm{S}}^{(s)}\times 1\right)}{\boldsymbol{y}^{(s)}} = \underset{\left(n_{\mathrm{S}}^{(s)}\times 1\right)}{\boldsymbol{0}} \tag{8}$$

where:

| | | | | | |
|---|---|---|---|---|---|
| $n_{\mathrm{S}}$ | Total number of sensor dofs | $n_{\mathrm{V}}$ | Number of Virtual Points | $(s)$ | Substructure index |
| $n_{\mathrm{aS}}$ | Number of additional sensor dofs | $N_S$ | Total number of substructures | | |

**Adaption of IBS Equations - Interface Equilibrium** Next, the interface equilibrium has to be written for the virtual point coordinates $\boldsymbol{q}^{(s)}$, i.e., the Lagrange multipliers $\boldsymbol{\lambda}_m$ are applied to the virtual point as virtual forces:

$$\boldsymbol{m}_\lambda^{(s)} = \boldsymbol{B}_m^{(s)\,\mathrm{T}} \boldsymbol{\lambda}_m \tag{9}$$

To calculate a response to these forces as required for the IBS scheme, they must be transformed into the basis of the impulse response functions $\boldsymbol{H}^{(s)}$, namely the physical space. This is done by inserting the transformation eq. (6), i.e., $\boldsymbol{T}_f^{\mathrm{T}}\boldsymbol{m}$, into eq. (9):

**Interface Equilibrium on Virtual Point**

$$\underset{\left(n_{\mathrm{I}}^{(s)}\times 1\right)}{\boldsymbol{f}_\lambda^{(s)}} = \underset{\left(n_{\mathrm{I}}^{(s)}\times\left(n_{\mathrm{aI}}^{(s)}+n_{\mathrm{V}}\cdot 6\right)\right)}{\boldsymbol{T}_f^{(s)\mathrm{T}}} \underset{\left(\left(n_{\mathrm{aI}}^{(s)}+n_{\mathrm{V}}\cdot 6\right)\times(n_{\mathrm{V}}\cdot 6)\right)}{\boldsymbol{B}_m^{(s)\mathrm{T}}} \underset{((n_{\mathrm{V}}\cdot 6)\times 1)}{\boldsymbol{\lambda}_m} \tag{10}$$

where:

| | | | | | |
|---|---|---|---|---|---|
| $n_{\mathrm{I}}$ | Total number of impact locations | $n_{\mathrm{V}}$ | Number of Virtual Points | $(s)$ | Substructure index |
| $n_{\mathrm{aI}}$ | Number of additional impact locations | $N_S$ | Total number of substructures | | |

**Adaption of IBS Equations - Lagrange Multiplier Calculation** Lastly, the calculation of the Lagrange multipliers must be adjusted. Using the discretization used in [4], this is given as:

$$\left(\sum_{s=1}^{N_S} \boldsymbol{B}_y^{(s)}\boldsymbol{H}^{(s)}[0]\boldsymbol{B}_f^{(s)\mathrm{T}}\right)\boldsymbol{\lambda}_f[k-1] = -\sum_{s=1}^{N_S}\boldsymbol{B}_y^{(s)}\tilde{\boldsymbol{y}}^{(s)}[k]\cdot\frac{1}{\Delta t} \tag{11}$$

On the right-hand side, the compatibility condition is exchanged with the one written for VPT coupling (eq. (8)). On the left-hand side, the interface forces $\boldsymbol{B}_f^{(s)\mathrm{T}}\boldsymbol{\lambda}_f = \boldsymbol{f}_\lambda^{(s)}$ are exchanged with the expression in eq. (10). As the left-hand side would then yield a response due to the Lagrange multipliers in the real space, a premultiplication by the transformation from eq. (4) is necessary to be consistent with the right-hand side, representing the gap in the virtual point coordinates. This yields the new Lagrange multiplier calculation with VPT coupling:

**Calculation of Lagrange Multipliers on VPT Interface**

$$\left(\sum_{s=1}^{N_S} \boldsymbol{B}_q^{(s)}\boldsymbol{T}_y^{(s)}\boldsymbol{H}^{(s)}[0]\boldsymbol{T}_f^{(s)\mathrm{T}}\boldsymbol{B}_m^{(s)\mathrm{T}}\right)\boldsymbol{\lambda}_m[k-1] = -\sum_{s=1}^{N_S}\boldsymbol{B}_q^{(s)}\boldsymbol{T}_y^{(s)}\tilde{\boldsymbol{y}}^{(s)}[k]\cdot\frac{1}{\Delta t} \tag{12}$$

**Discretized IBS Equations with Virtual Point Coupling** The only remaining step is to substitute the projection of the VPT interface forces (eq. (10)) into the predictor and corrector step of the IBS scheme. To summarize, the adapted IBS equations with virtual point coupling are given by:

**IBS with VPT Predictor Step**

$$\tilde{\boldsymbol{y}}^{(s)}[k] = \left(\sum_{i=0}^{k-2}\boldsymbol{H}^{(s)}[k-(i+1)]\left(\boldsymbol{f}^{(s)}[i] + \boldsymbol{T}_f^{(s)\mathrm{T}}\boldsymbol{B}_m^{(s)\mathrm{T}}\boldsymbol{\lambda}_m[i]\right)\Delta t\right) + \boldsymbol{H}^{(s)}[0]\,\boldsymbol{f}^{(s)}[k-1]\,\Delta t \tag{13}$$

**Calculation of Lagrange Multipliers on VPT Interface**

$$\left(\sum_{s=1}^{N_S} \boldsymbol{B}_q^{(s)}\boldsymbol{T}_y^{(s)}\boldsymbol{H}^{(s)}[0]\boldsymbol{T}_f^{(s)\mathrm{T}}\boldsymbol{B}_m^{(s)\mathrm{T}}\right)\boldsymbol{\lambda}_m[k-1] = -\sum_{s=1}^{N_S}\boldsymbol{B}_q^{(s)}\boldsymbol{T}_y^{(s)}\tilde{\boldsymbol{y}}^{(s)}[k]\cdot\frac{1}{\Delta t} \tag{14}$$

**IBS with VPT Corrector Step**

$$\boldsymbol{y}_\lambda^{(s)}[k] = \boldsymbol{H}^{(s)}[0]\boldsymbol{T}_f^{(s)\mathrm{T}}\boldsymbol{B}_m^{(s)\mathrm{T}}\boldsymbol{\lambda}_m[k-1]\,\Delta t \tag{15}$$

**Sensor and Impact Consistency Criterion** When applying the VPT for measurements, various issues could exist, e.g., some impacts are too soft (bad signal-to-noise ratio), or sensor positions, orientations, or calibrations are wrong. Also, the rigid interface assumption will not hold for higher frequencies. Such issues within FBS applications can be detected using the established performance indicators of sensor and impact consistency [5]. Therefore, an adaption for IBS is highly motivated.

The sensor consistency is defined using the response to a unit load case at any excitation dof, which in the frequency domain is represented by a magnitude of one at all frequencies. This essentially extracts the respective column of the FRF matrix [5]. In the time domain, the equivalent of the unit load case is given by a Dirac impulse. Then, the convolution of the IRF matrix with a Dirac impulse at dof $i$ also extracts the $i$-th column $\boldsymbol{H}_i$ of the IRF matrix $\boldsymbol{H}$. The filtered (time) response $\bar{\boldsymbol{y}}$, i.e., transformed onto the virtual point and back to the physical space, can then be found by substituting eq. (4) in eq. (3):

$$\bar{\boldsymbol{y}} = \boldsymbol{R}_y \boldsymbol{T}_y \boldsymbol{y} = \boldsymbol{R}_y \boldsymbol{T}_y \boldsymbol{H}_i \tag{16}$$

Similarly, the impact consistency uses the response at one sensor channel to all performed impacts, equivalent to one row of the FRF matrix [5], respectively $\boldsymbol{H}_{1i}$ of the IRF matrix. The filtered force $\bar{\boldsymbol{f}}$ is found by substituting eq. (5) in eq. (6):

$$\bar{\boldsymbol{f}} = \boldsymbol{T}_f^{\mathrm{T}} \boldsymbol{R}_f^{\mathrm{T}} \boldsymbol{f} = \boldsymbol{T}_f^{\mathrm{T}} \boldsymbol{R}_f^{\mathrm{T}} \boldsymbol{H}_{1i} \tag{17}$$

In the frequency domain, the criteria for sensor consistency $\rho_y^2$ and impact consistency $\rho_f^2$ can be evaluated using a formulation similar to the Modal Assurance Criterion (MAC), giving a value for each frequency [5]:

$$\begin{aligned} \rho_y^2 &= \mathrm{MAC}\left(\bar{\boldsymbol{y}}, \boldsymbol{H}_i\right) \\ \rho_f^2 &= \mathrm{MAC}\left(\bar{\boldsymbol{f}}, \boldsymbol{H}_{1i}\right) \end{aligned} \qquad \text{with} \qquad \mathrm{MAC} = \frac{\left(a^{\mathrm{H}} b\right)\left(b^{\mathrm{H}} a\right)}{\left(a^{\mathrm{H}} a\right)\left(b^{\mathrm{H}} b\right)} \tag{18}$$

However, this evaluation is not as straightforward in the time domain, i.e., in the context of IBS. While applying an FFT and using eq. (18) would be possible, the question is how representative this frequency domain result is. For instance, when only short time signals are used, e.g., only $50\,\mathrm{ms}$, like for the IRFs here, the frequency resolution will be very coarse, and the leakage introduced by the cut-off tremendous. Therefore, a comparison in the time domain should be favored. Nevertheless, the criteria are not further developed here due to the complexity of meaningfully comparing time domain signals of mechanical systems.

## Experimental Setup

The substructuring benchmark structure depicted in fig. 4 is used as a test case. The measurements are performed for two physical configurations, once for the assembled system A+B (fig. 4), connected by a bolt with washers torqued to $20\,\mathrm{N\,m}$, and once for the two unconnected substructures A and B (fig. 4). A closer view of the interface is also given in fig. 4. For all configurations, the respective structures are suspended from rubber ropes and can swing freely.

The systems are excited using an impact hammer with a vinyl tip. A total of ten impacts are performed per impact location. Accelerations are measured using *KISTLER 8688A50T* triaxial sensors that are glued to the structure. The signals are acquired with a *Müller-BBM PAK MKII* system with a sample rate of $f_S = 102.4\,\mathrm{kHz}$. The sensor as well as impact locations are identical between configurations and are summarized in table 1. The impact locations are described as needed in the following.

**Table 1** Summary of utilized sensor and impact locations on the substructuring benchmark structure

| | **Substructure A** | | **Substructure B** | |
|---|---|---|---|---|
| **Grouping** | Internal | Interface | | Internal |
| **Sensors** | Sensor 1 & 2 | Sensor 3 - 5 | Sensor 6 - 8 | Sensor 9 - 11 |
| **Impacts** | H1 - H4<br>H31 - H33 | H5 - H14 | H15 - H25[1] | H26 - H30<br>H34 - H38 |

[1]Note: Impact H21 could not be performed but is retained for consistent numbering

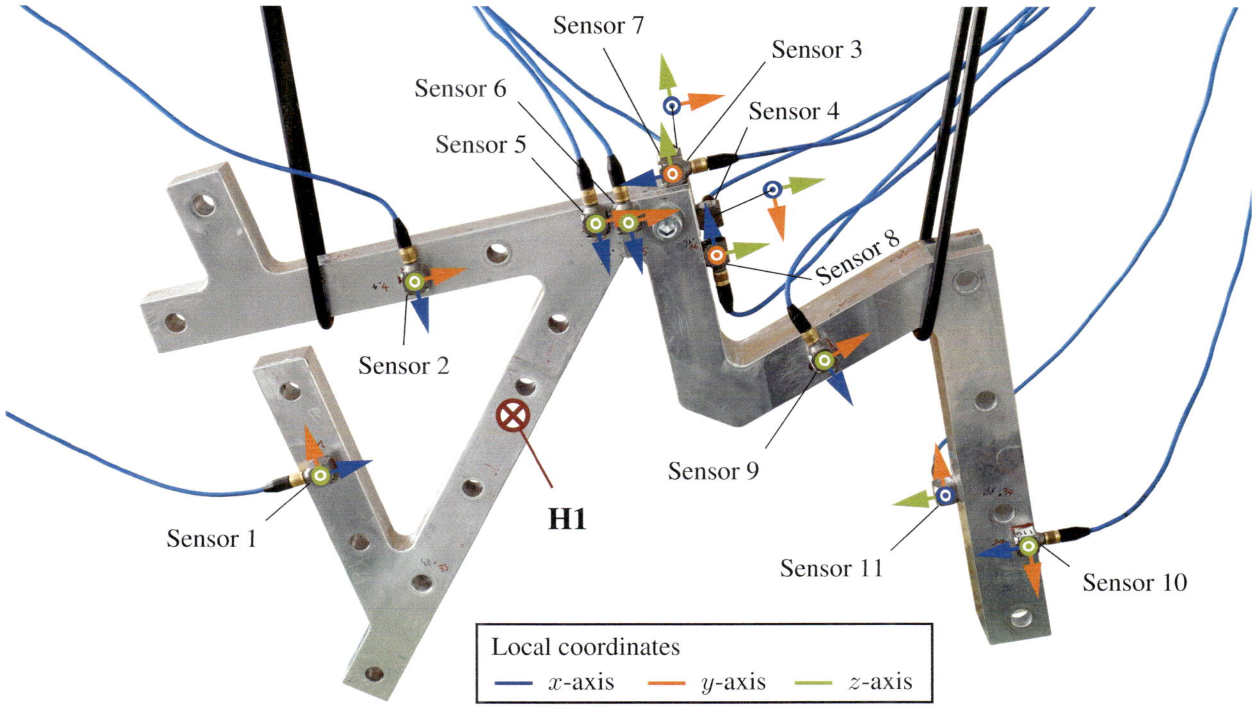


**Fig. 4** Experimental setup for the reference measurement of assembled system A+B

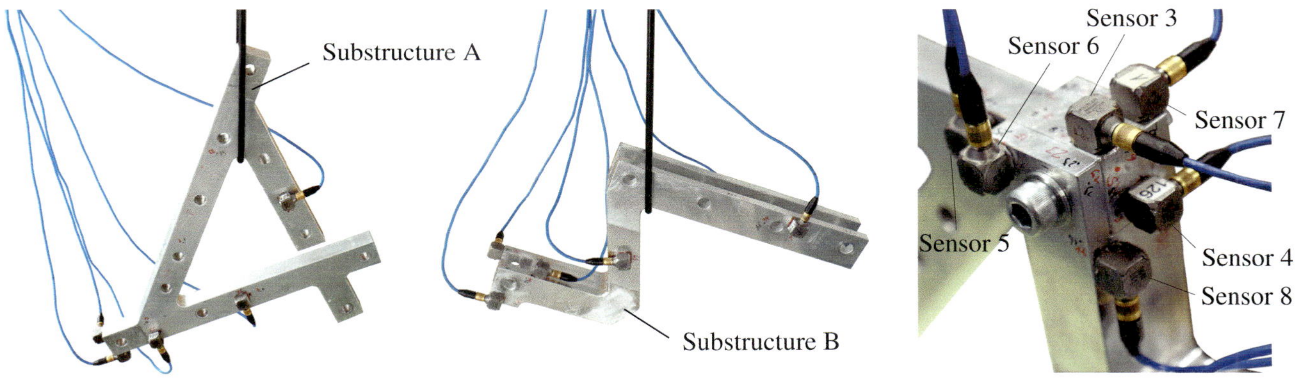


**Fig. 5** Experimental setup of individual substructures (left) and interface close-up (right)

IRFs are calculated from the measured signals as described in [4] with a length of $h_{\text{len}} = 50\,\text{ms}$. The required transformation matrices $\boldsymbol{T}_y^{(s)}$ and $\boldsymbol{T}_f^{(s)}$ are generated from sensor and impact positions using *pyFBS* [10]. All measured time series as well as an Excel sheet containing all sensor and impact positions are available for download at: https://doi.org/10.14459/2024mp1754069 [11]. Additionally, a Python script to retrieve the VPT matrices using *pyFBS* is provided.

## Experimental Results

For the test setup and sample rate of $f_S = 102.4\,\text{kHz}$ used here, the IBS scheme with VPT immediately becomes unstable, manifesting in the responses and Lagrange multiplier growing toward infinity very quickly. This is because the system excitation is limited and does not excite the whole measurement bandwidth, as already discussed in fig. 2, leading to erroneous high amplitudes in the IRFs at higher frequencies. For illustration, fig. 6 shows the FFT of one averaged experimental IRF.

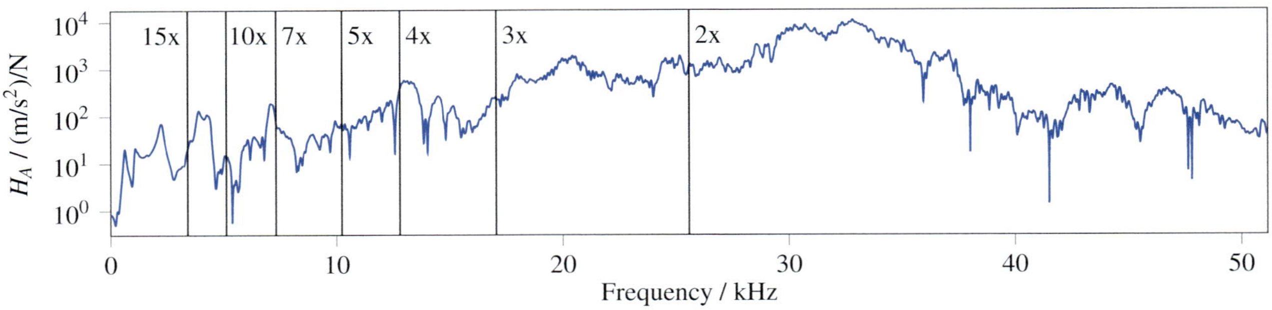


**Fig. 6** Frequency domain view of averaged S1+Z IRF with impact at H1 of substructure A. Vertical lines indicate the cut-off point for a downsampling factor as annotated

As can be seen, the amplitudes around 30 kHz, where basically no excitation above the noise floor exists, are almost two orders of magnitude higher than the peaks in the low-frequency region. Since there is no excitation, the high peaks are most likely only noise that dominates the time domain signal. Because for FBS the substructure assembly is per frequency line, i.e., independent of the FRF content at other frequencies [12], the noisy part does not influence the result at lower frequencies. However, every frequency component is present for IBS at any time (as long as the oscillation did not fully decay). Therefore, this frequency region should not be considered in the IBS scheme and must be removed.

Here, the frequency content is truncated by downsampling (with low-pass filter) the measured signals before estimating the IRFs. Note, however, that the same results could be achieved by reducing the sample rate during the measurements. The IBS results of one combination of sensor channel and impact H1 (see fig. 4) for various downsampling factors are shown in fig. 7.

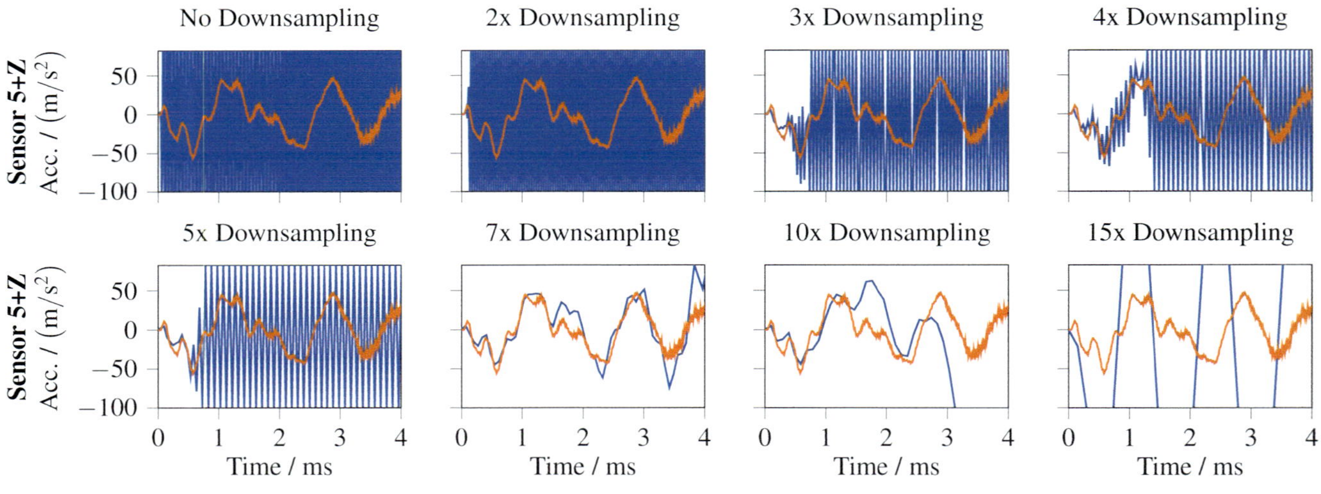


**Fig. 7** IBS results of the experimental test case with **excitation at H1** and **response at S5+Z** for various downsampling factors as indicated. Plot limits are fixed around reference measurement amplitudes, — IBS Result, — Measurement

Since the limits of the shown plots are fixed around the reference measurement, the actual amplitudes of the IBS results cannot be seen from fig. 7; however, they grow to very high values as soon as instability occurs. The instability usually manifests in an oscillation with growing amplitudes between positive and negative numbers.

As can be seen, with an increase of the downsampling factor, the instability of the IBS scheme generally decreases. One exception to this is the change from 4x to 5x, where the unstable, high frequent oscillations start to occur sooner than with the lower downsampling factor. The best result can be seen for 7x downsampling, where the IBS result closely matches the reference measurement and then slowly starts to diverge. A further increase of the downsampling factor leads to an earlier divergence, and for 15x, an unstable oscillation can be seen immediately at the start. Using a downsampling factor of 6x or 8x did not produce better results than with 7x downsampling. Therefore, this factor is used in the following.

The IBS results for the other sensors attached to substructure A, specifically, the channels aligned with the impact direction, are shown in fig. 8. For approximately the first millisecond, the IBS results are close to the reference, with a tendency toward overestimation. After that, all responses except the one of Sensor 2+Z show oscillations not present in the measured reference. The response of Sensor 2+Z also becomes unstable and shows growing amplitudes.

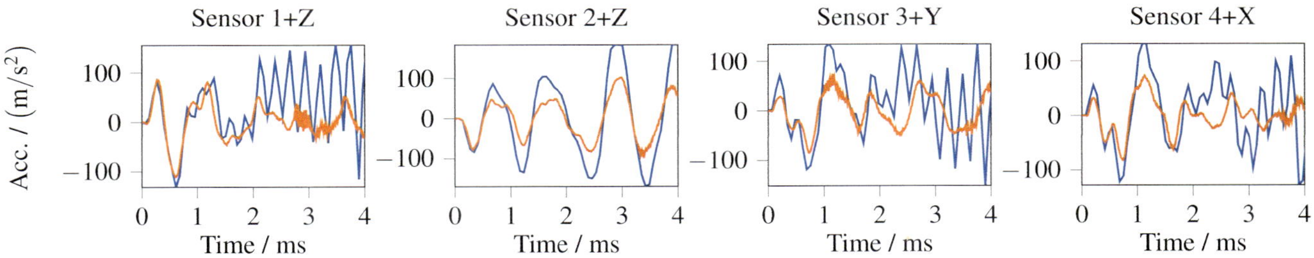


**Fig. 8** IBS results of the experimental test case with **excitation at H1** and responses of sensors attached to **substructure A**. Only sensor channels aligned with impact direction are shown. Plot limits fixed, — IBS Result, — Measurement

Responses of some sensors attached to substructure B, i.e., on the other interface side compared to the impact location, are shown in fig. 9; again, only responses in the impact direction are shown. Here, basically, the same observations can be made, where the IBS prediction of the acceleration response is decent in the beginning and then starts to diverge from the reference, which is accompanied by high frequent oscillations for Sensor 8+Y.

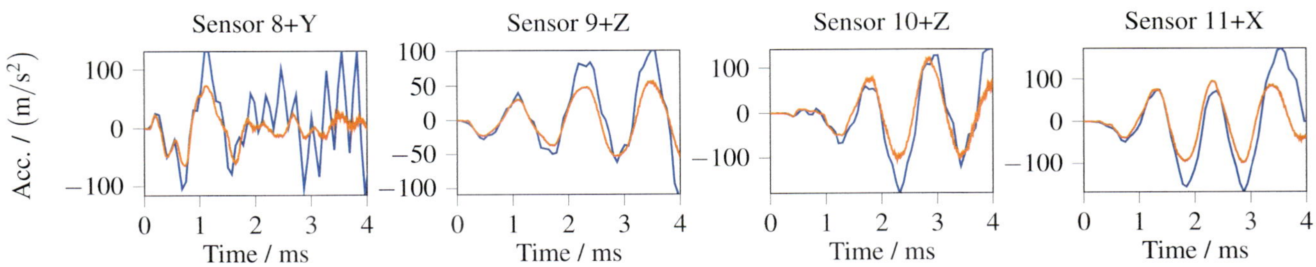


**Fig. 9** IBS results of the experimental test case with **excitation at H1** and responses of sensors attached to **substructure B**. Only sensor channels aligned with impact direction are shown. Plot limits fixed, — IBS Result, — Measurement

The full IBS results, including all sensor channels and the Lagrange multipliers over time, for impact location H1, as well as five other impact locations, can be found in the appendix. For all impacts, there are responses predicted by IBS that match the reference measurement well in the beginning, while other responses do not match the reference or are immediately unstable. Instability manifests in slowly growing amplitudes, high frequent oscillations, or in some cases, e.g., Sensor 1 for impact H26, the responses shoot off toward infinity. In most cases, at least one Lagrange multiplier also steadily grows in amplitude.

Here, it is impossible to identify the exact origin of the stability issues. Certainly, errors are introduced by noise, varying impacts, inaccuracies in the IRF estimation, and other typical experimental issues. Since the IBS scheme is based on convolution products, any error will accumulate over time. This implies that at some point, due to unavoidable experimental errors, the IBS scheme will become unstable. Further, any errors in the estimated IRFs will be amplified due to the coupling with Lagrange multipliers; more details are provided in [4]. Using the VPT, the hard enforcement of the interface compatibility is weakened, which should alleviate some of the associated issues already.

## Conclusion

In this paper, the theory of the Virtual Point Transformation was incorporated into the discretized equations of the Impulse-Based Substructuring scheme, allowing for an experimental application of IBS on three-dimensional structures. Using time domain deconvolution for estimating the Impulse Response Functions, combined with a reduction of the considered frequency bandwidth using downsampling with a low-pass filter, it was possible to experimentally apply IBS on a substructuring benchmark structure, where the initial acceleration response could be predicted successfully.

Due to some experimental errors, e.g., noise, the IBS scheme becomes unstable after some time. In future works, an improvement of the IRF estimation techniques could potentially lead to a longer amount of the predicted response being stable, i.e., better following the expected response. Nevertheless, the work presented here provides a framework for an experimental application of IBS using the VPT and enables further investigations to improve experimental IBS applications.

## References

[1] Gordis, J. H. "Integral equation formulation for transient structural synthesis". In: *AIAA Journal* 33.2 (1995), pp. 320–324. ISSN: 0001-1452. DOI: 10.2514/3.12454.

[2] Rixen, D. J. "Substructuring using Impulse Response Functions for Impact Analysis". In: *Proceedings of the IMAC-XXVIII* (2011). DOI: 10.1007/978-1-4419-9834-7_56.

[3] Rixen, D. J. and Van der Valk, P. L. C. "An Impulse Based Substructuring approach for impact analysis and load case simulations". In: *Journal of Sound and Vibration* 332.26 (2013), pp. 7174–7190. ISSN: 0022-460X. DOI: 10.1016/j.jsv.2013.08.004.

[4] Zobel, O., Trainotti, F., and Rixen, D. "Enabling Experimental Impulse-Based Substructuring Through Time Domain Deconvolution and Downsampling". In: *Experimental Techniques* (2024). ISSN: 1747-1567. DOI: 10.1007/s40799-024-00756-3.

[5] Van der Seijs, M., Bosch, D. van den, Rixen, D., and Klerk, D. *An improved methodology for the virtual point transformation of measured frequency response functions in dynamic substructuring*. 2013. DOI: 10.7712/120113.4816.C1539.

[6] Van Der Seijs, M., Van Der Valk, P., Van Der Horst, T., and Rixen, D. "Towards Dynamic Substructuring Using Measured Impulse Response Functions". In: *Proceedings of the IMAC-XXXII* (2014). DOI: 10.1007/978-3-319-04501-6_6.

[7] de Klerk, D., Rixen, D. J., Voormeeren, S., and Pasteuning, P. "Solving the RDoF problem in experimental dynamic substructuring". In: *Proceedings of the IMAC-XXVI* (2008).

[8] *Virtual Point Transformation - pyFBS documentation*. URL: https://pyfbs.readthedocs.io/en/latest/examples/fbs/virtual_point_transformation.html.

[9] Haeussler, M. and Rixen, D. J. "Optimal Transformation of Frequency Response Functions on Interface Deformation Modes". In: *Dynamics of Coupled Structures, Volume 4*. Cham: Springer International Publishing, 2017, pp. 225–237. ISBN: 978-3-319-54930-9. DOI: 10.1007/978-3-319-54930-9_20.

[10] Bregar, T., Mahmoudi, A. E., Kodrič, M., Ocepek, D., Trainotti, F., Pogačar, M., Göldeli, M., Čepon, G., Boltežar, M., and Rixen, D. J. "pyFBS: A Python package for Frequency Based Substructuring". In: *Journal of Open Source Software* 7.69 (2022), p. 3399. ISSN: 2475-9066. DOI: 10.21105/joss.03399.

[11] Zobel, O. M., Trainotti, F., and Rixen, D. J. *3D Experimental Impulse-Based Substructuring AM Benchmark Structure Measurement Data*. Dataset. 2024. DOI: 10.14459/2024mp1754069.

[12] Allen, M. S., Rixen, D., Seijs, M. van der, Tiso, P., Abrahamsson, T., and Mayes, R. L. *Substructuring in Engineering Dynamics - Emerging Numerical and Experimental Techniques*. 1st ed. Springer Cham, 2020. ISBN: 978-3-030-25532-9. DOI: 10.1007/978-3-030-25532-9.

## Appendix: Full IBS Results for Selected Impact Locations

In the following, the full experimental results of the substructuring benchmark structure for selected impacts and 7x down-sampling are shown on double pages, where the responses of substructure A are depicted on the left-hand side and the ones of substructure B on the right-hand side. Additionally, the sensors and the used impact location are shown on the top left of the double pages, and the Lagrange multipliers are on the bottom right.

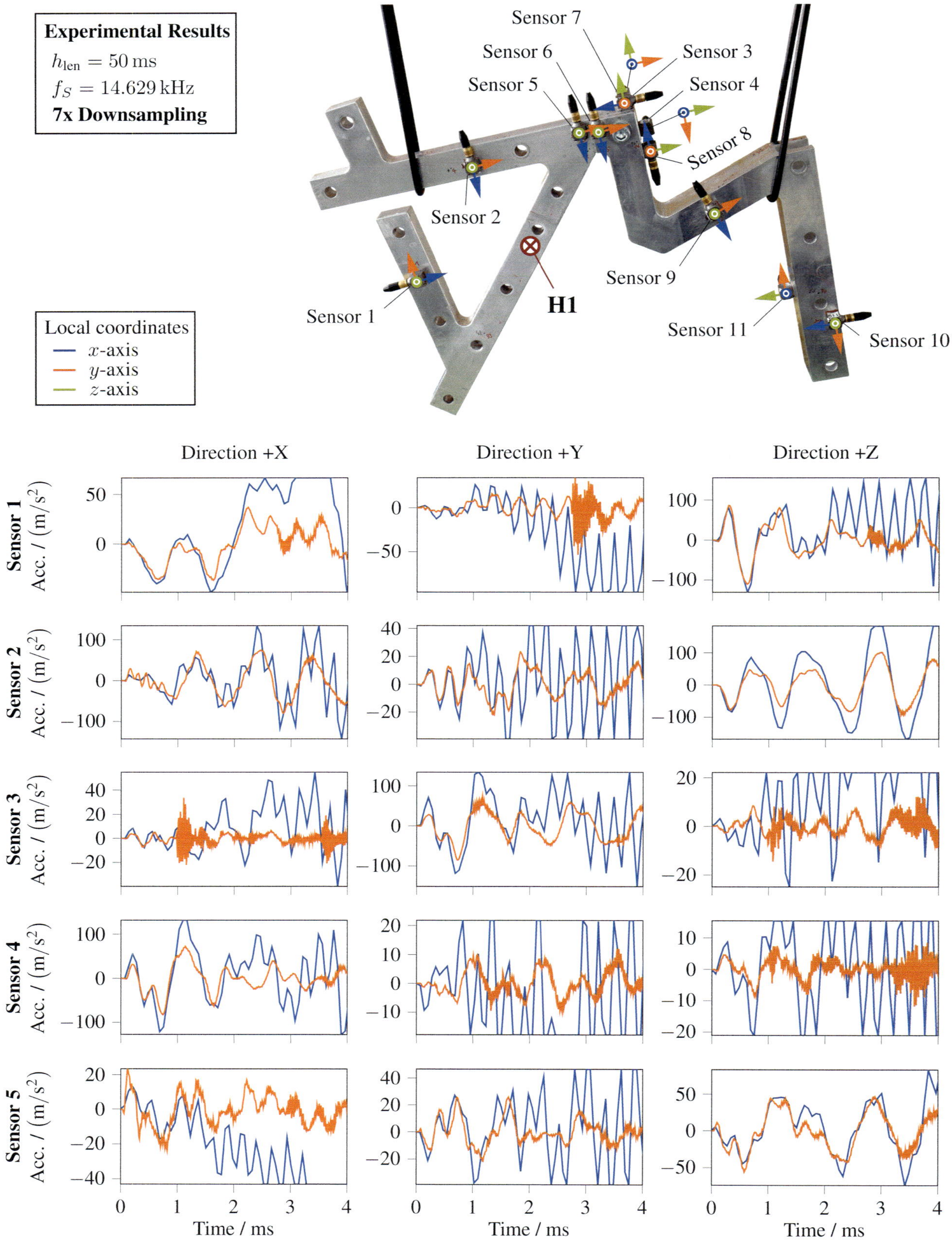


**Fig. 10** Experimental IBS results of **Substructure A** sensor dofs for IBS force applied at **impact location H1**, $f_S = 14.629\,\text{kHz}$ (x7 Downsampling), — IBS Result, — Measurement

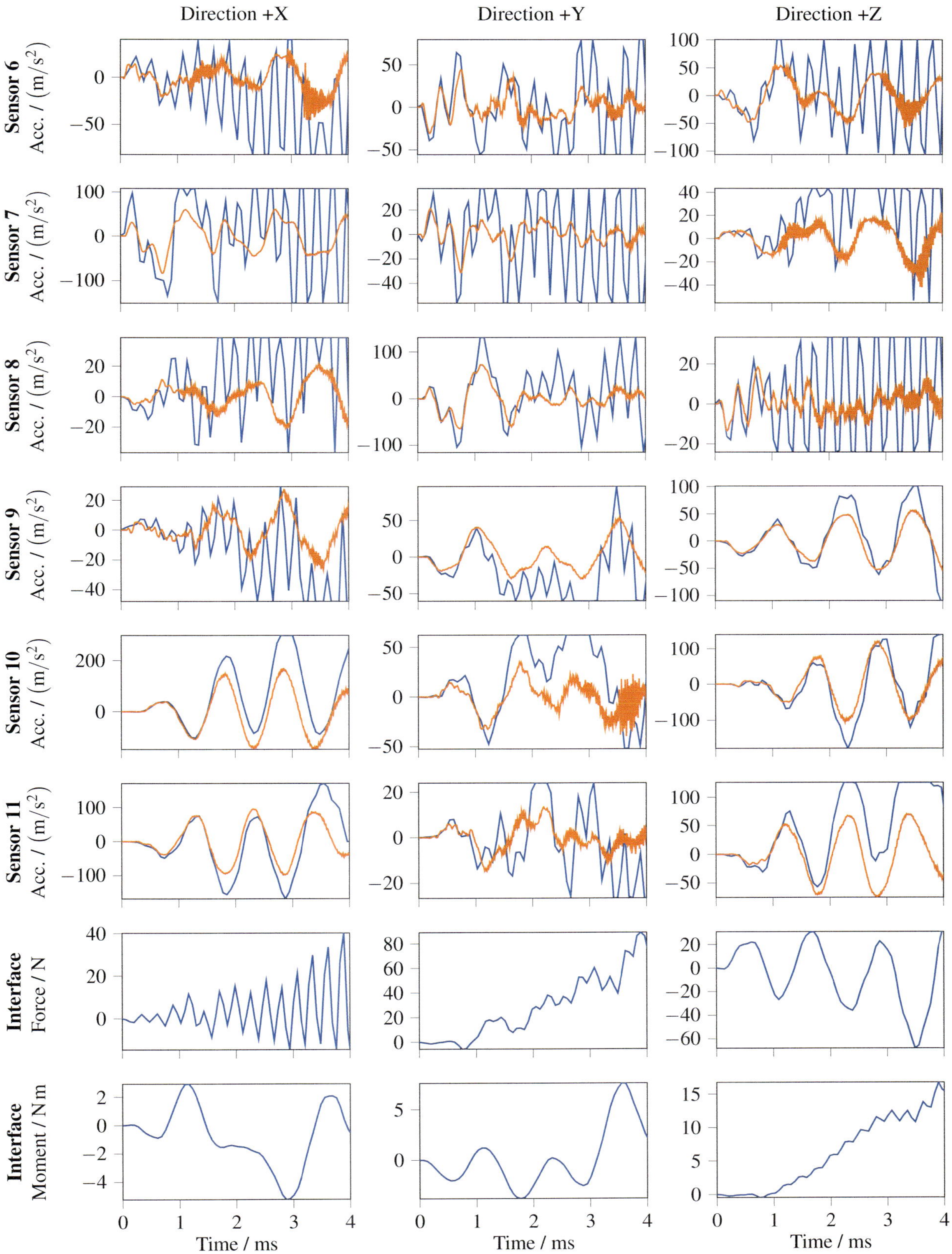


**Fig. 11** Experimental IBS results of **Substructure B** sensor dofs and Lagrange multipliers for IBS force applied at **impact location H1**, $f_S = 14.629\,\text{kHz}$ (x7 Downsampling), — IBS Result, — Measurement

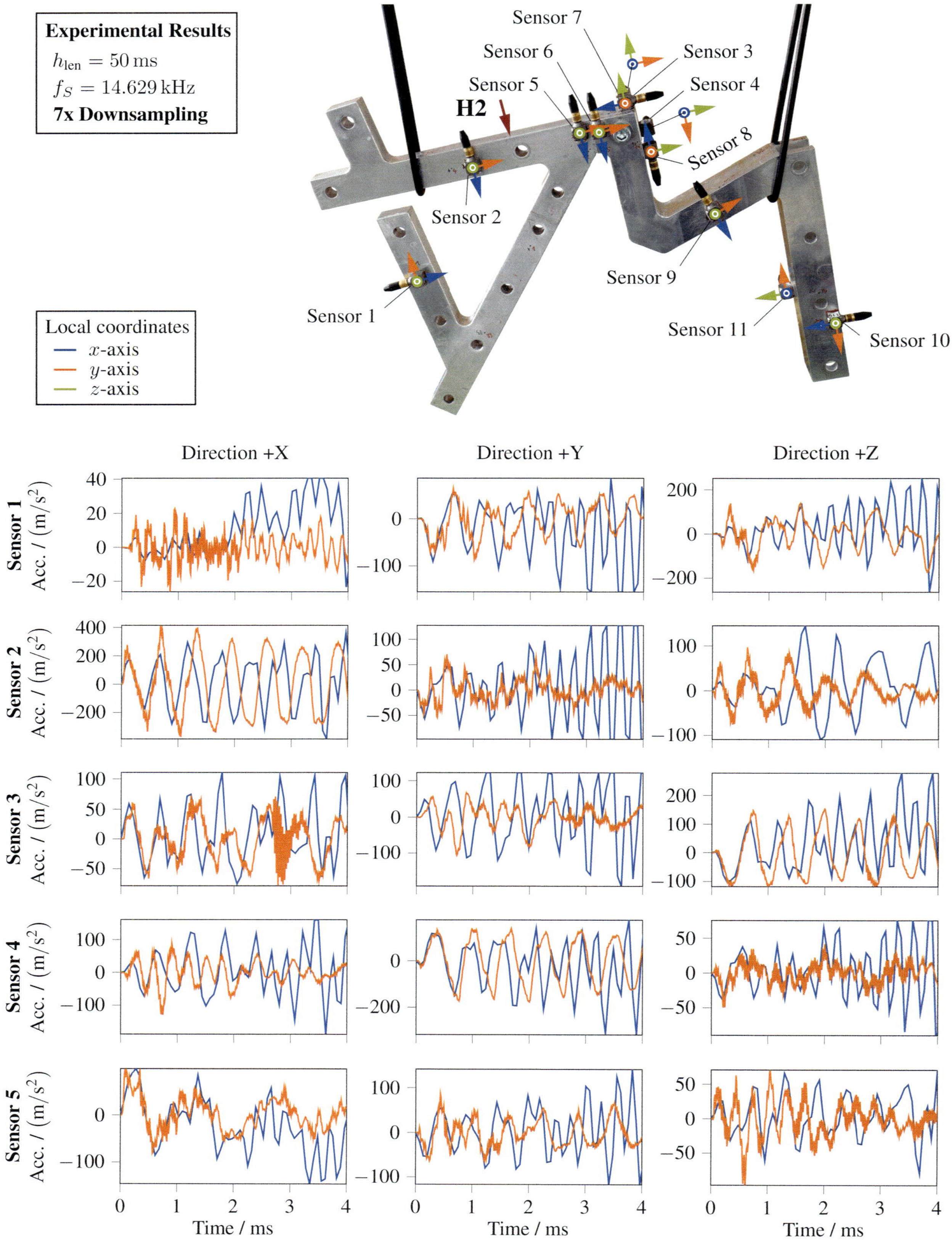


**Fig. 12** Experimental IBS results of **Substructure A** sensor dofs for IBS force applied at **impact location H2**, $f_S = 14.629$ kHz (x7 Downsampling), — IBS Result, — Measurement

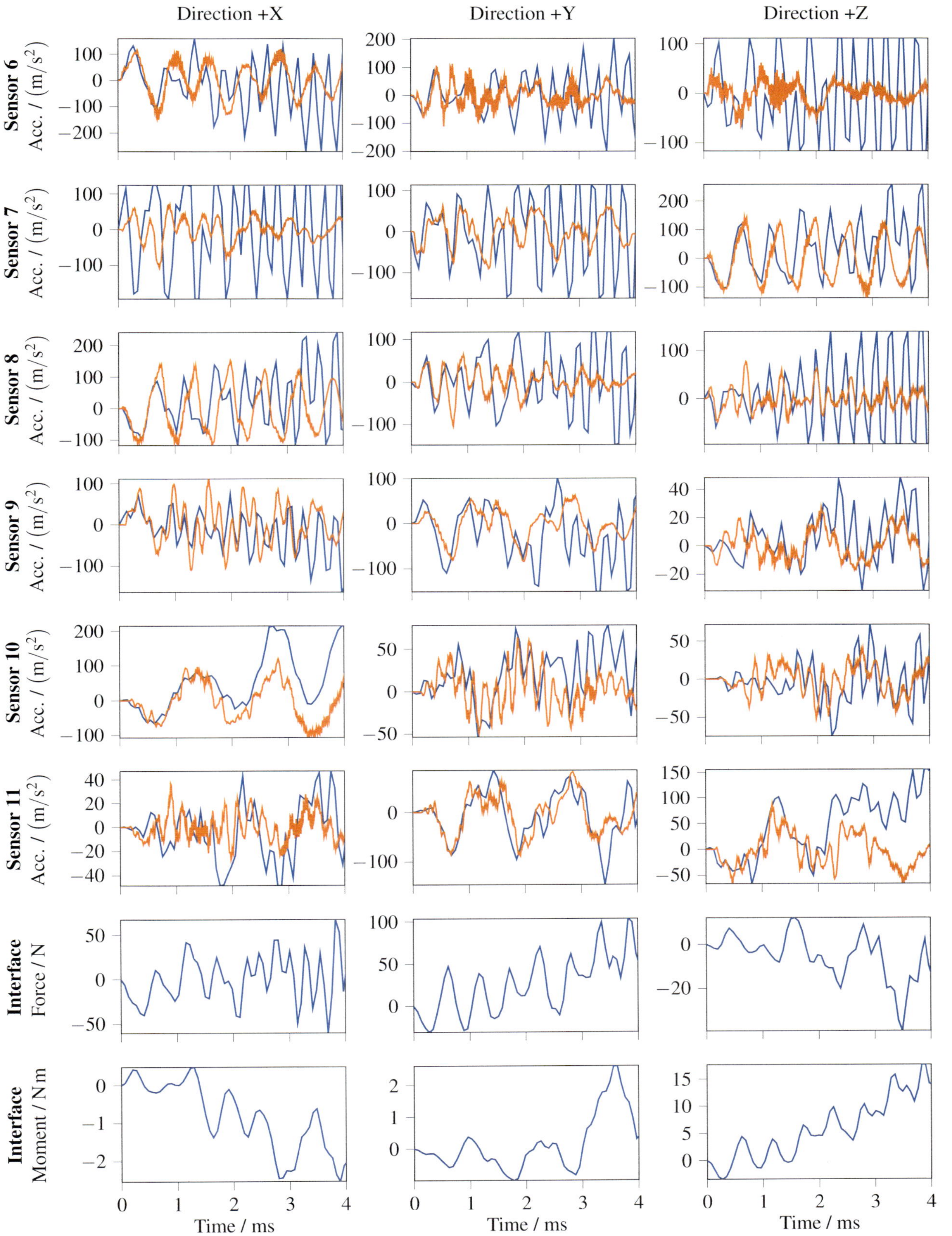


**Fig. 13** Experimental IBS results of **Substructure B** sensor dofs and Lagrange multipliers for IBS force applied at **impact location H2**, $f_S = 14.629\,\text{kHz}$ (x7 Downsampling), — IBS Result, — Measurement

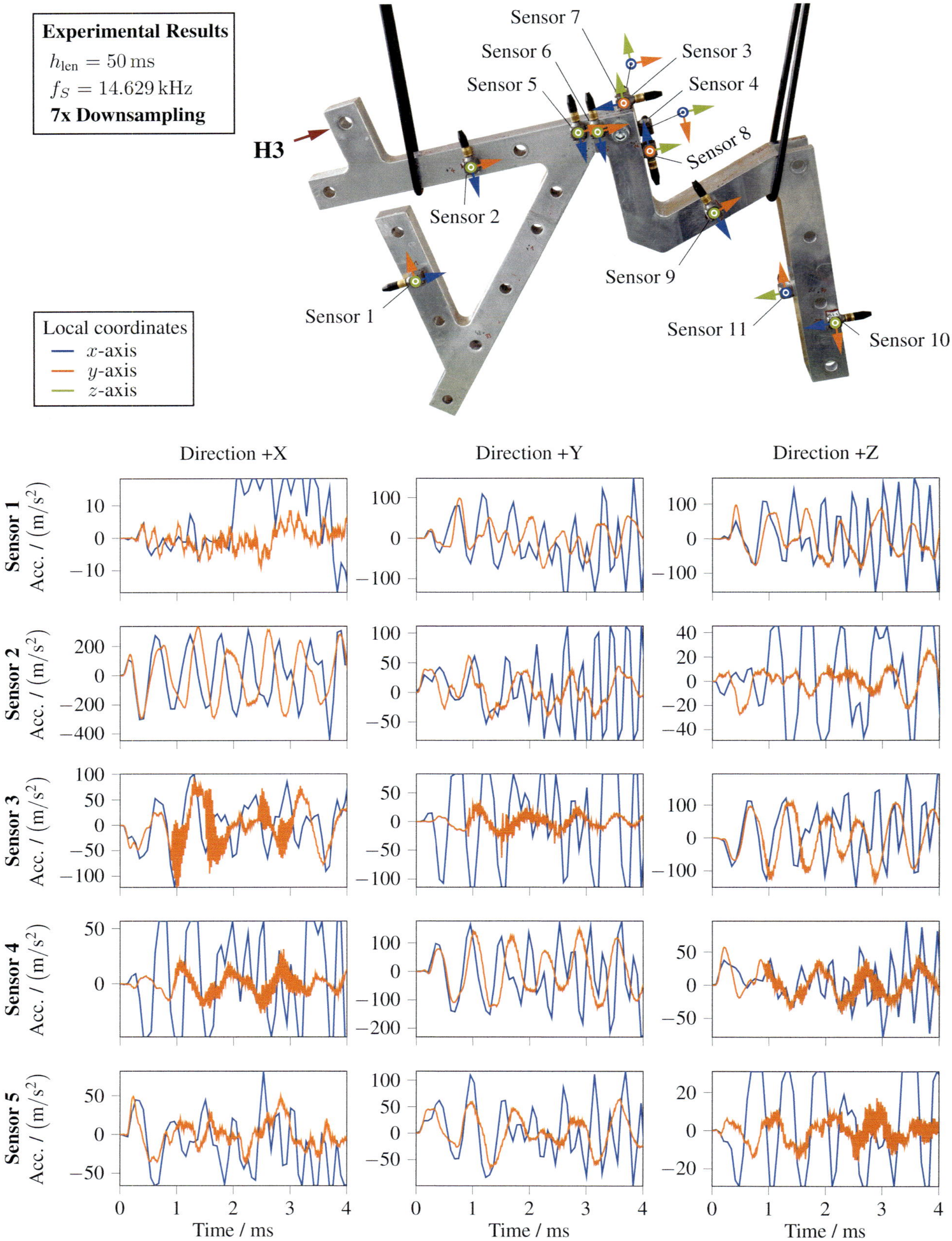


**Fig. 14** Experimental IBS results of **Substructure A** sensor dofs for IBS force applied at **impact location H3**, $f_S = 14.629\,\text{kHz}$ (x7 Downsampling), — IBS Result, — Measurement

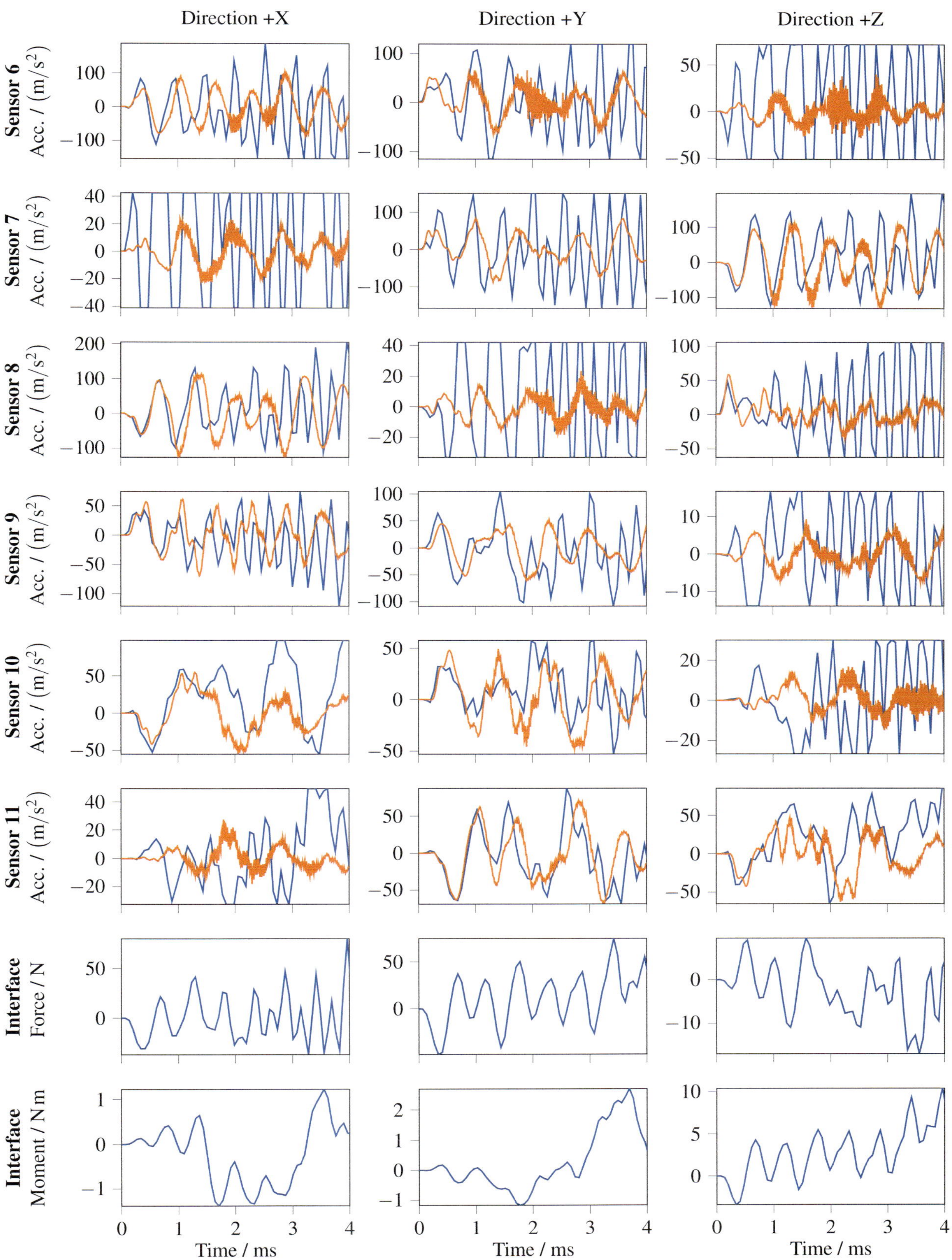


**Fig. 15** Experimental IBS results of **Substructure B** sensor dofs and Lagrange multipliers for IBS force applied at **impact location H3**, $f_S = 14.629\,\text{kHz}$ (x7 Downsampling), — IBS Result, — Measurement

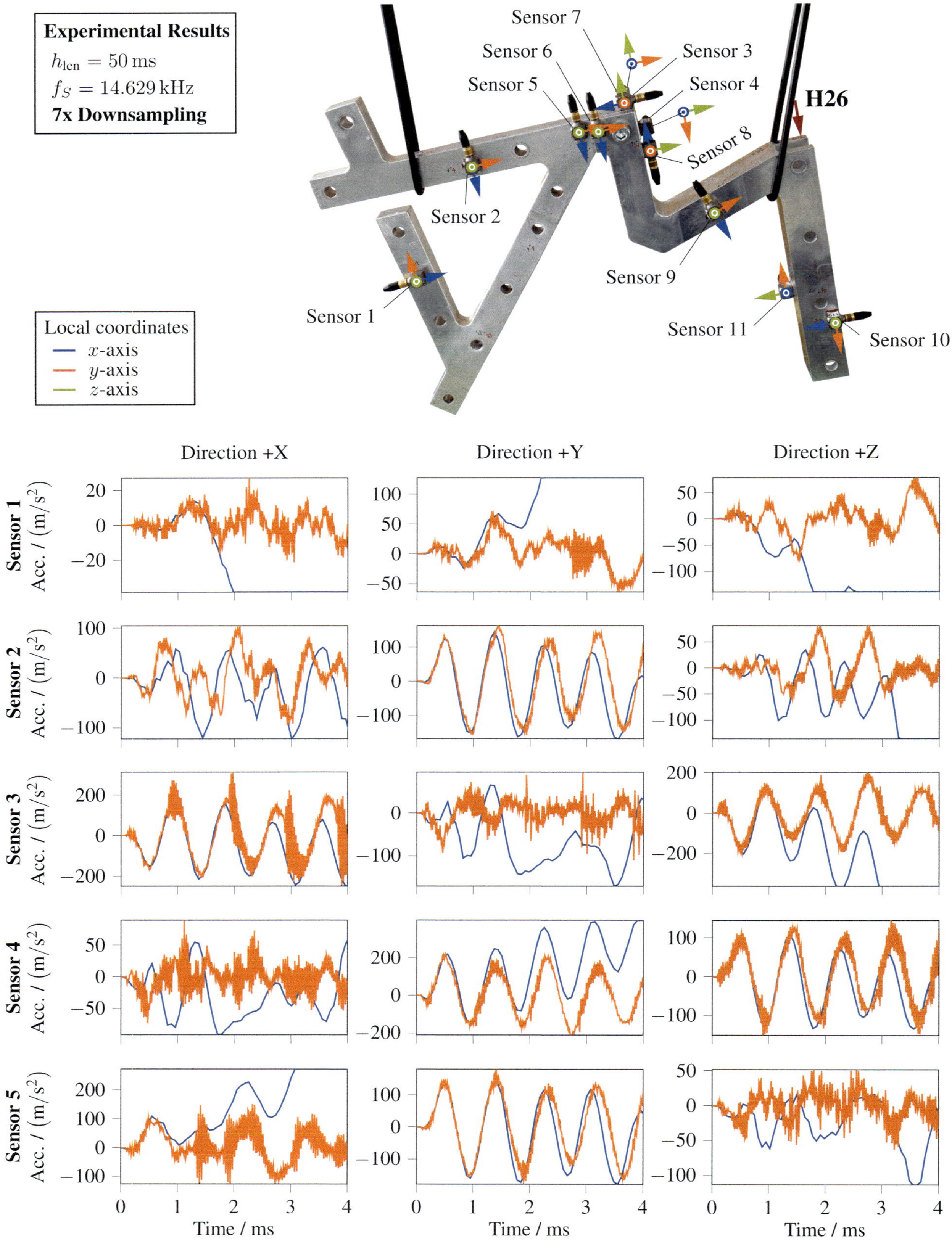


**Fig. 16** Experimental IBS results of **Substructure A** sensor dofs for IBS force applied at **impact location H26**, $f_S = 14.629\,\text{kHz}$ (x7 Downsampling), — IBS Result, — Measurement

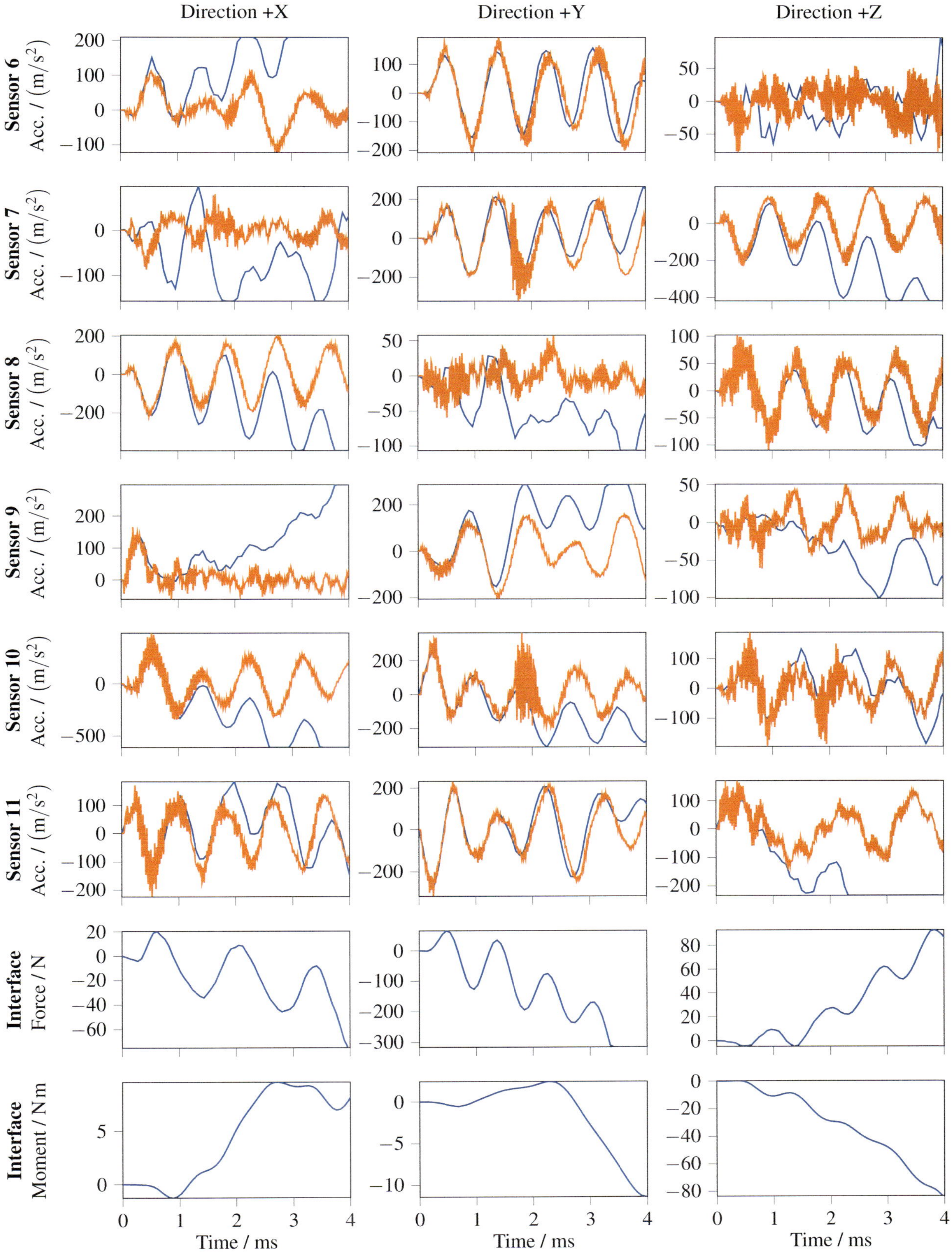


**Fig. 17** Experimental IBS results of **Substructure B** sensor dofs and Lagrange multipliers for IBS force applied at **impact location H26**, $f_S = 14.629\,\text{kHz}$ (x7 Downsampling), — IBS Result, — Measurement

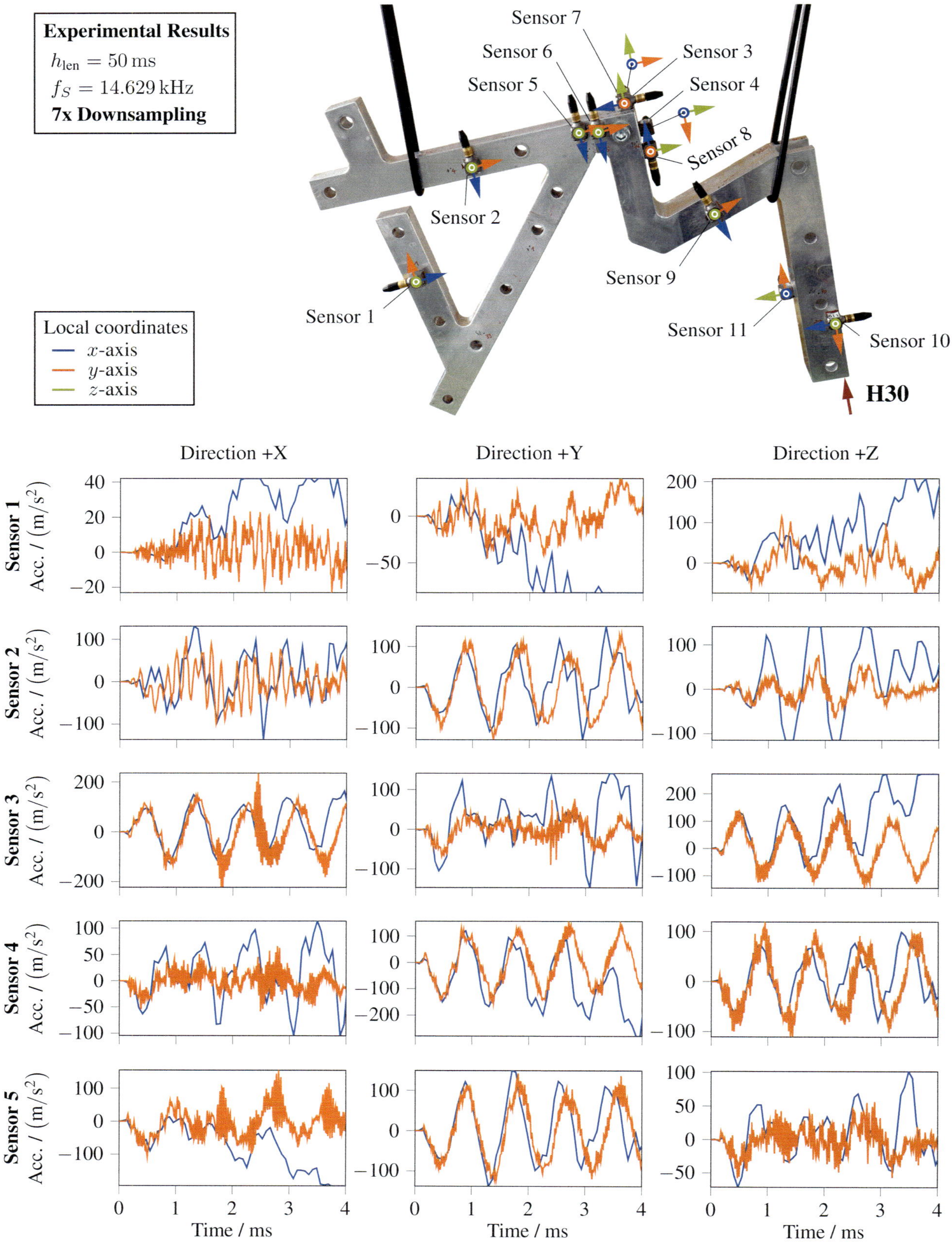


**Fig. 18** Experimental IBS results of **Substructure A** sensor dofs for IBS force applied at **impact location H30**, $f_S = 14.629$ kHz (x7 Downsampling), — IBS Result, — Measurement

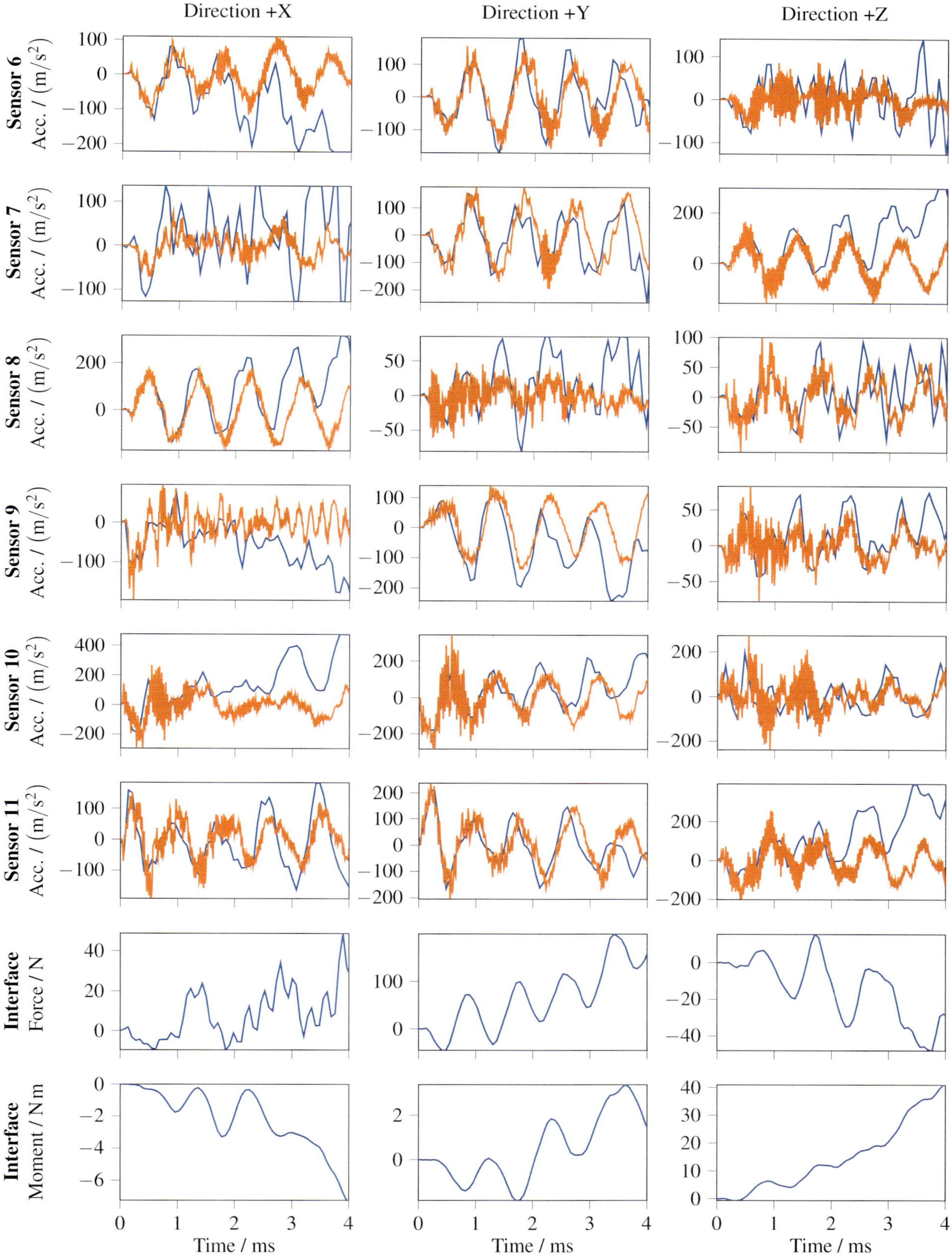


**Fig. 19** Experimental IBS results of **Substructure B** sensor dofs and Lagrange multipliers for IBS force applied at **impact location H30**, $f_S = 14.629\,\text{kHz}$ (x7 Downsampling), — IBS Result, — Measurement

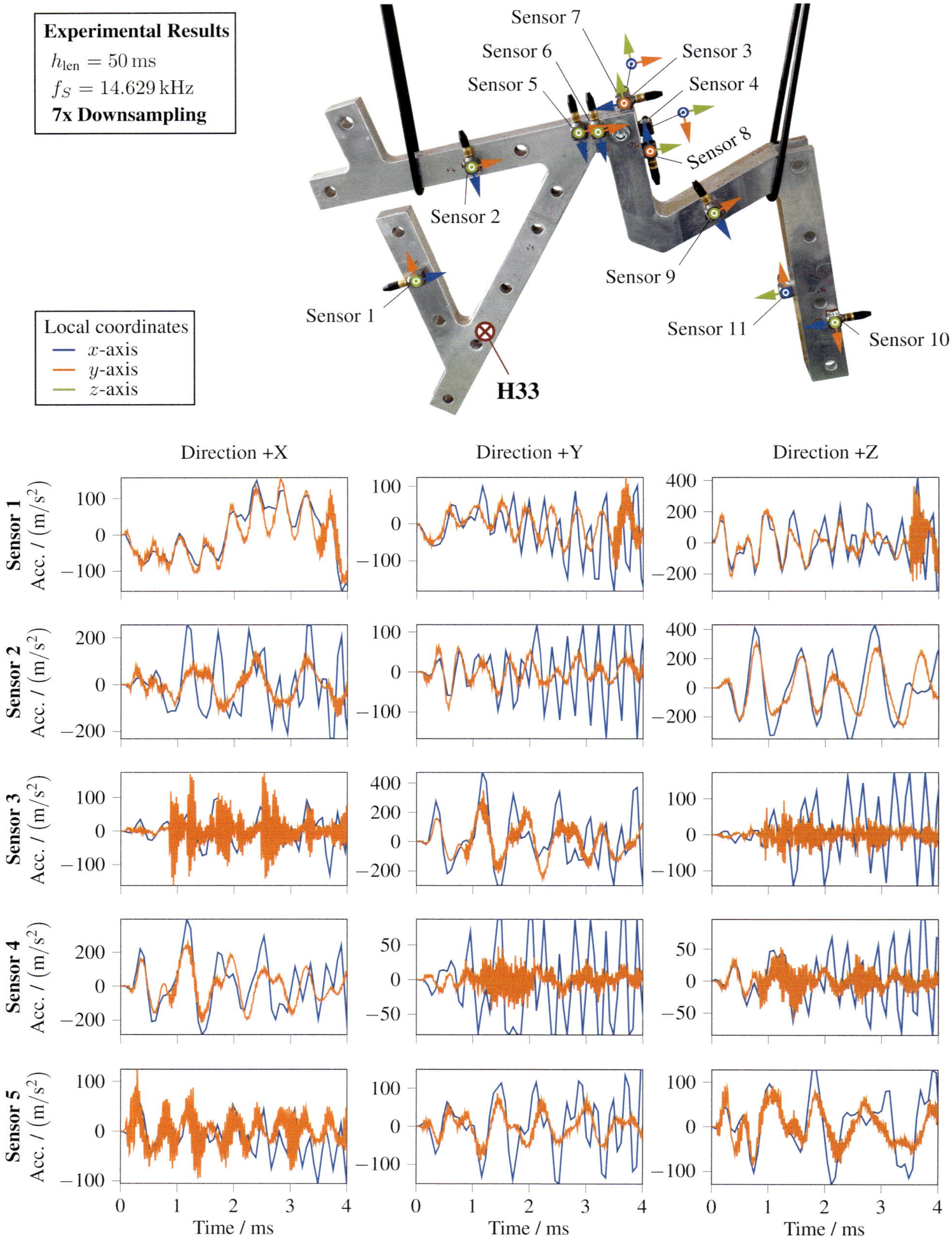


**Fig. 20** Experimental IBS results of **Substructure A** sensor dofs for IBS force applied at **impact location H33**, $f_S = 14.629\,\text{kHz}$ (x7 Downsampling), — IBS Result, — Measurement

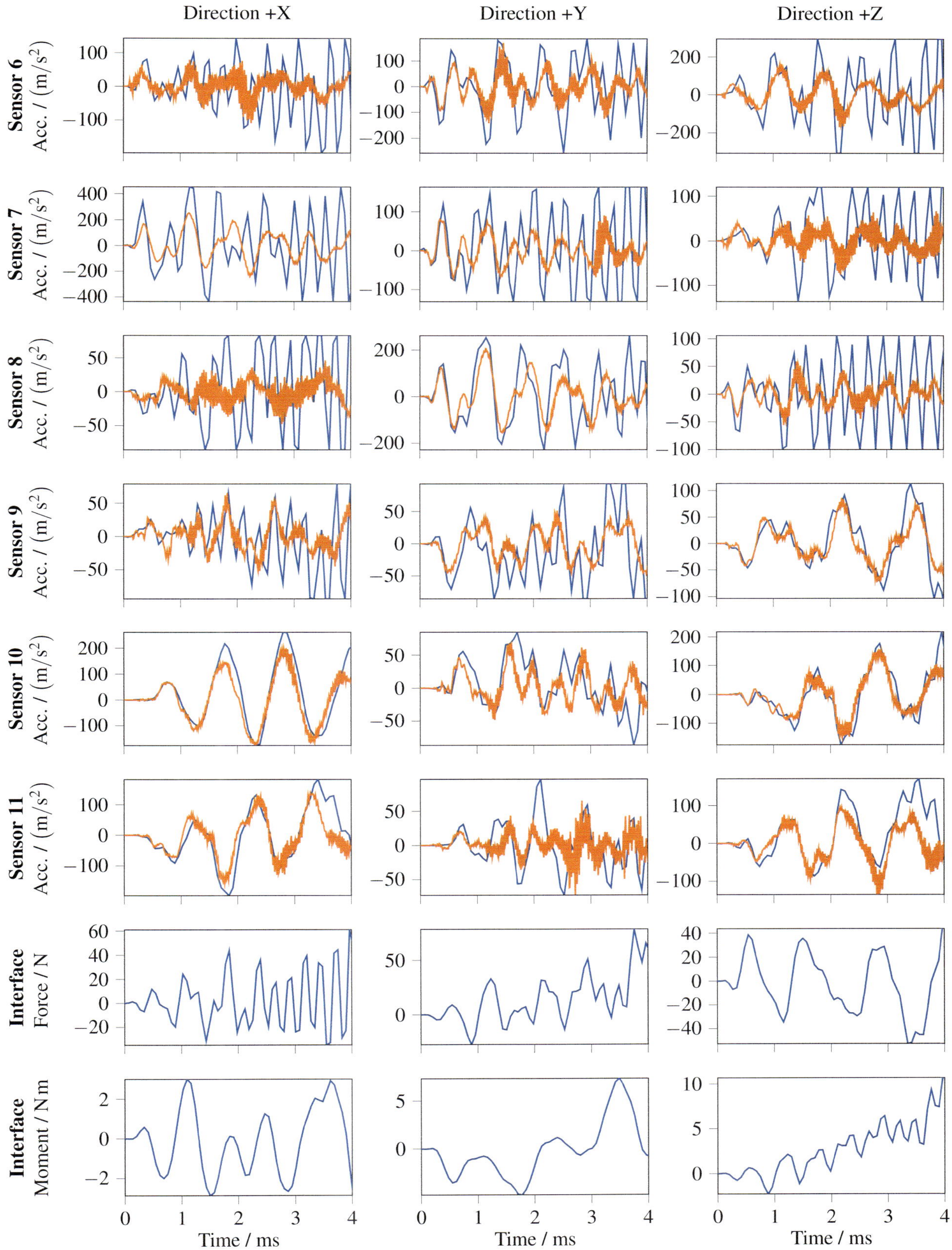


**Fig. 21** Experimental IBS results of **Substructure B** sensor dofs and Lagrange multipliers for IBS force applied at **impact location H33**, $f_S = 14.629\,\text{kHz}$ (x7 Downsampling), — IBS Result, — Measurement

# Chapter 3
# A Look at Overfitting in Source Estimation Problems

**Steven Carter**

**Abstract** Inverse source estimation (ISE) is a fundamental technique that is widely studied in transfer path analysis (TPA) and multiple-input/multiple-output (MIMO) vibration testing. It is commonly accepted that errors in the frequency response functions (FRFs) and target responses can lead to overfitting errors, which can present as over-predictions of the source amplitudes. These overfitting errors are generally attributed to the ISE problem being "ill-posed", but there seems to be limited discussion in the literature about how specific errors (such as modeling errors) lead to overfitting. This work describes a study that was done to gain a deeper understanding of the causes of overfitting by evaluating how the FRF and response errors impact the ISE problem differently. It will evaluate how common techniques that are used to evaluate and mitigate overfitting (such as cross-validation and regularization) are influenced by the different types of error and will share the lessons that were learned over the course of the study.

**Keywords** TPA · MIMO Testing · Force Reconstruction

## Introduction

Inverse source estimation (ISE) is a fundamental technique in noise and vibration engineering that is widely studied in force reconstruction [1], transfer path analysis (TPA) [2], and multiple-input/multiple-output (MIMO) vibration testing [3]. The most common ISE technique uses the frequency response function (FRF) equation of motion, as shown in (1) where $\{X\}$ is a vector of response linear spectra, $[H]$ is a matrix of FRFs, and $\{F\}$ is a vector of source linear spectra (forces for force reconstruction and TPA, shaker drives for MIMO vibration testing).

$$\{X\} = [H]\{F\} \tag{1}$$

The sources are recovered from a set of target responses and FRFs with (2) where $[\bullet]^{\dagger}$ indicates the pseudo-inverse of the FRF matrix.

$$\{F\} = [H]^{\dagger}\{X\} \tag{2}$$

While force reconstruction and TPA commonly utilize the linear spectra formulation, MIMO vibration testing typically uses the cross-power spectral density (CPSD) formulation that is shown in (3) and (4). In these equations $[G_{xx}]$ is a (square) matrix of response CPSDs, $[G_{ff}]$ is a (square) matrix of source CPSDs, and $[\bullet]^{*}$ indicates the conjugate transpose of the given matrix.

$$[G_{xx}] = \{X\}\{X\}^{*} = \{[H]\{X\}\}\{[H]\{X\}\}^{*} = [H][G_{ff}][H]^{*} \tag{3}$$

$$[G_{ff}] = [H]^{\dagger}[G_{xx}][H]^{\dagger^{*}} \tag{4}$$

Source estimation by FRF matrix inversion is an ill-posed process, which can lead to overfitting errors [4, 5], resulting in erroneously high source amplitudes since the inverse problem is fitting errors in the FRFs or responses. There has been significant research into methods that can be used to mitigate the overfitting errors, either through regularization [6–9], transformations [10–12], or thoughtful selection of response degrees of freedom (DOFs) [13, 14]. These methods have been shown to dramatically reduce overfitting, but there has been little discussion (in the ISE literature) on the root causes for the overfitting errors and why the mitigation techniques work. As such, a study was performed to better understand overfitting. This paper details the results of that study, which attempted to address the following questions:

Steven Carter
Sandia National Laboratories, P.O. Box 5800 – MS0557, Albuquerque, NM, 87185
e-mail: spcarte@sandia.gov



Walter D'Ambrogio, et al. (eds.), *Dynamic Substructuring & Transfer Path Analysis, Vol. 4*, Conference Proceedings of the Society for Experimental Mechanics Series, https://doi.org/10.13052/9788743801498-3

- What is the theoretical mechanism for the overfitting error in ISE problems?
- How do the different sources of ISE error (i.e., modeling errors or measurement noise in the target responses) align with the theoretical mechanism for overfitting?
- Does cross-validation work as a method for identifying an overfit solution? This question is being asked since previous experience [4] has shown that cross-validation may be insufficient for identifying overfitting.
- How does regularization improve the source estimation, based on the theoretical mechanism for overfitting?

The remainder of this paper is structured as follows. First, a theoretical model for overfitting will be introduced. Second, the theoretical model for overfitting will be applied to cross-validation and regularization to gain a deeper understanding of the methods. Third, various test cases will be performed on a numerical model to evaluate hypotheses related to the theory of overfitting, how regularization works to mitigate overfitting, and when cross-validation works to identify overfitting. Lastly, there will be conclusions on the results of the study.

## Theoretical Model for Overfitting in Source Estimation

The fundamental sources of error (excluding model form error) in an ISE problem should be determined before a theoretical model for overfitting can be developed. In this work, it is assumed that the potential errors can be modeled as additive errors to the FRFs or target responses, as shown in (5) and (6), where $\varepsilon$ is an error term for the quantity that is described in the superscript. Note that the error term for the FRFs will be referred to as "modeling errors" and the error term for the target responses will be referred to as "response errors" for the remainder of this paper. Also note that these errors could be applied to either the linear or power spectra version of the ISE equations but are only being shown for linear spectra to simplify the algebra.

$$\{X\} = \left[[H] + \left[\varepsilon^{FRF}\right]\right]\{F\} \tag{5}$$

$$\{\{X\} + \{\varepsilon^{response}\}\} = [H]\{F\} \tag{6}$$

One can start to develop an intuitive understanding of the overfitting error if (5) and (6) are solved to estimate the sources, as shown in (7) and (8).

$$\{F\} = \left[[H] + \left[\varepsilon^{FRF}\right]\right]^{\dagger}\{X\} \tag{7}$$

$$\{F\} = [H]^{\dagger}\{\{X\} + \{\varepsilon^{response}\}\} = [H]^{\dagger}\{X\} + [H]^{\dagger}\{\varepsilon^{response}\} \tag{8}$$

A high-level reason for overfitting is easily seen in (8) since the response error term turns into an additive error in the sources. Similar logic for overfitting can be applied when there are modeling errors, as shown in (7). However, the additive error in the FRFs does not become an additive error in the sources since the pseudo-inverse inseparably combines the FRFs with the modeling errors. Further engineering intuition can be developed on overfitting when the inverse problem is viewed through the singular components of the FRF matrix, as shown in (9) and (10) where $[U]$ is a rectangular matrix of left singular vectors, $[S]$ is a diagonal matrix of singular values, and $[V]$ is a square matrix of right singular vectors.

$$[H] = [U][S][V]^{*} \tag{9}$$

$$\{F\} = [V][S]^{\dagger}[U]^{*}\{X\} \tag{10}$$

The first two matrix operations (from the right) in (10), which are isolated in (11), can be viewed as a response transformation [15] that has been scaled by the inverse of the singular values. Response transformations have been widely studied in structural dynamics and there are several theoretical formulations depending on the use case. Examples for the different types of response transformations include the virtual point transformation (VPT) in TPA and substructuring [10, 16, 17], output transformations in MIMO vibration testing [18], and modal filtering [19].

$$\{F\} = [V]\left\{X^{Transformed}\right\} \rightarrow \left\{X^{Transformed}\right\} = [S]^{\dagger}[U]^{*}\{X\} \tag{11}$$

The modal filtering perspective on the transformation seems to be the most appropriate for this case since the left singular vectors of the FRF matrix are combinations of the mode shapes of the system [20]. Consequently, the transformation can be described similarly to a modal filtering operation. In this case, the physical responses are transformed into "singular responses" through a least squares projection onto the left singular vectors of the FRF matrix. The term singular response is being used to describe the transformed response, instead of the term "modal responses" (which is used in modal filtering), because the transformation basis is singular vectors instead of mode shapes.

Recent work in modal filtering [21] has highlighted two main errors that could occur in the transformation when there are response or modeling errors in the ISE problem, which could lead to mismatches between the left singular vectors of the FRF matrix and the deflection shapes in the target responses:

1. It may be impossible for the left singular vectors of the FRF matrix to be combined in a way to fit the deflection shapes in the target responses, eliminating some information from the inverse problem. For example, a test set-up will never be able to excite a particular mode of a system if the excitation locations are at a nodal line of that mode.
2. Alternatively, it may be possible for the left singular vectors of the FRF matrix to be combined in a way to fit the deflection shapes in the target response, even if it is inappropriate to do so. For an example, a complicated set of singular vectors may be able to fit random measurement noise in the target responses even though the measurement noise represents a "non-physical" response that does not match the structural dynamics of the structure.

It is hypothesized that the second transformation error that was described above is the root cause for overfitting in ISE problems because spurious combinations of singular vectors fit the target responses in an erroneous manner, causing the singular responses to have excessive amplitudes. The singular response errors in-turn will cause excessively high source amplitudes. It is important to note that the condition number FRF matrix should also be considered in overfitting, since the amplitude of the sources is directly related to the amplitude of the inverted singular values, as shown in (11). However, the author argues that the root cause for overfitting is the same, regardless of if the overfitting is significant to the ISE problem.

## Cross Validation as a Method to Identify Overfitting

Cross validation (CV) is typically used in ISE to test if a set of sources can recreate the "true" response at all locations on a structure, not just the locations that were used to compute the sources. The typical process for CV in ISE (based on the authors experience) is to split the full set of target response DOFs into a "training set" and "validation set". The training set is used to compute the sources while the validation set is used to assess the predictive capability of the sources. This assessment is done by computing the responses at the validation DOFs with the sources that were estimated with the training dataset. The recomputed and target responses at the validation DOFs are then compared to each another to assess the predictive capability of the estimated sources, which is rated depending on how well the responses match.

The basic process for CV is relatively straight-forward and seems to provide a reasonable check for overfitting at first glance. However, there is an important requirement that the training and validation datasets be independent from one another [22]. From a modal perspective, this requirement means that the training and validation DOF sets should describe unique portions of the deflection shapes.

Essentially, the CV check (as described above) is checking to see if the estimated sources can reconstruct the deflection shapes that are described by both the training and validation DOFs sets. Note that this check is different than validating that the estimated sources can reconstruct the correct deflection shape all over the structure, for two reasons. First, significant portions of the overall deflection shapes of the structure may be neglected, if the training and validation DOF sets are poorly chosen (e.g., a critical component is left out). Second, the validation DOF set may not describe a unique portion of the deflection shapes of the structure, compared to the training DOF set, meaning that the CV check would not provide new information. Consequently, it is hypothesized that the CV check may not be sensitive to overfitting unless the training and validation DOFs are carefully chosen. Further, it is hypothesized that it may be difficult to obtain a meaningful CV check in cases where there are several target response DOFs that provide a good description of the deflection shapes of the structure.

It is important to note that the above hypothesis about CV should *not* be interpreted to say that CV checks are not useful. Rather, it is pointing to the fact that practitioners should carefully design validation experiments, so they are significant for the problem at-hand. For example, a CV check could be used in a way to validates that the sources are sufficient to recreate the responses at any point, on one version of a structure (i.e., the goal in a MIMO test). However, this same CV check may not be able to validate the sources are generic to any version of a structure (i.e., the goal in component based TPA), meaning that a different type of check (that spans multiple versions of the structure) should be used.

## Regularization as a Method to Mitigate overfitting

Regularization is commonly used to mitigate overfitting in ISE problems [6-9]. The most common regularization methods for ISE are the truncated singular value decomposition (TSVD) and Tikhonov regularization. The equations for these methods are shown in (12) and (13), where $N$ is a tunable parameter for the number of singular values to retain in the TSVD (the remaining values are set to zero) and $\lambda$ is a tunable regularization parameter for Tikhonov regularization.

$$\{F\} = [V]\,[S\,(1:N)]^{\dagger}[U]^{*}\,\{X\} \tag{12}$$

$$\{F\} = [V]\,\frac{[S]}{[S]^{*}\,[S]+\lambda}[U]^{*}\,\{X\} \tag{13}$$

It is useful to partition the regularized ISE equations in the same manner as (11), as shown in (14) and (15) for the TSVD and Tikhonov regularization, respectively.

$$\{F\} = [V]\,\{X^{Transformed}\} \rightarrow \{X^{Transformed}\} = [S\,(1:N)]^{\dagger}[U]^{*}\,\{X\} \tag{14}$$

$$\{F\} = [V]\,\{X^{Transformed}\} \rightarrow \{X^{Transformed}\} = \frac{[S]}{[S]^{*}\,[S]+\lambda}[U]^{*}\,\{X\} \tag{15}$$

Both (14) and (15) show that both the TSVD and Tikhonov regularization tend to reduce the amplitude of the estimated sources, which is an effect that is known as shrinkage in the statistical literature [22]. The TSVD reduces the amplitudes of the estimated sources by eliminating the small singular values of the FRF matrix, which would become large when inverted. Tikhonov regularization works similarly to the TSVD, but it increases the amplitude of the small singular values (reducing their inverted value), rather than eliminating them (and the associated singular vectors) from the inverse problem.

Comparing the effect of the TSVD and Tikhonov regularization to the hypothesis for overfitting, it is evident that both methods assume that the small singular values (and associated singular vectors) are the primary cause of overfitting. Put more plainly, both methods assume that the significant left singular vectors (associated with the large singular values) in the FRFs are a good match to the primary deflection shapes in the target responses. However, there may be mismatches between the minor deflection shapes in the FRFs and target responses, which are associated with the small singular values of the FRF matrix. As previously described, the mismatches could lead to spurious singular responses, so the singular responses that are associated with the small singular values should be eliminated or made less significant to the ISE problem, which is how the TSVD and Tikhonov regularization mitigate overfitting.

Other forms of regularization, such as the least absolute shrinkage and selection operator (LASSO) and the elastic net [9, 23, 24], also exist for ISE problems. These methods perform both a shrinkage function (similar to the TSVD and Tikhonov regularization) and a subset selection function [22, 25], where source DOFs may be automatically set to zero if they are considered mathematically insignificant to the ISE problem. These methods are mentioned here for completeness but are considered outside of the scope of this work since they are not commonly used ISE and are significantly more mathematically complex than the TSVD and Tikhonov regularization.

## Example Problem for the Overfitting Study

The previously mentioned hypotheses were tested on numerical models of the round robin system for the substructuring technical division of the Society of Experimental Mechanics [26, 27]. An image of the reduced modal model of the system is shown in Figure 1. Two versions of the model were used in this study, a calibrated finite element model of the system and an experimental modal model of the system, where the experimental modal model was used for the finite element (FE) model calibration. Both of these models were developed in the work described in [5]. These models were used for the overfitting study because they are reasonably similar, allowing for a version of the model that represents the truth and a version of the model with minor modeling errors. Note that the FE model was used as the truth model in this case, since it could allow for more in-depth analysis than the experimental modal model, if necessary. The FE model will be referred to as the "truth model" and the experimental modal model will be referred to as the "test model" for the rest of this paper.

The differences between the truth and test models are shown via a modal assurance criterion (MAC) comparison (for the elastic modes) in Figure 2 and a frequency difference comparison between high MAC modes (MAC values greater than 0.9) is shown in Figure 3. The truth and test shapes generally have high MAC values, with a single match for most mode except for mode three in the truth model, which matches two test modes. Lower MAC values are more common at higher frequencies, with the lowest MAC value being ~0.38. The frequency difference comparison shows that the high MAC modes (with a MAC value greater than 0.9) have a good frequency match, with the highest difference being ~3%. Note that the modal damping ratio for all modes in both the truth and test models was set to 0.005, which is why only the shapes and frequencies were compared to show the similarities and differences in the models.

Three sets of triaxial response DOFs were selected on the wing to use as the target responses in the overfitting study. The reason for having three sets of response DOFs was to have two DOF sets that could be used as training sets in the ISE and one

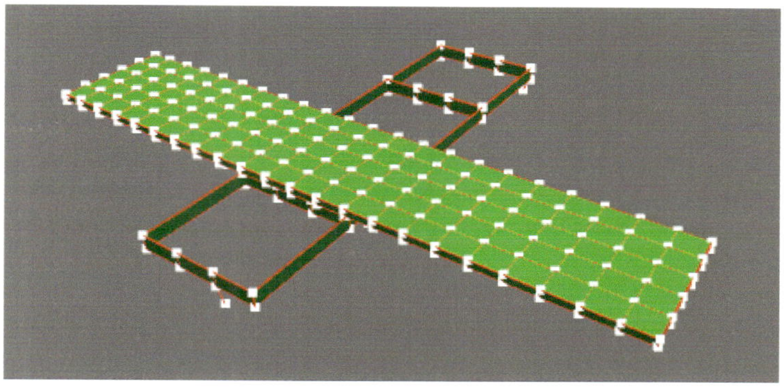

**Fig. 1** The reduced modal model of the round robin system that was used to test the hypotheses in the overfitting study.

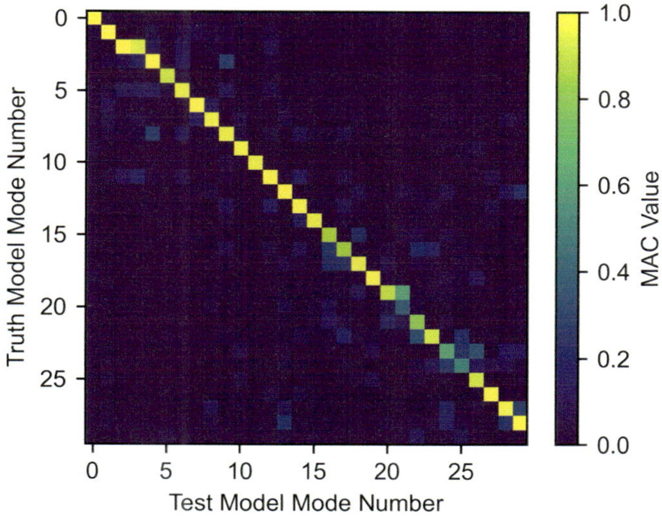


**Fig. 2** MAC comparison between the modes in the truth and test models for the overfitting study.

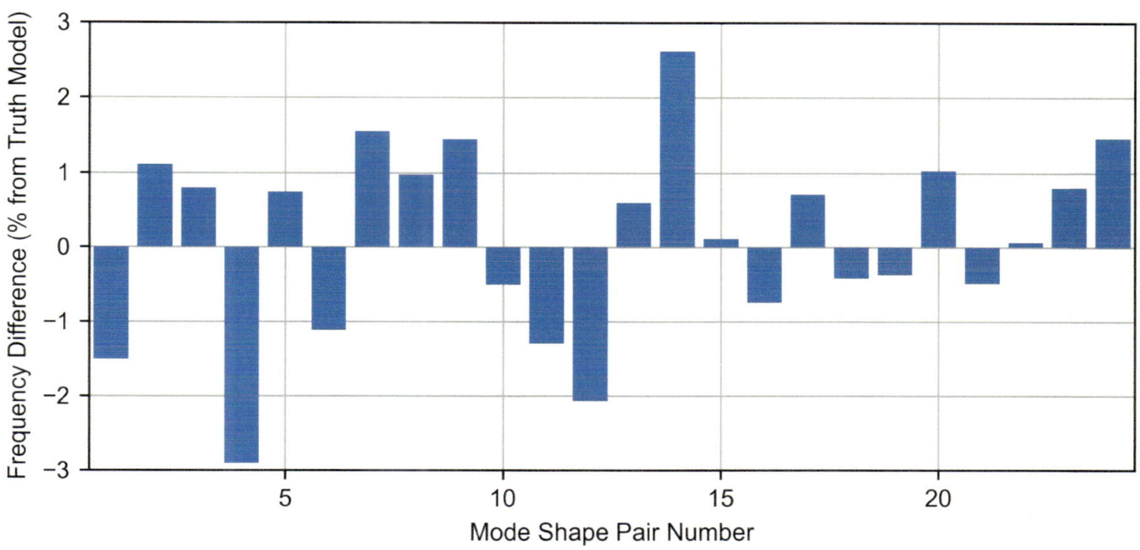


**Fig. 3** Frequency difference comparison between high MAC modes in the truth and test models for the overfitting study.

validation DOF set that could be used for a CV check. One training set contained "good" response DOFs that describe the response deflection shapes well, and the other training set contained "bad" response DOFs that poorly describe the response deflection shapes. Consequently, it is assumed that the validation DOF set will *not* add independent information to the ISE problem when the good DOFs are used in the training set (since the deflection shapes are well defined by the good DOFs) but will add independent information when the bad DOFs are used in the training set (since the bad DOFs poorly describe the deflection shapes). As a result, the different DOF sets make it possible to test the previously described hypothesis on CV checks, depending on if the validation DOFs can identify overfitting when the sources are estimated with responses at the good and bad DOFs (where comparisons between the estimated and truth sources are the ultimate check).

The DOF sets were objectively selected with an effective independence (EFI) [28] algorithm to eliminate engineering bias from the results of the study. Both the good and bad training DOF sets used thirty DOFs at ten triaxial response locations and the validation DOF set used twelve DOFs at four triaxial response locations. Note that the same validation DOF set was used regardless of the training DOF set. The good DOF set was selected with the typical EFI process and metric for triaxial accelerometers, the selected (good) locations are shown in Figure 4. The bad DOF set also used the typical EFI process for triaxial accelerometers, but the EFI metric was inverted so the worst sensor locations were chosen, the selected (bad) locations are shown in Figure 5. The validation DOF set was also selected with the standard EFI process and metric (for triaxial accelerometers), but the algorithm was initialized with the good and bad DOF sets, the selected (validation) locations are shown in Figure 6.

The "truth" target responses for the overfitting study were computed by applying broadband random excitation to triaxial input DOFs at the corners of the frame (twelve DOFs total) in the truth model. The input DOFs are shown in Figure 7.

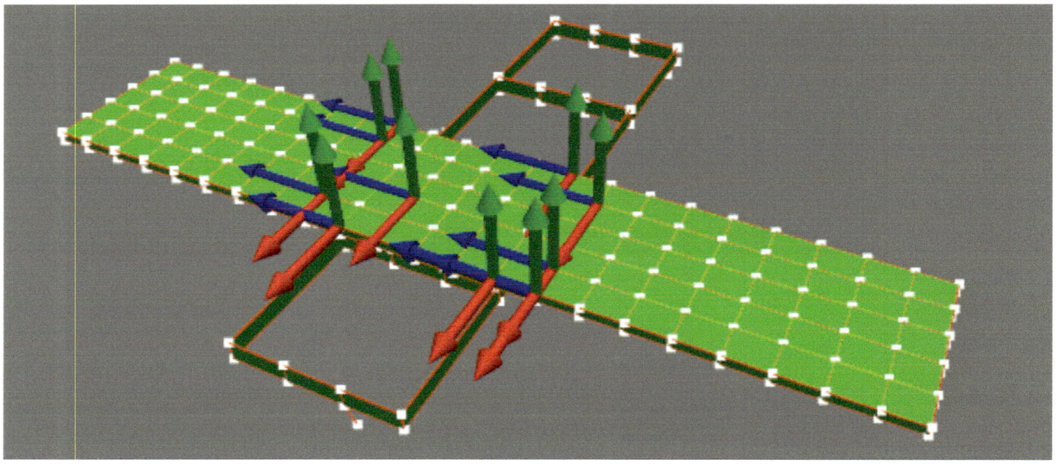

**Fig. 4** The good set of response DOFs for the overfitting study.

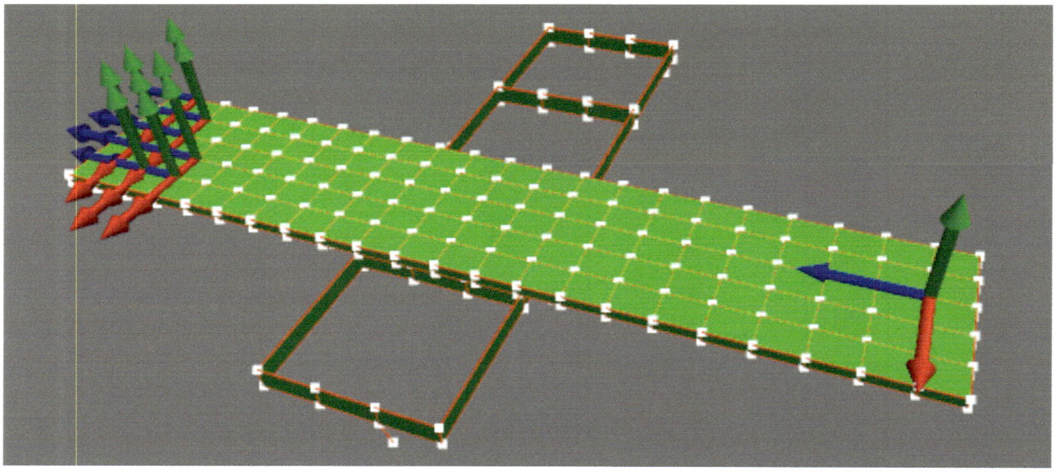

**Fig. 5** The bad set of response DOFs for the overfitting study.

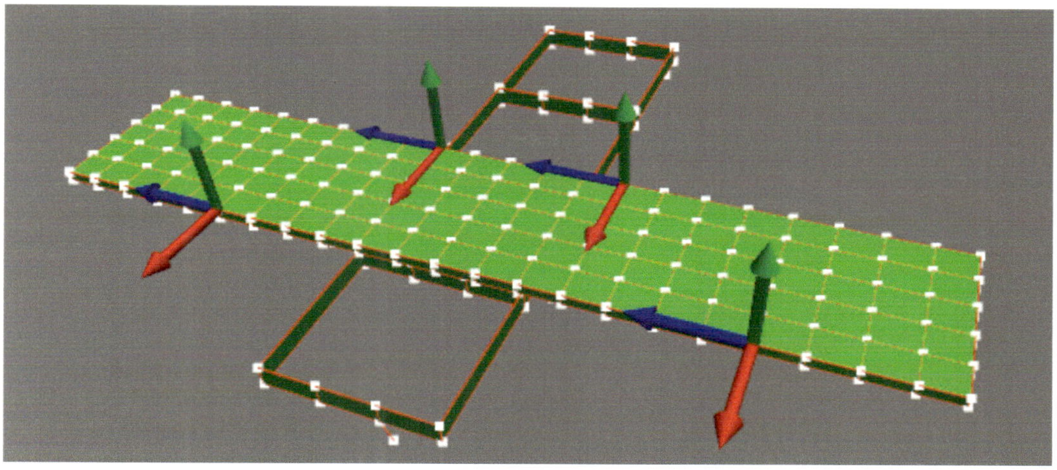

**Fig. 6** The validation set of response DOFs for the overfitting study.

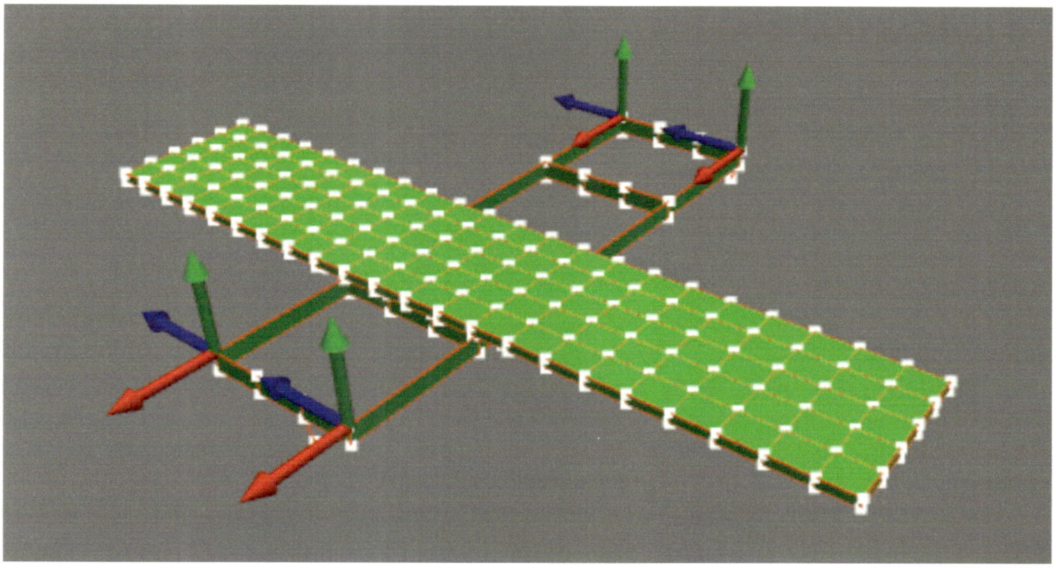

**Fig. 7** The input DOFs for the overfitting study.

The spectral responses and inputs were computed as phase referenced spectra, using a force DOF as the reference and the signal processing parameters that are shown in Table 1. The responses with errors, which will be referred to as the "noised responses", were computed by adding broadband random noise to the truth response spectra, prior to phase referencing and averaging. Note that the signal-to-noise ratio (SNR) for the noised responses was tuned so an ISE with modeling errors or response errors would result in similar levels of overfitting, as determined by the amplitude errors in the estimated sources, while still having a typical SNR for vibration measurements. The DOF averaged SNR and summed (truth and noised) response power spectral density (PSD) for all the response DOFs in the overfitting study are shown in Figure 8. This plot shows that the noised response has a good signal to noise ratio and that the noise is not dramatically contaminating the truth responses.

The models and synthesized response data for the overfitting study were combined three different ways to match the different types of error that could contaminate an ISE problem. All the model/data combinations were used with both the good and bad training response DOFs, depending on the hypothesis that was being tested, and will be referred to accordingly. The model combinations were:

1. The truth model and truth response data were combined to create the baseline ISE problem
2. The truth model and noised response data were combined to create the response error ISE problem
3. The test model and truth response data were combined to create the modeling error ISE problem

**Table 1** Signal processing parameters for spectra computation.

| Parameter | Setting |
|---|---|
| Sampling Frequency | 4,000 Hz |
| Frequency Resolution | 1 Hz |
| Number of Averages | 100 |
| Window Type | Hanning |

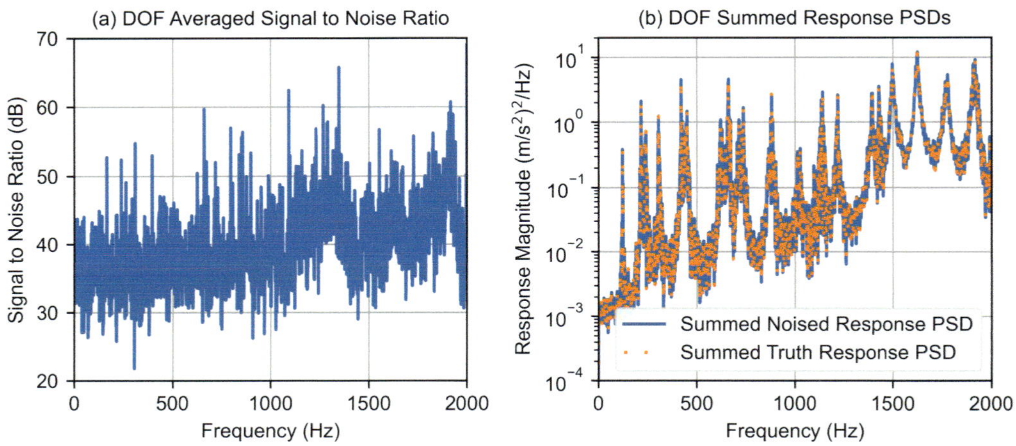


**Fig. 8** DOF averaged signal to noise ratio for summed (truth and noised) response for all the response DOFs in the overfitting study.

The objective of the ISE problem in this example was to recover the "truth" sources that were used to compute truth responses at the input DOFs that are shown in Figure 7. Since the truth sources are known in this case, a comparison between the summed truth and estimated source PSDs can be used as a simple metric to show how significant the overfitting is for the different model/data combinations. The comparison between the truth and estimated source PSDs for both the good and bad training DOF sets is shown in Figure 9 for the baseline ISE problem, Figure 10 for the response error ISE problem, and Figure 11 for the modeling error ISE problem. These plots show that either training set can accurately recover the sources when there are not response or modeling errors in the ISE problem, but that there are significant errors when

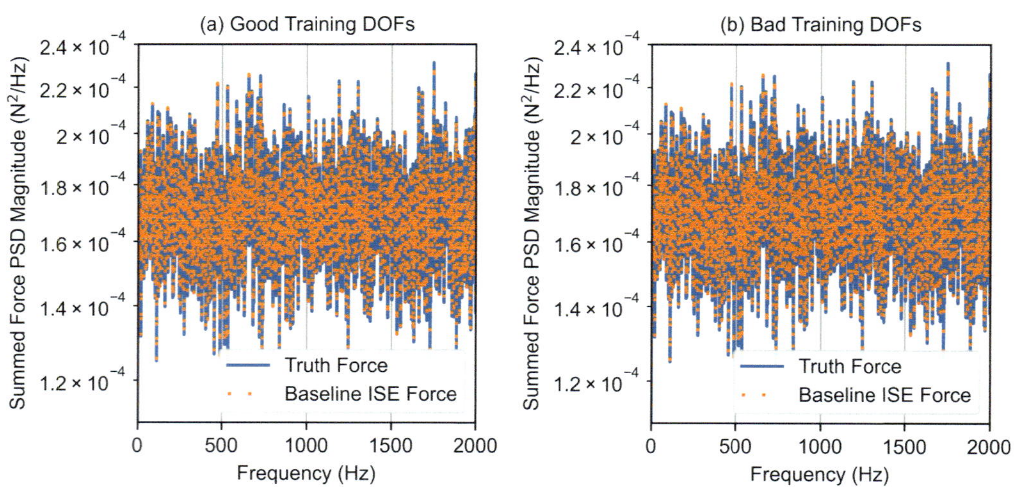


**Fig. 9** Summed truth and estimated source PSDs for the baseline ISE problem for the different training DOF sets.

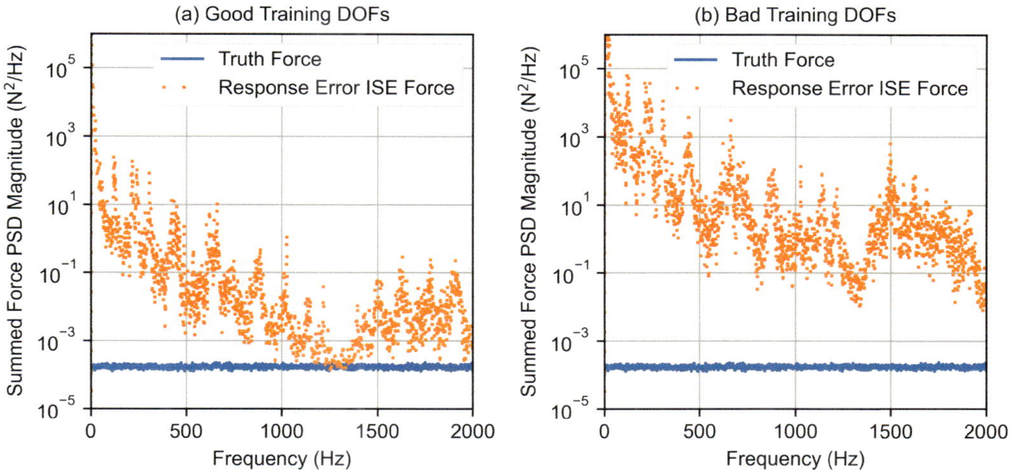


**Fig. 10** Summed truth and estimated source PSDs for the response error ISE problem for the different training DOF sets.

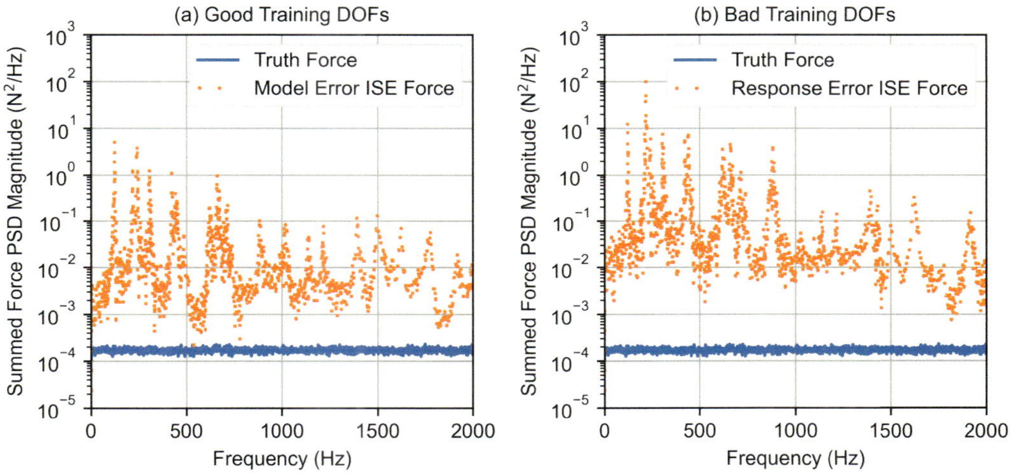


**Fig. 11** Summed truth and estimated source PSDs for the modeling error ISE problem for the different training DOF sets.

there are response or modeling errors. Further, these plots show that the bad training DOFs result in worse source estimation errors than the good training DOFs (as expected). Lastly, the overfitting is slightly more significant when there are response errors compared to modeling errors. However, the reason for this increased sensitivity is unclear and was not explored in this work.

## Testing The Hypothesis For Overfitting

The hypothesis for the root cause of overfitting was tested by comparing the singular responses (computed with the transformed response that is shown in (11)) from the ISE problems with response and modeling errors to the singular responses from the baseline ISE problem, which represents the truth singular responses. These comparisons can be seen in Figure 12 and Figure 13 for the ISE problem with response and modeling errors, respectively. The same comparisons were made for the singular responses that were not scaled by the inverted singular values (by removing $[S]^{\dagger}$ from the equation), as shown in Figure 14 and Figure 15 for the ISE problem with response and modeling errors, respectively. Note that these comparisons were made with the singular responses that were computed with the good training DOFs only.

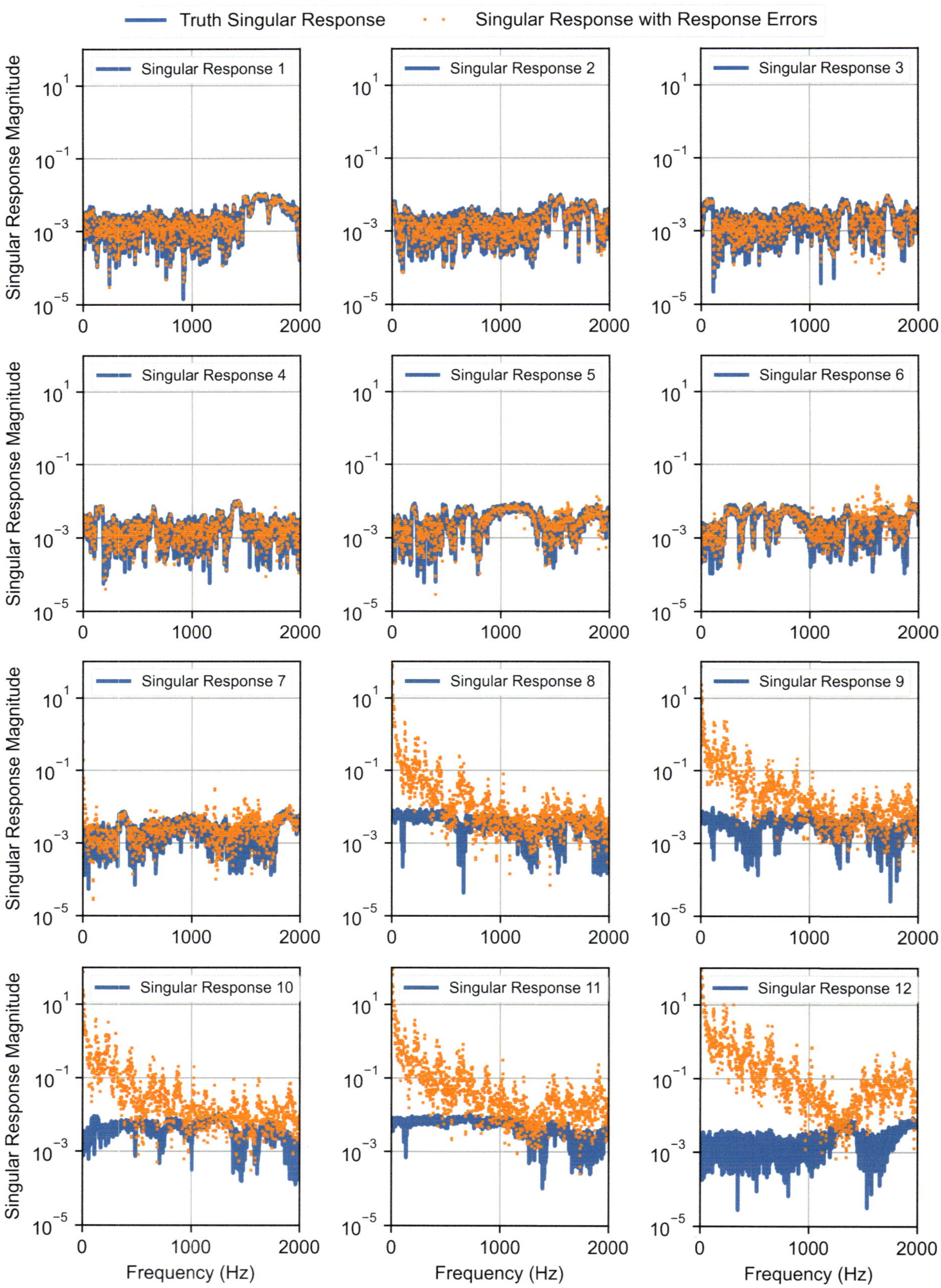


**Fig. 12** The singular responses (scaled the inverted singular values) for the response error and baseline ISE problems.

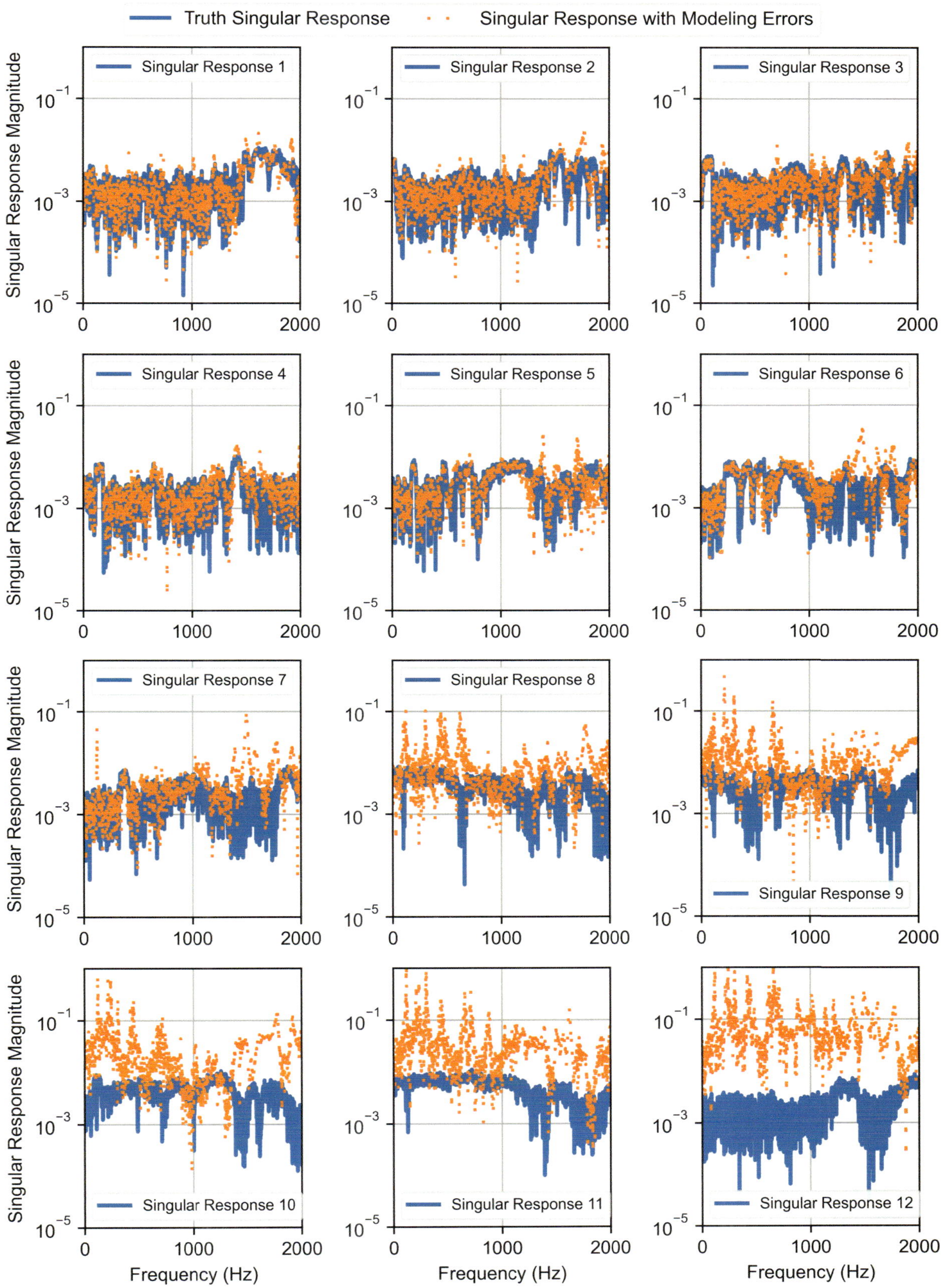


**Fig. 13** The singular responses (scaled the inverted singular values) for the modeling error and baseline ISE problems.

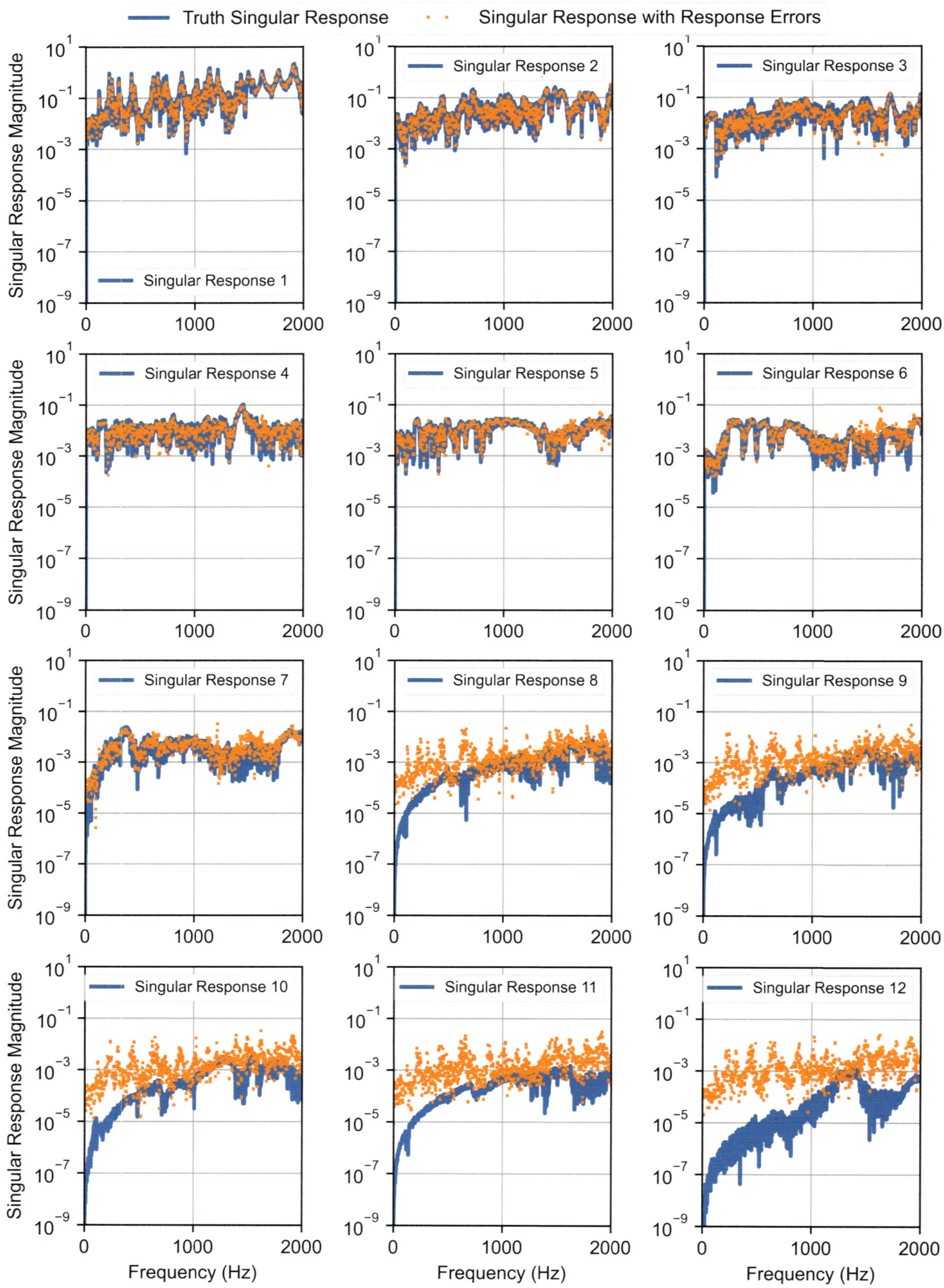


**Fig. 14** The singular responses (not scaled by the inverted singular values) for the response error and baseline ISE problems.

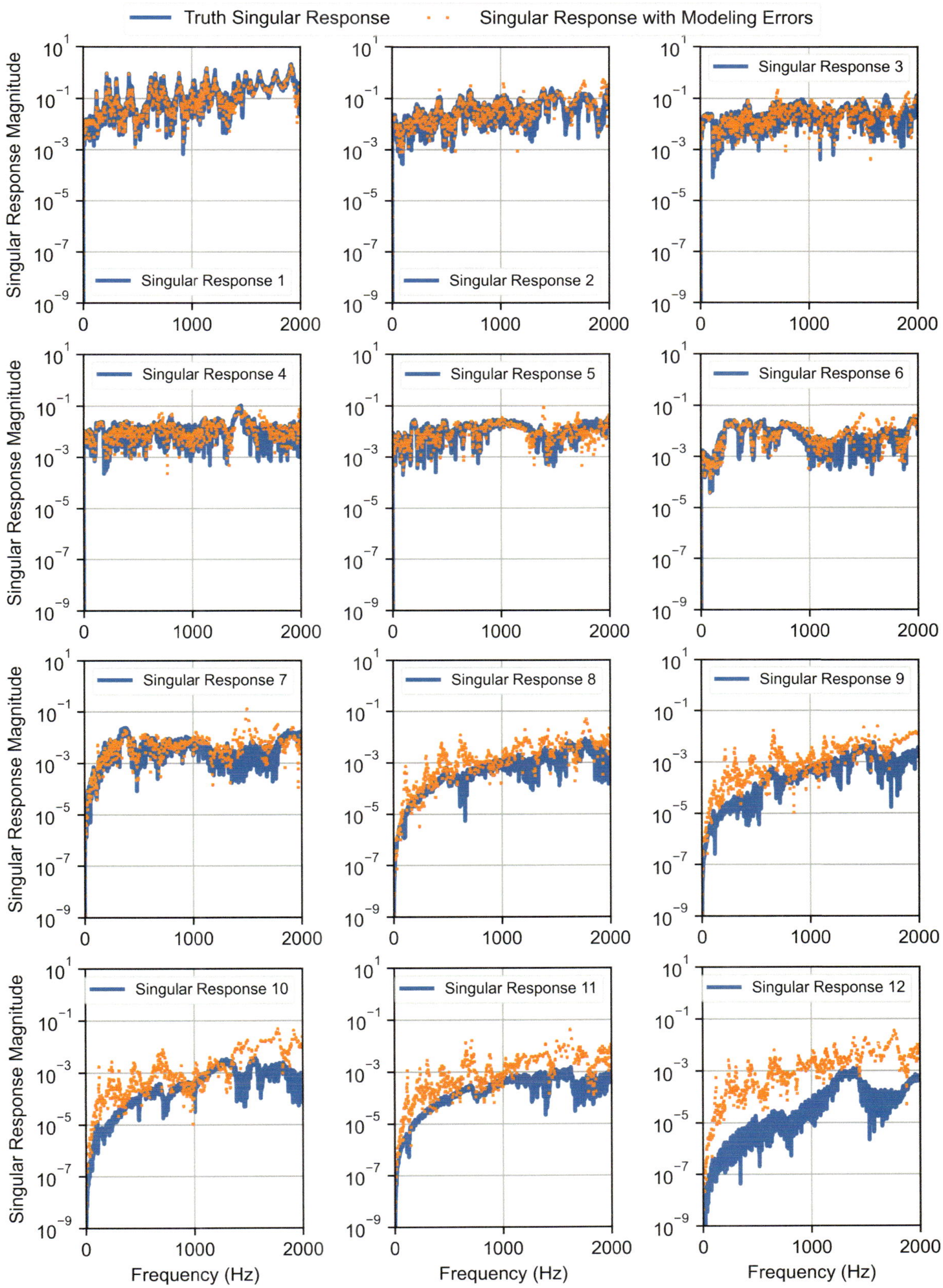


**Fig. 15** The singular responses (not scaled by the inverted singular values) for the modeling error and baseline ISE problems.

The singular responses are significantly overpredicted for both the response and modeling error ISE problems, regardless of if singular responses scaled by the inverted singular value. However, it can clearly be seen that scaling the singular responses by the inverted singular values increases the significance of the singular response errors, given on the relative amplitude of the errors compared to the amplitude of the truth singular responses. These comparisons seem to confirm the hypothesis that overfitting is caused by transformation errors when the responses are projected onto the left singular vectors of the FRF matrix, and that the significance of the overfitting is controlled by the relative amplitudes of the singular values of the FRF matrix (i.e., the condition number). Interestingly, the singular response errors seem to result in the same spectrum shape as the errors in the sources that were shown Figure 10 and Figure 11, providing further evidence that the transformation errors are likely the primary cause of overfitting in ISE.

The most significant singular response errors are seen in the singular responses that are associated with the small singular values (9-12 in this case) for both the response and modeling error ISE cases. This finding seems confirm the assumption for how the TSVD and Tikhonov regularization mitigate overfitting. However, it is interesting to note that moderate singular response errors are seen in the singular responses that are associated with the large singular values, when there are modeling errors in the ISE problem. But not when there are response errors. This makes intuitive sense because the modeling errors will likely contaminate all the singular vectors of the FRF matrix, since there are fundamental errors in the mode shapes that are described by the FRFs. On the other hand, the response errors in this example contaminated the response deflection shapes with random noise, which likely can't be fit by simplistic deflection shapes in the FRFs. As such, the response errors likely map to the complicated singular vectors in the FRF matrix, which are associated with the small singular values.

## Evaluating How Regularization Mitigates Overfitting

The Tikhonov regularized singular responses, as formulated in (15), were computed to demonstrate how regularization can be used to mitigate overfitting. In this case, the regularization parameter was automatically selected using the L-curve method (on a frequency line by frequency line basis) that is described in [9, 29]. Note that only the good training DOFs were used here, for brevity. The regularized source estimates were compared to the truth sources to evaluate how well the regularization resolved the overfitting issue as shown in Figure 16 and Figure 17 for the response and modeling error ISE problems, respectively. The regularization dramatically reduces the overfitting issue and actually underestimates the sources. It is unclear if further tuning could increase accuracy of the source estimation, especially if the truth forces are unknown. Regardless, this example seemed sufficient to demonstrate the effects of regularization and further tuning was not attempted since it was outside the scope of this work.

The comparisons between the truth and regularized singular responses (scaled by the inverted singular values) are shown in Figure 18 for the response error ISE problem and Figure 19 for the modeling error ISE problem. These comparisons clearly demonstrate that Tikhonov regularization is mitigating the errors in the singular responses that are associated with the small singular values, as expected. It is interesting to note that while the singular responses are generally underpredicted,

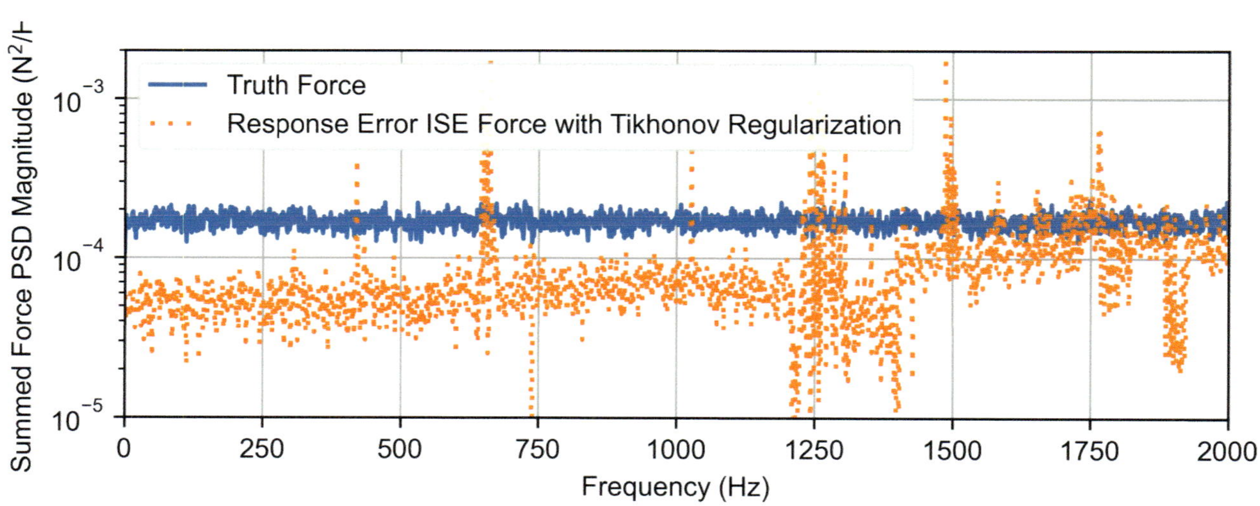


**Fig. 16** Summed truth and estimated source PSDs for the response error ISE problem with the good training DOFs.

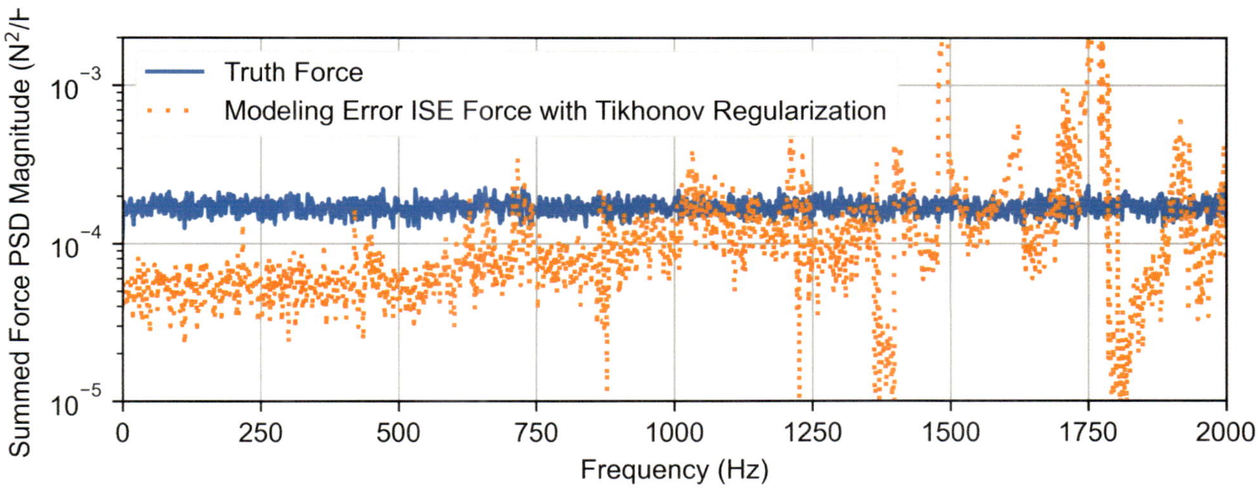


**Fig. 17** Summed truth and estimated source PSDs for the modeling error ISE problem with the good training DOFs.

the errors in the singular responses do not seem to have a similar spectrum shape to the errors that are seen in the estimated sources, which contradicts the finding from the overfit case that was shown in Figure 12 and Figure 13. However, it is unclear what (if anything) this contradiction means for the hypothesis on the cause of overfitting, given the extreme flexibility in selecting the regularization parameter.

## Testing the CV Hypothesis

The efficacy of CV checks was evaluated using the different DOF sets that were described above. First, the sources were estimated with both the good and bad training DOFs and then the estimated sources were then used to reconstruct the responses at the training and validation DOFs. The reconstructed responses (for both the training and validation DOFs) were compared to the target responses to see if the responses at either the training or validation DOFs could predict overfitting. For brevity, the average dB error for all the DOFs in either the training or validation sets was used as metric for the error in the reconstructed responses. The formula for this error metric is shown in (16), where $\{X\}$ is a vector of response PSDs for either the reconstructed or target responses, depending on the superscript, and $n$ is the number of response DOFs in the given set.

$$\varepsilon^{reconstructed} = 10 * \log_{10}\left(\frac{\sum\left(\{X^{reconstructed}\} - \{X^{target}\}\right)}{n}\right) \tag{16}$$

The reconstructed response error for the training and validation DOFs when the sources are estimated with both the good and bad training DOFs is shown in Figure 20 and Figure 21 for the ISE problems with response and modeling errors, respectively. Both these plots show that the validation and training DOFs show similar levels of reconstructed response error when the good training DOFs are used to estimate the sources. Conversely, the validation DOFs show significantly higher reconstructed response error than training DOFs when the bad training DOFs are used to estimate the sources. This seems to confirm the previously described hypothesis that CV checks may be insensitive to overfitting errors if the training DOFs describe the response deflection shapes well, but further work should be done to see if this is a general finding.

It should be noted that the CV check seems to be more sensitive to the overfitting error when there are responses errors in the ISE problem vs. when there are modeling errors in the ISE problem. This could be due to the fact that the responses errors were essentially random perturbations of the response deflection shapes, which could accentuate the differences in the deflection shapes that are described by the training and validation DOFs (increasing the sensitivity of the CV check). However, the increased sensitivity in the CV check could simply be due to the fact that the overfitting is much worse in the response error ISE problem with the bad training DOFs than the modeling error ISE problem with the bad training DOFs, as evidenced source estimation errors in Figure 10 and Figure 11. As such, further study should be done to understand how CV checks apply to overfitting.

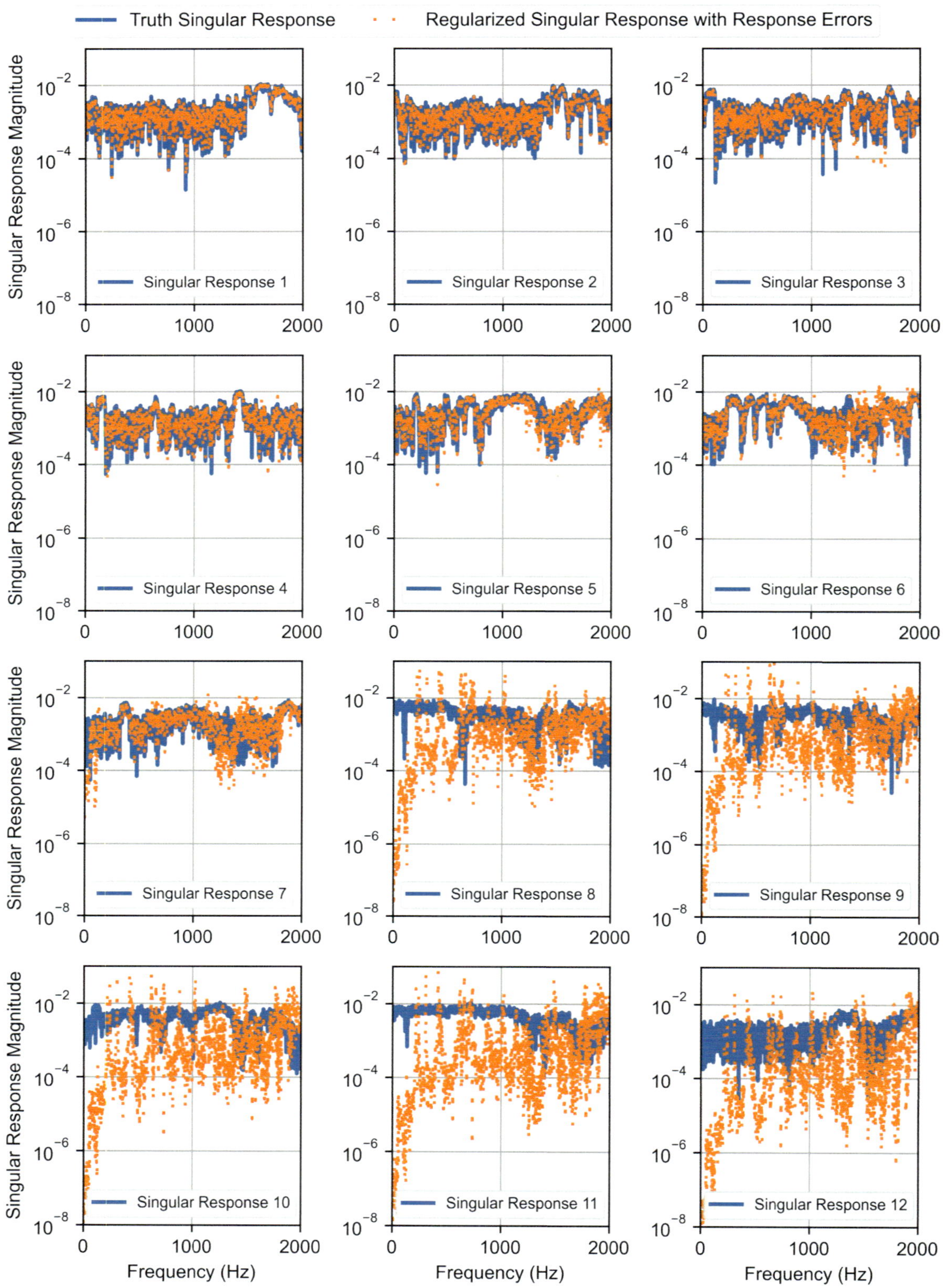


**Fig. 18** Regularized singular responses for the response error ISE problem compared to the unregularized baseline ISE problem.

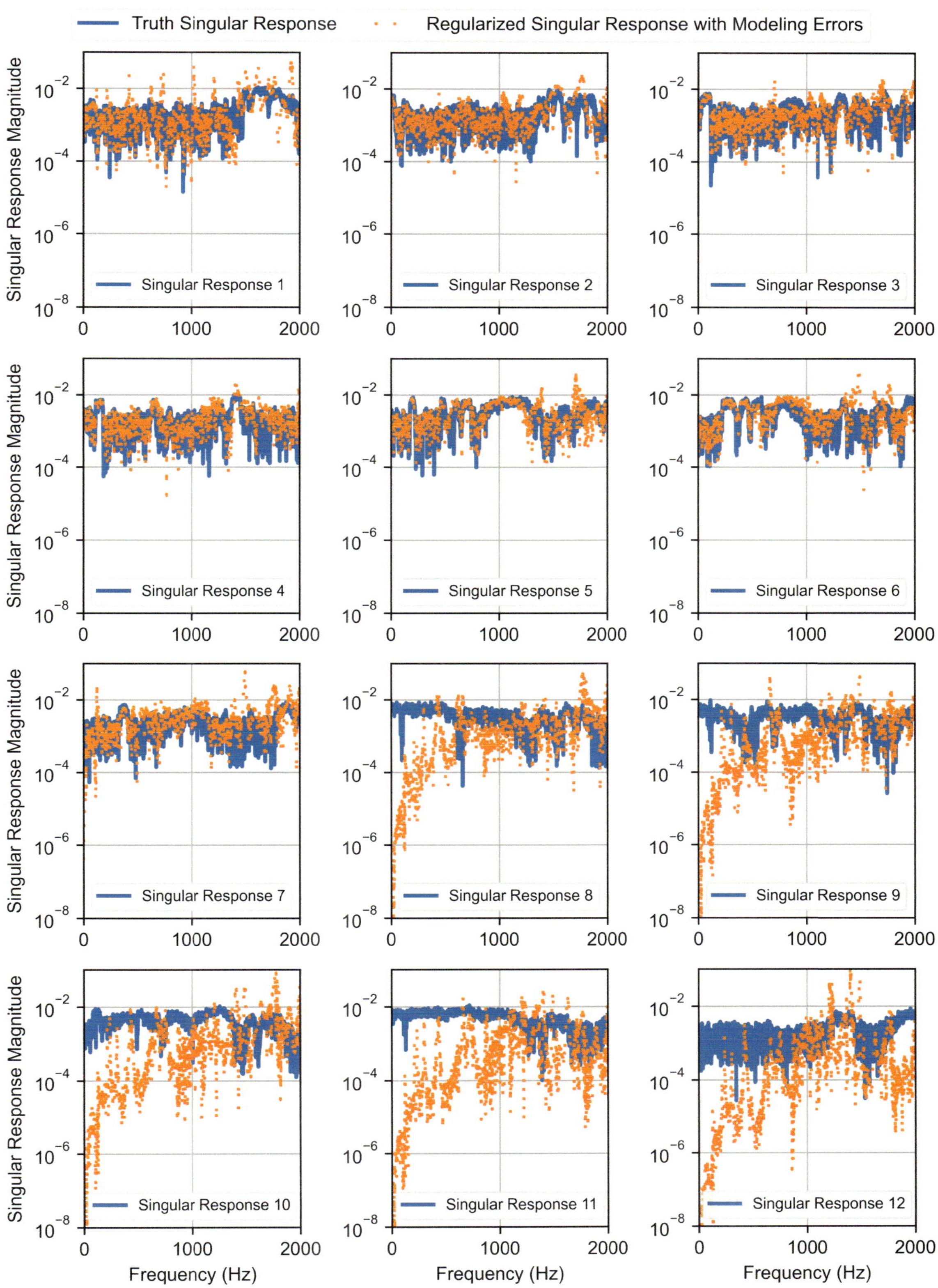


**Fig. 19** Regularized singular responses for the modeling error ISE problem compared to the unregularized baseline ISE problem.

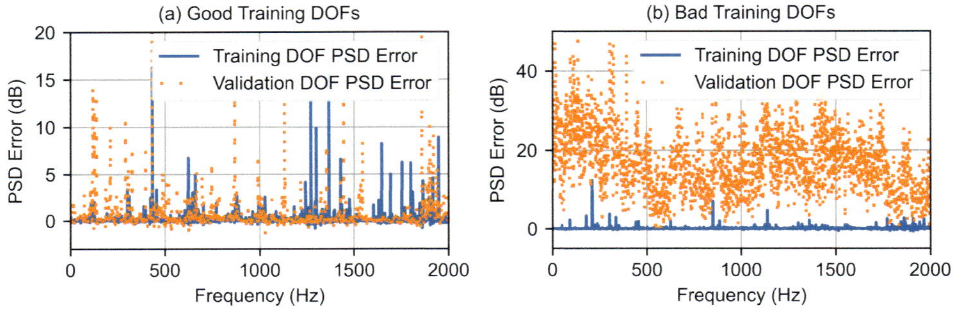


**Fig. 20** Reconstructed response error for the training and validation DOFs in the response error ISE problem from sources that were estimated with both the good and bad training DOFs.

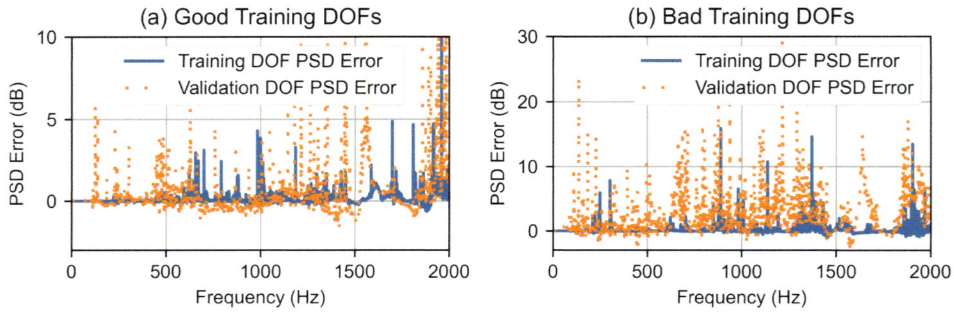


**Fig. 21** Reconstructed response error for the training and validation DOFs in the modeling error ISE problem from sources that were estimated with both the good and bad training DOFs.

## Conclusion

This paper developed a theoretical understanding for overfitting in ISE, which led to a hypothesis that overfitting is fundamentally caused by transformation errors when the target response is projected onto the left singular vectors of the FRF matrix, where the relative amplitudes of the singular values (i.e., the condition number) determine the significance of the overfitting problem. This hypothesis was tested and seemingly confirmed on two different ISE problems, where there were errors in the target responses and errors in the FRFs. The overfitting hypothesis was also applied to regularization and CV checks to better understand how those methods can be used to mitigate and identify overfitting errors. It was shown that CV checks may be insensitive to overfitting if high quality response DOFs are used in the training set for the ISE, since the validation set may not add any information to the ISE problem.

The concepts in this paper were described in a qualitative manner and future work should attempt to describe overfitting in a mathematically rigorous fashion. Future work should also include the development of a framework for overfitting in ISE problems so the different error, mitigation methods, and data analysis methods can be discussed using common terminology. Further work should also be done to understand the efficacy of CV checks in ISE so better methods for identifying overfitting errors can be developed. Lastly, work should be done to see if other ISE techniques, such as the sum of weighted acceleration technique (SWAT) or Kalman filtering, that don't rely on FRF matrix inversion have a similar root cause for overfitting.

**Acknowledgments** *This article has been authored by an employee of National Technology & Engineering Solutions of Sandia, LLC under Contract No. DE-NA0003525 with the U.S. Department of Energy (DOE). The employee owns all right, title and interest in and to the article and is solely responsible for its contents. The United States Government retains and the publisher, by accepting the article for publication, acknowledges that the United States Government retains a non-exclusive, paid-up, irrevocable, world-wide license to publish or reproduce the published form of this article or allow others to do so, for United States Government purposes. The DOE will provide public access to these results of federally sponsored research in accordance with the DOE Public Access Plan* https://www.energy.gov/downloads/doe-public-access-plan.
*This paper describes objective technical results and analysis. Any subjective views or opinions that might be expressed in the paper do not necessarily represent the views of the U.S. Department of Energy or the United States Government.*

## References

1. J. Sanchez and H. Benaroya, "Review of force reconstruction techniques," Journal of Sound and Vibration, vol. 333, no. 14, pp. 2999–3018, 2014/07/07/ 2014, doi:https://doi.org/10.1016/j.jsv.2014.02.025.
2. M. V. van der Seijs, D. de Klerk, and D. J. Rixen, "General framework for transfer path analysis: History, theory and classification of techniques," Mechanical Systems and Signal Processing, vol. 68-69, pp. 217–244, 2016/02/01/2016, doi:https://doi.org/10.1016/j.ymssp.2015.08.004.
3. P. Daborn, "Smarter dynamic testing of critical structures," PhD dissertation, Aerospace Department, University of Bristol, 2014.
4. S. Carter and B. Owens, "Errors using six degree-of-freedom force estimates to translate environments system-to- system," in ISMA, Leuven, BE, 2022.
5. S. Carter and B. Owens, "Using Component Based TPA to Translate Vibration Environments Between Versions of the Round Robin Structure with FRFs Derived from Analytical Models," in Proceedings of IMAC XLI, the 41st International Modal Analysis Conference, Austin, 2023.
6. A. N. Thite and D. J. Thompson, "The quantification of structure-borne transmission paths by inverse methods. Part 1: Improved singular value rejection methods," Journal of Sound and Vibration, vol. 264, no. 2, pp. 411–431, 2003/07/03/ 2003, doi:https://doi.org/10.1016/S0022-460X(02)01202-6.
7. A. N. Thite and D. J. Thompson, "The quantification of structure-borne transmission paths by inverse methods. Part 2: Use of regularization techniques," Journal of Sound and Vibration, vol. 264, no. 2, pp. 433–451, 2003/07/03/2003, doi:https://doi.org/10.1016/S0022-460X(02)01203-8.
8. R. Schultz, "A Demonstration of Force Estimation and Regularization Methods for Multi-Shaker Testing," 2020, pp. 229-243.
9. S. Carter, "Thoughts on Using Sparse Inverse Solutions in Transfer Path Analysis," in Proceedings of IMAC XLII, the 42nd International Modal Analysis Conference, Orlando, 2024.
10. M. Van der Seijs, D. van den Bosch, D. Rixen, and D. Klerk, "An improved methodology for the virtual point transformation of measured frequency response functions in dynamic substructuring," in Proceedings of the 4th International Conference on Computational Methods in Structural Dynamics and Earthquake Engineering, Kos Island, 2013, pp. 4334–4347, doi:10.7712/120113.4816.C1539.
11. R. Schultz and P. Avitabile, "Shape-constrained Input Estimation for Efficient Multi-shaker Vibration Testing," Experimental Techniques, vol. 44, no. 4, pp. 409-423, 2020/08/01 2020, doi:10.1007/s40799-020-00361-0.
12. D. Ocepek, F. Trainotti, G. Cepon, M. Boltezar, and D. J. Rixen, "Different displacement reduction spaces for the use in admittance-based TPA methods," in Proceedings of IMAC XLII, the 42nd International Modal Analysis Conference, Orlando, 2024.
13. A. N. Thite and D. J. Thompson, "Selection of response measurement locations to improve inverse force determination," Applied Acoustics, vol. 67, no. 8, pp. 797–818, 2006/08/01/ 2006, doi:https://doi.org/10.1016/j.apacoust.2006.01.001.
14. C. Beale, R. Schultz, C. Smith, and T. Walsh, "Degree of Freedom Selection Approaches for MIMO Vibration Test Design," in Proceedings of IMAC-XL, the 40th International Modal Analysis Conference, Orlando, 2022, doi:10.1007/978-3-031-05405-1_10.
15. J. DeClerck, "Inverse Force Estimation for TPA," 2000.
16. M. Haeussler and D. Rixen, "Optimal transformation of frequency response functions on interface deformation modes," in Proceedings of IMAC XXXV, the 35th International Modal Analysis Conference, Los Angeles 2017.
17. F. Trainotti, T. Bregar, S. W. B. Klaassen, and D. J. Rixen, "Experimental decoupling of substructures by singular vector transformation," Mechanical Systems and Signal Processing, vol. 163, p. 108092, 2022/01/15/ 2022, doi:https://doi.org/10.1016/j.ymssp.2021.108092.
18. D. Rohe, R. Schultz, and N. Hunter, "Rattlesnake User's Manual," Sandia National Laboratories, 2021.
19. L. Meirovitch and H. Baruh, "The implementation of modal filters for control of structures," Journal of Guidance Control and Dynamics, vol. 8, pp. 707–716, 1985.
20. D. Otte, "Development and evaluation of singular value analysis methodologies for studying multivariate noise and vibration problems," Department of Mechanical Engineering, Katholieke Universiteit Leuven, 1994.
21. T. Schoenherr, "Using Modal Projection Error to Optimize Accelerometer Locations for a Modal Filter," in Proceedings of IMAC XLII, the 42nd International Modal Analysis Conference, Orlando, 2024.
22. T. Hastie, R. Tibshirani, and J. Friedman, The Elements of Statistical Learning: Data Mining, Inference, and Prediction, 2nd Edition ed. New York: Springer New York, 2017.
23. M. Aucejo and O. De Smet, "Bayesian source identification using local priors," Mechanical Systems and Signal Processing, vol. 66-67, pp. 120–136, 2016/01/01/ 2016, doi:https://doi.org/10.1016/j.ymssp.2015.05.004.
24. M. Aucejo and O. De Smet, "An optimal Bayesian regularization for force reconstruction problems," Mechanical Systems and Signal Processing, vol. 126, pp. 98–115, 2019/07/01/ 2019, doi:https://doi.org/10.1016/j.ymssp.2019.02.021.
25. T. Hastie, R. Tibshirani, and M. Wainwright, Statistical Learning with Sparsity: The Lasso and Generalizations (Monographs on Statistics and Applied Probability 143). Boca Raton, FL: Chapman & Hall/CRC, 2015.
26. D. Roettgen and A. Linderholt, "Substructuring excercises on combinations of steel and aluminum components of the benchmark structure of the technical division on dynamic substructures," in The 40th International Modal Analysis Conference, Orlando, 2022.
27. D. Roettgen, G. Lopp, A. Jaramillo, and B. Moldenhauer, "Experimental Substructuring of the Dynamic Substructures Round-Robin Testbed," 2022, pp. 119-123.
28. D. Kammer and M. Tinker, "Optimal placement of triaxial accelerometers for modal vibration tests," Mechanical Systems and Signal Processing, vol. 18, pp. 29–41, 01/31 2004, doi:10.1016/S0888-3270(03)00017-7.
29. H. G. Choi, A. N. Thite, and D. J. Thompson, "Comparison of methods for parameter selection in Tikhonov regularization with application to inverse force determination," Journal of Sound and Vibration, vol. 304, no. 3, pp. 894–917, 2007/07/24/ 2007, doi:https://doi.org/10.1016/j.jsv.2007.03.040.

# Chapter 4
# A Numerical Decoupling of a Wing Together with Two Nonlinear Radius-pylons, from the Technical Division's Substructuring Four-unit Frame Structure

**A. Linderholt, B. Moldenhauer and D. Roettgen**

**Abstract** To understand the structural dynamics of a complex mechanical assembly, it is often advantageous to study the dynamics of the individual components within the system. This has been utilized in numerical models for decades. An alternative is to represent the dynamics of these substructures using test data. Combining models based on vibrational test data with analytical/numerical models is known as experimental dynamic substructuring.

The Society of Experimental Mechanics' (SEM's) Technical Division on Dynamic Substructuring recognized a need for a geometrically simple yet conceptually challenging benchmark structure. A four-unit frame structure was designed to form its base. The substructuring collaboration was initially focused on linear substructuring but recent developments have led to an interesting nonlinear challenge problem.

Joints between substructures often cause nonlinear behavior in built-up systems; typical underlying physics including dry friction, varying contact areas, and other tribomechadynamic phenomena. Other causes are material nonlinearities, large deformation and contact in gaps/clearances. To include well-defined nonlinearities in the benchmark structure, the team from Sandia National Laboratories designed and manufactured a set of nonlinear subcomponents that serve as baseline examples. The design utilizes the four-unit frame as a basis and introduces wings with midspan lap joints as well as underwing pylons. The two pylon designs, the Gap Pylon (GP) and the Radius Pylon (RP), differ in the shape of the mounting blocks. The RP contains a smoothly varying contact with a metal strip while the GP features an abrupt change in the contact.

This study utilizes the four-unit Frame and rectangular jointed wing with attached RPs. Experimental dynamic substructuring is used to form a representation of the isolated nonlinear dynamics of the wing assembly by decoupling the dynamics of the Frame subsystem from the full structure. This process is completed at a range of excitation levels to characterize the nonlinear dynamics of the wing with two pylons as a function of response level. To avoid nonlinearities other than the one introduced by the RP, washers are inserted between the fuselage and the wing, between the wing parts in the lap joints, and between the wing and the upper pylon part. This work seeks to extend the boundaries of what is achievable with dynamic substructuring and demonstrate the process on a challenging use case.

**Keywords** Substructuring · Benchmark · Nonlinear · Decoupling

---

Sandia National Laboratories is a multi-mission laboratory managed and operated by National Technology & Engineering Solutions of Sandia, LLC (NTESS), a wholly owned subsidiary of Honeywell International Inc., for the U.S. Department of Energy's National Nuclear Security Administration (DOE/NNSA) under contract DE-NA0003525. This written work is authored by an employee of NTESS. The employee, not NTESS, owns the right, title and interest in and to the written work and is responsible for its contents. Any subjective views or opinions that might be expressed in the written work do not necessarily represent the views of the U.S. Government. The publisher acknowledges that the U.S. Government retains a non-exclusive, paid-up, irrevocable, world-wide license to publish or reproduce the published form of this written work or allow others to do so, for U.S. Government purposes. The DOE will provide public access to results of federally sponsored research in accordance with the DOE Public Access Plan.

---

A. Linderholt
Department of Mechanical Engineering, Linnaeus University
e-mail: andreas.linderholt@lnu.se

B. Moldenhauer · D. Roettgen
Sandia National Laboratories
e-mail: bmolden@sandia.gov; drroett@sandia.gov



Walter D'Ambrogio, et al. (eds.), *Dynamic Substructuring & Transfer Path Analysis, Vol. 4*, Conference Proceedings of the Society for Experimental Mechanics Series, https://doi.org/10.13052/9788743801498-4

## Introduction

The Society of Experimental Mechanics' (SEM's) Technical Division on Dynamic Substructuring has established a challenging yet practical benchmark structure for experimental-analytical substructuring exercises. A team with members from many research institutes set out from several desirable properties and a unit-frame structure was designed to form the base. The benchmark structure is built up by a frame and different kinds of plates. Together, they can represent structures from a variety of fields; automotive chassis, wing-fuselage structure, and building floors are examples of assemblies that would be possible to represent by the benchmark structure [1]. The frame is made as a one-piece structure consisting of four square units with 10/32 threaded inserts that can be used to attach other components. Additional 10/32 tapped holes are included on the side of the frame to mount impedance heads or force transducers.

Structural joints commonly cause nonlinear behavior due to their inherent physics such as frictional slip and varying contact areas. Understanding these nonlinear phenomena are a significant and increasingly important challenge when attempting to identify and model the dynamic characteristics of complex structural assemblies such as aircraft. To initiate this challenge, the team from Sandia National Laboratories designed and manufactured a set of nonlinear subcomponents for the four-unit frame that serve as a baseline example of what hardware could be utilized [2]. To facilitate collaboration and involvement, the technical drawings and CAD models of these subcomponents are freely available to any interested researchers on the Dynamic Substructuring Wiki [3]. This design is largely based on that used in [4], and originally proposed in [5]. The pylons are effectively pendulums in which a thin strip of metal is clamped at the top via two mounting blocks and has two masses adhered to the bottom. There are two designs that differ only in how the metal strip contacts the top mounting blocks. The two options, deemed the Gap Pylon and Radius Pylon, feature either an abrupt transition or a smooth curvature, respectively, between where the metal strip is clamped and the base of the mounting blocks.

The main objectives in this study are to measure the nonlinear characteristics of the four-unit frame together with the jointed wing and pylons and to conduct a numerical decoupling exercise. A major part in this work has been to devise a test setup enabling accurate measurements of the frame, wing and pylons. Another part has been to measure the dynamics of the structure at different excitation levels. During the presentation at the conference, results will be shown to demonstrate how the substructuring decoupling performed at different amplitude levels of the benchmark system.

## Testing

The test structure offers many possible causes of nonlinear dynamics. It also enables many ways to define the substructures. Additionally, test setups, tests and decoupling can each be performed in many ways. Here, the jointed wing with two radius pylons constitutes the nonlinear substructure and the four-unit frame fuselage constitutes the substructure to be subtracted from the assembly. The decoupling operation is performed using virtual points where the two subsystems are assembled and results in linearized eigenmodes for different excitation levels.

To avoid nonlinearities other than the one in the pylon, washers were included between the fuselage and the wing, between the wing parts in the lap joints, and between the wing and the upper pylon part. The test structure was suspended by fish lines in a "free-free" condition. The excitation for the assembly was provided by a Modal Shop Modal Shaker Model 2025 attached to the corner of the frame and a chirp signal was used as the source definition. A PCB 288D01 impedance head was used to estimate the drive point accelerance. Finally, a Polytec 3D Scanning Laser Doppler Vibrometer (SLDV) was used for the motion measurements. This first part in the study, i.e. establishing a test setup, at Linnaeus University (LNU), that can be used to conduct accurate vibrational tests on this nonlinear structure, resulted in the test setup shown in Figure 1. Additional testing will be performed at Sandia National Laboratories to test a nominally identical nonlinear benchmark system to gauge the similarity and repeatability of the assembly and setup.

## Modal Analysis

First, a low excitation level of about 1 Newton was utilized to generate a baseline linear response. Some of the lowest elastic eigenmodes are described in Table 1 and shown in Figure 2. These modes will be monitored at increasing excitation levels regarding their observable nonlinear behavior. For each excitation level, decoupling of the fuselage from the wing and pylons will be conducted using experimental substructuring. Tracking the frequency and dampingchange in the modes will be essential for understanding the performance of dynamic substructuring techniques.

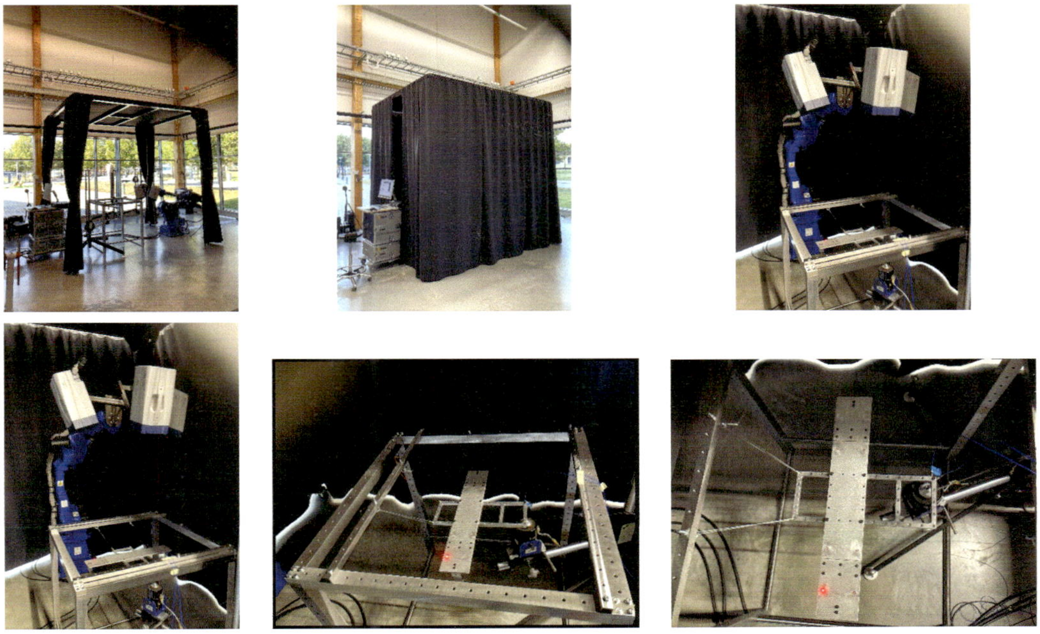

**Fig. 1** The test setup at LNU of the fuselage and the wing together with two Radius Pylons. A modal shaker was used for excitation and a Polytec 3D Scanning Laser Doppler Vibrometer was used for non-contact measurements.

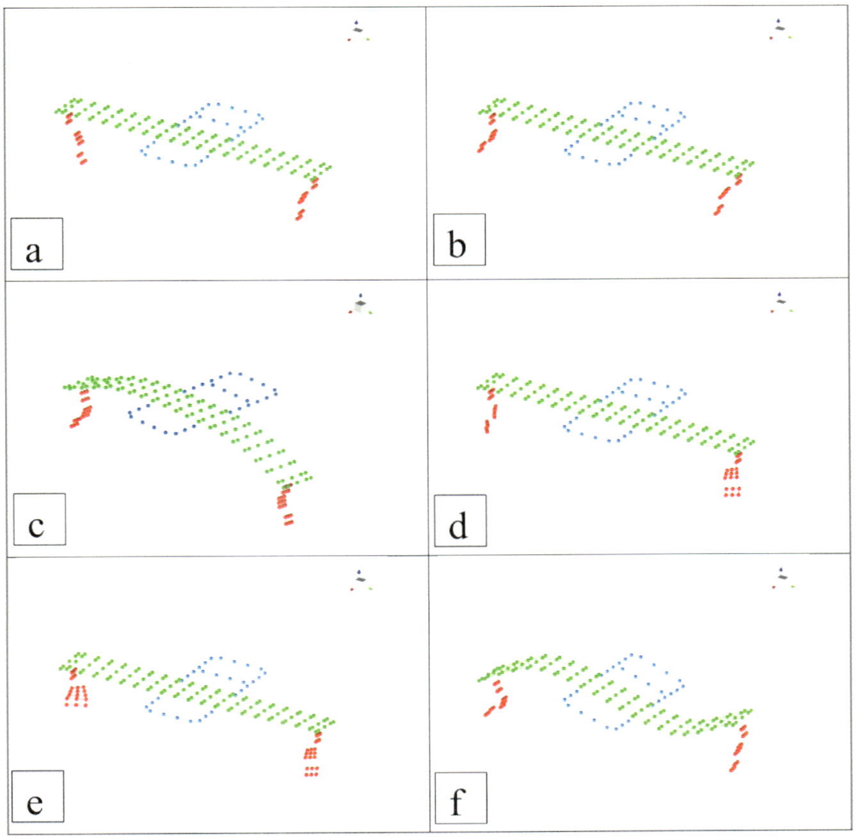


**Fig. 2** Six elastic mode shapes at a low excitation level. a) symmetric bending of the pylons, b) anti-symmetric bending of the pylons, c) first symmetric wing bending, d) symmetric torsion of the pylons, e) anti- symmetric torsion of the pylons, f) first anti-symmetric wing bending.

**Table 1** Six elastic modes estimated from measurements with a low excitation level.

| Natural Frequency [Hz] | Mode Shape Description |
|---|---|
| 14.3 | symmetric bending of the pylons |
| 14.9 | anti-symmetric bending of the pylons |
| 49.9 | first symmetric wing bending |
| 96.3 | symmetric torsion of the pylons |
| 97.0 | anti-symmetric torsion of the pylons |
| 109.4 | first anti-symmetric wing bending |

## Conclusion

The four-unit frame together with the lap-jointed wing and two Radius Pylons are used in a test campaign in which chirp signals supplied to a shaker were used to excite the structure. Linearized eigenmodes were extracted for increasing excitation levels. Some of these modes are monitored as a function of excitation level. A main part of the work was to achieve a test set-up enabling accurate vibration tests of the structure. From the first tests, it can be concluded that the structure has a high modal density and should provide an enlightening challenge problem for dynamic substructuring decoupling.

## References

1. D. Roettgen, G. Loop, A. Linderholt, P. Avitabile, J. Harvie and S. Klaassen, Technical Division Benchmark Structure for Dynamic Substructuring, 2020. W309W10080
2. D. Roettgen, B. Moldenhauer and A. Linderholt, "Nonlinear Subcomponent Attachments for the Technical Division on Dynamic Substructuring Benchmark Structure," in International Modal Analysis Conference, 2024. W309W10080
3. "Dynamic Substructures Wiki - Round Robin Frame Structure," [Online]. Available: https://sem.mywikis.wiki/wiki/Round_Robin_Frame_Structure.W309W10080
4. Cooper, Manzato, Borzacchiello, Bregant, Peeters and Barthorpe, "Investigating Nonlinearities in a Demo Aircraft Structure Under Sine Excitation," in IMAC - Model Validation and Uncertainty Quantification, Volume 3, Orlando, 2019. W309W10080
5. Platten, Wright, Cooper and Dimitriadis, "Identification of a Nonlinear Wing Structure Using an Extended Modal Model," Journal of Aircraft, vol. 46, no. 5, pp. 1614–1626, 2009. W309W10080

# Chapter 5
# Finite Element Modeling and Substructuring to Simulate Shock Plate Testing

**Harrison C. Denning, Scott Tuley, Matthew S. Allen, Jeffrey R. Hill, and Daniel R. Roettgen**

**Abstract** Aerospace structures such as rockets can experience various shock events during flight such as stage separation or impacts. These shock events can expose components inside the rocket to large accelerations and stresses which in turn can damage the parts. Shocks are often simulated in the lab using a resonant plate test. This work explores modeling of a 1kHz resonant plate. A shell finite element model is constructed both with and without simple parts attached. Modal transient simulations are used to investigate how many modes are needed to accurately reproduce the response of the plate and part, and the shock response spectrum (SRS) experienced. Then a component, modeled as a single degree-of-freedom system is attached to the center of the plate. The component's mass and fixed-base natural frequency are varied to understand how its SRS is affected.

**Keywords** Substructuring · Finite Elements · Shock Response

## Introduction

Resonant plate and bar tests measure the response of a device under test to shock. The test involves mounting a device onto a large metal plate and striking the opposite side with a projectile to induce vibration. Resonant plate and resonant bar testing have been used to qualify and test aerospace components since their first implementation in the 1980's and 1990's by Davie and Bateman of Sandia National Laboratories [1]. The responses obtained are often represented in shock-response spectra (SRS). While the test is traditionally used to measure the response along a single axis, in more recent years research has looked into how to extend the tests into multiple axis of vibration by changing the location of impact or how the test objects are mounted onto the plate [2, 3, 4, 5].

Most shock tests are designed empirically, with test engineers using a trial and error approach to adjust the parameters of the test (e.g. addition/location of damping bars, location of the test article, programmer material, etc...) until the desired SRS has been obtained. While this is often challenging even for single-axis shock tests, multi-axis testing makes the empirical

Sandia National Laboratories is a multimission laboratory managed and operated by National Technology and Engineering Solutions of Sandia, LLC., a wholly owned subsidiary of Honeywell International, Inc., for the U.S. Department of Energy's National Nuclear Security Administration under contract DE-NA-0003525.

Harrison C. Denning · Scott Tuley
Undergraduate Students, Department of Mechanical Engineering, Brigham Young University, Provo, UT, USA
e-mail: hc.denning22@gmail.com; sttuley@byu.edu

Matthew S. Allen
Professor, Department of Mechanical Engineering, Brigham Young University, Provo, UT, USA
e-mail: matt.allen@byu.edu

Jeffrey R. Hill
Associate Professor, Department of Mechanical Engineering, BYU, Provo, UT, USA
e-mail: dr.jeff.hill@byu.edu

Daniel R. Roettgen
Member of Technical Staff, Sandia National Laboratories, Albuquerque, NM, USA
e-mail: drroett@sandia.gov



Walter D'Ambrogio, et al. (eds.), *Dynamic Substructuring & Transfer Path Analysis, Vol. 4*, Conference Proceedings of the Society for Experimental Mechanics Series, https://doi.org/10.13052/9788743801498-5

approach a daunting task. Finite element analysis (FEA) has been used to model shock tests and shock loading [6, 7, 8] and has been used in several recent works to model resonant plate tests [3, 4, 9, 10, 11]. Efforts have been made to use predict the SRS of a component in a resonant plate shock test, showing promising results [11, 5]. This work explores the feasibility of using finite element analysis and dynamic substructuring to predict the shock response of a component in a resonant plate shock test.

Experimental dynamic substructuring has received increasing attention in recent years in order to accelerate test and analysis work [12]. Dynamic substructuring allows for accurate modeling of a large system by breaking it into component parts. Each subsystem may be modeled using traditional finite element models or a model may be derived experimentally, depending on the application of interest [13]. Subsystem models can then be assembled to accurately predict the response of the total system, avoiding the need to perform a test on the assembly. Jacobson et al. [5] appear to have been the first to use substructuring to predict the response of a component in a resonant plate shock test. They used LaGrange-Multiplier Frequency Based Substructuring (LM-FBS) to predict and optimize the response of a metal block in a multi-axis shock test. Substructuring allowed them to evaluate 110 possible configurations of their test and select the one that would theoretically produce the best multi-axis SRS. However, they also noted that the substructuring predictions under-predicted most of the resonances, as well as the knee frequency in the SRS, and showed that this was due to the way in which the interface was modeled.

This work seeks to build on those prior works by exploring a few questions. First, how many modes must be accurately modeled to predict the SRS in a resonant plate test? Shock has the potential to excite many modes, and it is unclear how many modes of each subcomponent would need to be measured or modeled in order for substructuring predictions to have adequate accuracy. Second, how sensitive is the SRS of a component to the dynamics of the component itself, as opposed to the dynamics of the resonant plate? It is hoped that the case studies presented here can lay the foundation needed to use modal substructuring to predict the SRS of an arbitrary component in a resonant plate test.

This paper will first review the construction of an accurate model of a resonant plate shock response using FEA. Abaqus will be used to create all FE models presented in this work. A model is constructed and the modes of the plate are computed. A study is performed to determine the mesh density is required to achieve convergence. A MATLAB mathematical model is then constructed to derive the response and hence the shock response spectrum (SRS) of the plate. The model is then interrogated to determine the appropriate number of modes to use in this analysis. This reveals that the SRS is dominated by the symmetric bending modes of the plate, as these are these are the modes most strongly excited by the shock pulse. Higher modes have a minor effect unless the damping is very light.

We then explore the influence of the device under test (DUT) on the shock response spectrum. A single degree-of-freedom (SDOF) system is coupled to the resonant plate at its center using substructuring and the fixed-base resonant frequency of the DUT is varied to see how that affects the SRS. The mass of the DUT is also varied, while keeping the resonant frequency constant, to explore its effect. In general, the SRS on the plate (i.e. at the base of the DUT) changes very little as the parameters of the DUT are varied.

The following section reviews the methods used, in particulars of the finite element model and the analytical solution for the response of a system to an impulsive load. Section  presents the results of the various case studies.

## Methods

Finite Element Analysis (FEA) software has been used increasingly in the last several years to aid in predicting the shock response of resonant plate and resonant bar tests [3, 9]. For our research, the Abaqus package was utilized to simulate the modes of a 1000Hz resonant plate and these modes were imported into MATLAB to compute the plate's response. The model was then augmented by adding a device under test (DUT), attached to the plate at its center, and the response and SRS computed as various parameters were varied.

### *Developing the Abaqus Model*

A resonant plate test is used in the testing and qualification of parts. Its has proved especially useful in replicating pyroshock environments for aerospace applications. The test is performed by striking a heavy metal plate with a projectile. The force pulse input by the projectile is shaped by placing a "programmer" material between the projectile and plate. The modes of the plate also serve to accentuate certain frequencies in the response. This research is concerned only with the standard case where the impact and the DUT are centered on opposite sides of the plate. Several steps were taken to create an accurate model of this test.

Our plate model was created first without any device under test attached to verify the expected behavior of the plate when subjected to the impact. The model was setup using a standard $508 \times 508 \times 50.8$ mm plate made of 6061 aluminum [4]. No damping bars or other attachments were added and the damping ratio of the plate was set to 0.02 as this appears to be typical [14]. The model was meshed with linear reduced integration (S4R) shell elements.

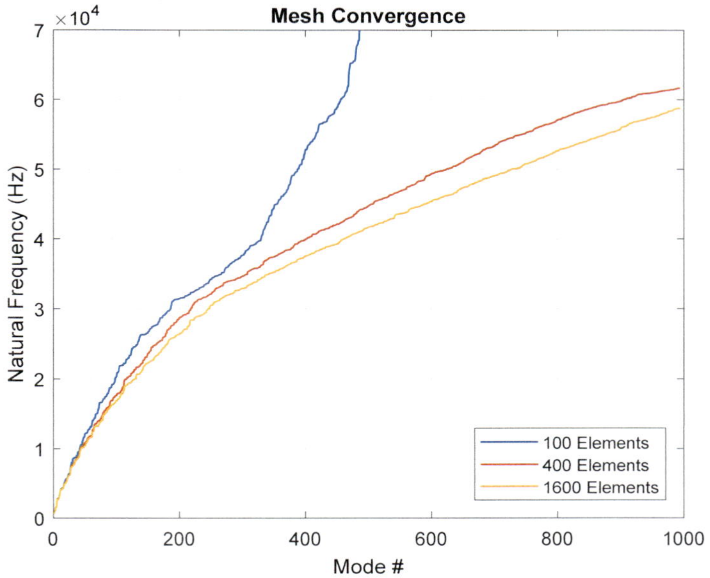


**Fig. 1** The natural frequencies computed by Abaqus for three different mesh densities

The appropriate mesh density was determined by comparing the calculated mode frequencies returned by the finite element solver as the mesh density varied. In all cases the square plate was meshed with a uniform grid of hexagonal elements with 10, 20 or 40 elements on a side. Figure 1 shows the frequencies of the first 1000 modes for a plate meshed with 100, 400, and 1600 elements and Figure 2 shows the meshes. The resolution was such that a plate modeled with only 100 elements deviated from the other models after only 50 modes, with an exponential divergence from the other models after 300 modes. When modeled with 400 and 1600 elements, the calculated modes showed a far lesser divergence, although the natural frequencies still differed by about $10^3$ Hz after 400 modes or 40kHz. That is most likely beyond the frequency range of interest. All models returned similar values over the first few dozen modes or up to about 10kHz and the modes for the 400 and 1600 element models stayed within 100Hz of one another in that range. This was deemed an acceptable margin and the 400 element model is used for most of the analyses presented subsequently.

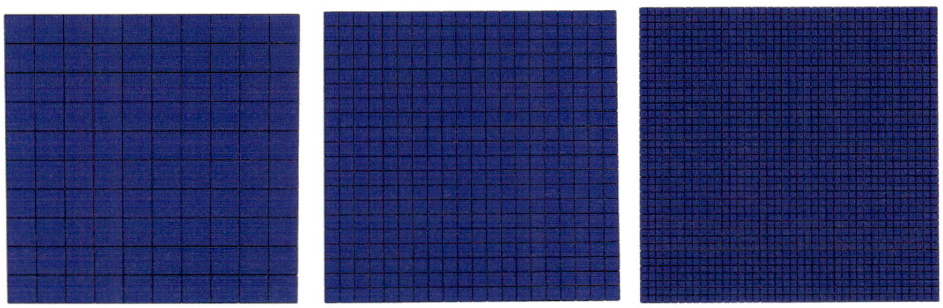

**Fig. 2** The three Abaqus models created with 100, 400, and 1600 shell elements

## *Response of Plate to Impulsive Loading*

Figure 3 shows a few of the lower frequency modes of the plate. Abaqus was only used for the initial calculation of mode shapes and frequencies, as it was faster to iterate on the time response in MATLAB. The first 1000 modes were found and imported into MATLAB where the response of the plate was calculated over time. The time response at the center of the plate was used to compute the Shock Response Spectrum (SRS).

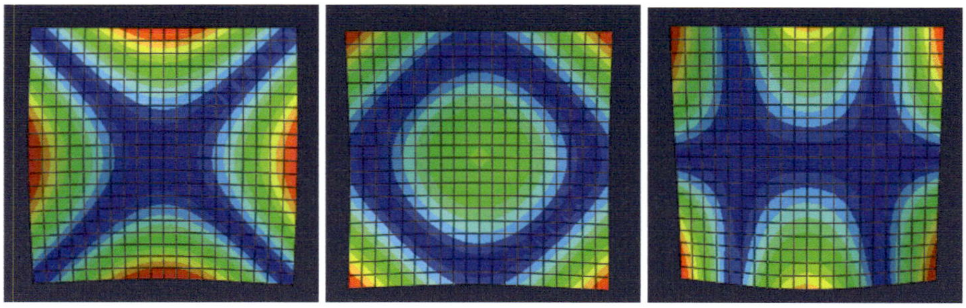

**Fig. 3** The Plate as modeled in Abaqus showing several modes, from the top left going clockwise: mode 2 at 912.5 Hz, mode 3 at 1125.4 Hz, and mode 4 at 1538.6 Hz are shown

Once the modes are imported from Abaqus the response of the plate can be solved for using standard solutions for ordinary second order differential equations. The problem is represented by Eq. (1)

$$\mathbf{M}\ddot{\mathbf{x}} + \mathbf{K}\mathbf{x} = \mathbf{F} f_p(t) \tag{1}$$

Where $\mathbf{M}$ and $\mathbf{K}$ are the mass and stiffness matrices, $\mathbf{F}$ represents a vector of zeros with a value of 1 where the force is applied and $f_p(t)$ is the amplitude of the force as a function of time. The force was applied at the center of the plate for all cases considered in this paper. Using the orthogonality properties in Eq. (2) these equations are transformed to the set of uncoupled modal equations in Eq. (3).

$$\phi_j^{\mathrm{T}}[\mathbf{M}]\phi_n = \begin{cases} 0 & \text{if } n \neq j, \\ \mu_j & \text{if } n = j. \end{cases} \qquad \phi_j^{\mathrm{T}}[\mathbf{K}]\phi_n = \begin{cases} 0 & \text{if } n \neq j, \\ \mu_j \omega_j^2 & \text{if } n = j. \end{cases} \tag{2}$$

$$\ddot{\eta}_r + 2\zeta_r\omega_r\dot{\eta}_r + \omega_r^2\eta_r = \boldsymbol{\phi}_r^{\mathrm{T}}\mathbf{F} f_p(t) \tag{3}$$

with $\phi_r^{\mathrm{T}}\mathbf{F}$ being a scalar value. This scalar equation can be solved for each mode of the plate. The solutions are then finally converted back to the original coordinates by using the following.

$$\mathbf{x} = [\boldsymbol{\Phi}]\boldsymbol{\eta} \tag{4}$$

This model was initially solved using a perfect impulse, i.e. $f_p(t) = F_0\delta(t)$ to excite the plate, with an amplitude of $F_0 = 100$ N-s. The solution for each mode is given in Eq. (5), where $\omega_{dr} = \omega_r\sqrt{1-\zeta_r^2}$.

$$\eta_r = \frac{F_0\boldsymbol{\phi}_r^{\mathrm{T}}\mathbf{F}}{\omega_{dr}} e^{-\zeta_r\omega_r t} Re(-ie^{i\omega_{dr}t}) \tag{5}$$

The response was calculated using several different numbers of modes and the SRS was plotted and is shown in Fig. 4. As can be seen, the SRS does not converge to a single SRS as the number of modes increased. As more modes are added the SRS increases and the knee frequency in the curves moves to higher and higher frequencies. This is not an accurate representation of the response that is measured in these types of resonant plate tests. In practice a certain resonant plate typically has a single knee frequency that is centered on the frequency of the first symmetric mode of the plate.

The prior simulations reveal that the shape of the input pulse, and in particular its frequency content are instrumental in setting the knee frequency in an SRS. Cisemore and Babuska [14] showed that sinusoidal and haversine pulses all exhibited a similar SRS, implying that the shape of the pulse was not as important as its bandwidth, so the input force was modeled as a half sine pulse of varying duration $f_p(t) = F_0 sin(\omega_p t)$ for $0 < t < \pi/\omega_p$, because it has a simple analytical solution [15].

The force was once again was applied at the center of the plate and the duration was set to $t_p = \pi/\omega_p$ where $\omega_p = 1000 * 2\pi$. The response of a single mode to this input is given in Eq. (6) for the $r$th mode. Note that this solution assumes that $\omega_p \neq \omega_r$. The authors actually found that it loses accuracy if $\omega_p \approx \omega_r$, so one must use care to avoid the case where any natural frequency is equal to the pulse frequency $\omega_p$.

$$\eta_r(t) = \frac{F_0\Phi^T\mathbf{F}}{((\omega_r^2-\omega_p^2)^2 + 4\zeta_r^2\omega_r^2\omega_p^2)}\mathrm{Re}[A_{1r}e^{i\omega_p t} + A_{2r}e^{-\zeta_r\omega_r t}e^{i\omega_{dr}t}] \tag{6}$$

where $A_{1r}$ and $A_{2r}$ are

$$A_{1r} = -2\zeta_r\omega_r - i(\omega_r^2 - \omega_p^2) \tag{7}$$

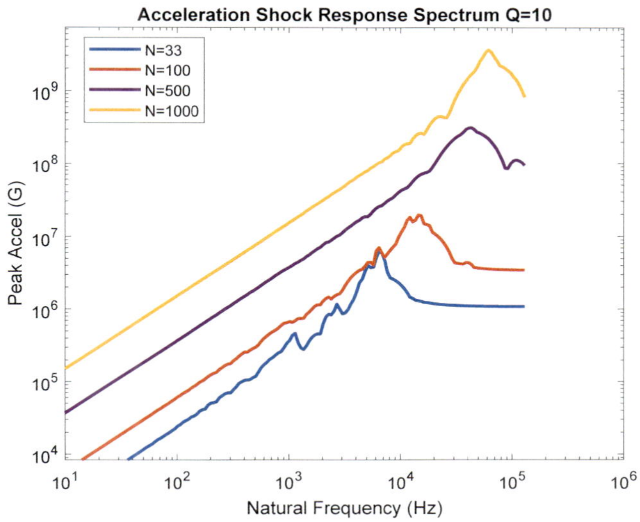


**Fig. 4** The SRS of the resonant plate in response to a pure impulse input $f_p(t) = F_0\delta(t)$ as calculated using different numbers of modes

$$A_{2r} = -2\zeta_r\omega_r\omega_p + i\frac{\omega_p(1-2\zeta_r^2)\omega_r^2 - \omega_p^2}{\omega_r\sqrt{1-\zeta_r^2}} \tag{8}$$

This solution holds for the entirety of the forced response or for $0 < t < t_p$, the final state of the forced response at $t = t_p$ is then used as the initial conditions ($\dot{\eta}_r(0)$ & $\eta_r(0)$) and the subsequent response is given by Eq. 9

$$\eta_r = e^{-\zeta_r\omega_r t}\text{Re}[A_{3r}e^{i\omega_{dr}t}] \tag{9}$$

where $A_{3r}$ is

$$A_{3r} = \eta_r(0) - i\frac{\dot{\eta}_r(0) + \zeta_r\omega_r\eta_r(0)}{\omega_{dr}} \tag{10}$$

Again the results are converted back to the original coordinate system using Eq. (4). The SRS was then computed and compared as the number of modes used and frequency of the half sine wave's input was varied. The results are shown in Fig. 5.

SRS plots have become standard in characterizing shock response and were generated for each test run. Their use began with Biot's application of SRS in the analysis of earthquake vibrations but they have since proved useful in analyzing shock and vibrations in many areas of research [16]. An SRS is the maximum acceleration response of an SDOF system whose base input is the shock signal. This is repeated for an SDOF system whose natural frequency is equal to each frequency in the spectrum of interest. Hence, it gives the worst case response of a family of components to the shock input [14]. All of the SRSs presented here where created using the tool provided by Irvine on https://www.vibrationdata.com/, which implements the Smallwood Algorithm with some modifications [17, 18]. For a more detailed mathematical derivation of the algorithm see [19].

When the input frequency of the pulse was kept at or below the plate's resonance a single knee frequency emerged, located at 1140Hz. In those cases (i.e. $\omega_p = 500$ or 1000 Hz), as more modes were added more peaks were observed after the knee, yet the SRS remained more or less at the same level. In contrast, when the higher frequency $\omega_p = 5000$ Hz input was applied, the SRS continued to rise after 1140 Hz and a new knee frequency emerged at 5120 Hz. The simulation failed to capture this effect when only 5 modes were used in the computation. This issue was found to become more prevalent as more modes were included in the computation or when a shorter (or higher frequency) pulse was applied to the system. While not shown here, these were also found to become more pronounced if the damping in the plate was lower.

These results, when taken in conjunction with those in Fig. 4, explain why the SRS did not converge when a perfect impulse was used. The response of the plate is governed by the relatively small number of modes that can be excited by an input at the center. Figure 6 shows the shapes of the first two symmetric modes of the plate; all of the other modes that appear below 5kHz for this plate are anti-symmetric and hence have node lines at the center. If the impulse has infinite bandwidth, then as more modes are added to the simulation the SRS continues to increase, with the knee frequency appearing near the

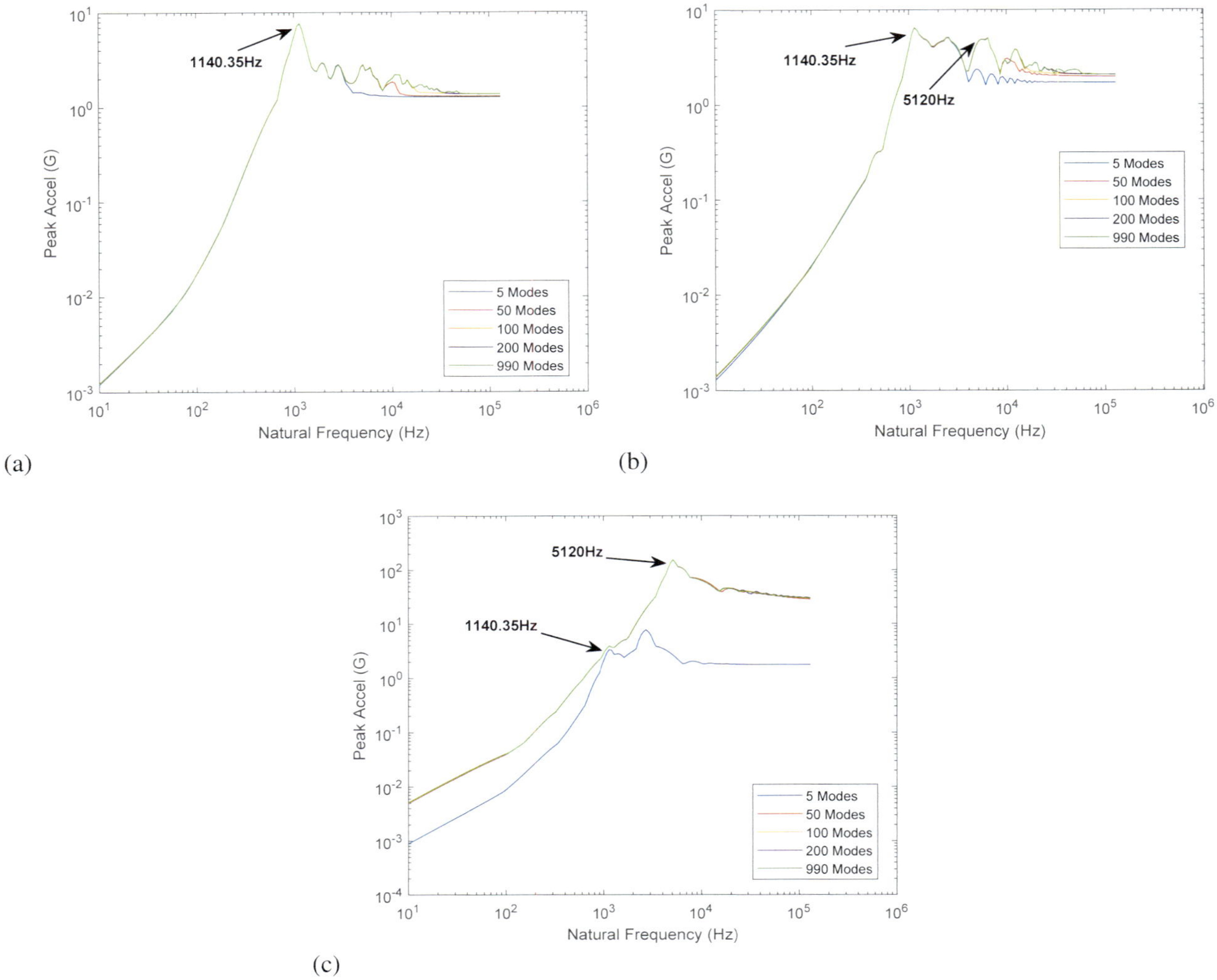


**Fig. 5** SRS of the center of the plate, computed using pulse frequencies $\omega_p$= 500, 1000, 5000 Hz in (a), (b) and (c) respectively and with various numbers of modes included in the computation.

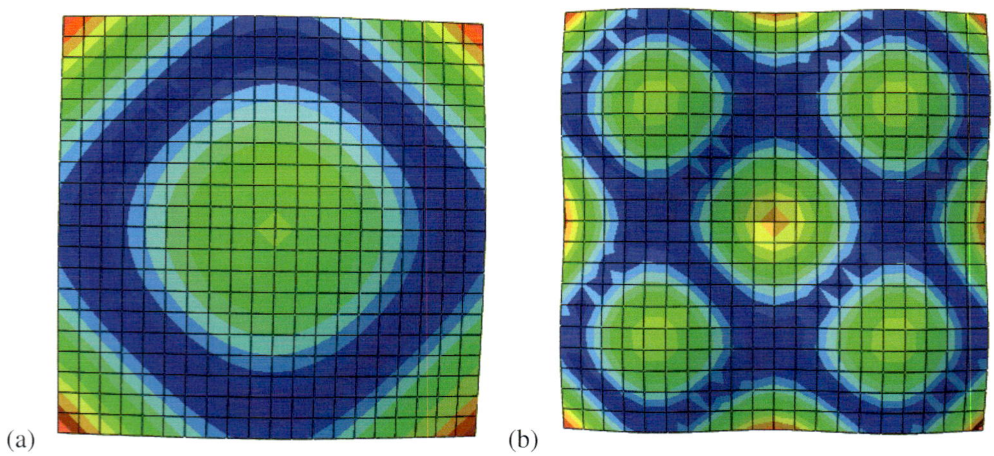


**Fig. 6** Mode Shapes of (a) Mode 2 at 1125.4 Hz and (b) Mode 19 at 5067.5Hz. These are the closest modes to the frequencies indicated in Fig. 5.

frequency of the highest frequency mode that was both excited and included in the simulation. In contrast, if the bandwidth of the pulse is limited, then it excites all modes up to a certain frequency and the highest frequency symmetric plate mode dominates the response and produces the knee in the SRS. This behavior explains why a resonant plate test is able to produce a consistent SRS. The "programmer" material on the back of the plate reduces the bandwidth of the input force, but it is difficult to control the pulse width precisely. These simulations show that the overall response is relatively insensitive to the force pulse width, because the plate accentuates the response, producing a knee near its fundamental breathing mode for a fairly wide range of input forces.

To investigate the effects of damping on the SRS of the plate, the simulations were repeated with varying levels of damping applied to the plate's modes and the results are shown in Figure 7. A higher damping in the plate has a tendency to smooth out the response beyond the knee frequency. This makes the simulations less sensitive to errors in the FE model. While Fig. 1 showed that the high freuqency modes were more sensitive to mesh density, these modes have little effect on the SRS if the damping is high. Once the response of the plate was verified, attention turned to the influence that a component would have on the plate.

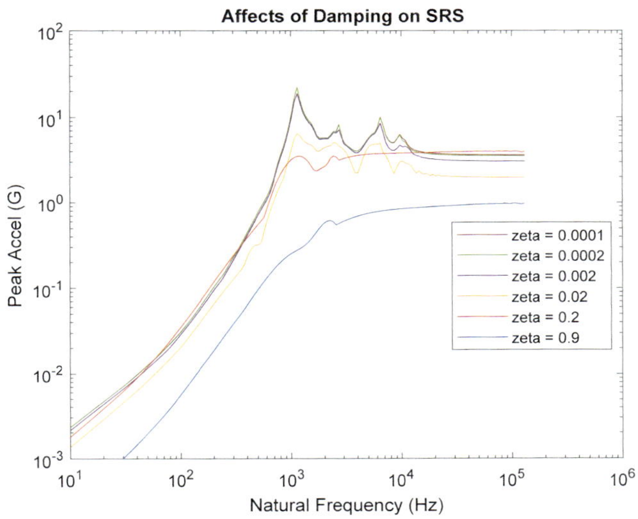


**Fig. 7** The SRS of a plate with different levels of damping when subjected to a 0.5msec pulse width. The same damping ratio was applied to all 50 modes used in the simulation.

## Results

Having resolved issues of convergence and establishing a workable model of the resonant plate, it was now possible to simulate more complex scenarios. In particular, it was desired to determine how the test article, or component mounted on the plate, changes its SRS. The simplest component could be modeled as a rigid mass, and its effect is already obvious; if the mass is large enough it would lower the natural frequency of the bubble mode and move the knee in the SRS to lower frequencies.

A more realistic component would contain both mass and stiffness. A simple component was modeled as a spring-mass system and was affixed to the plate using the modal model that was imported into MATLAB. The addition of the spring-mass system was achieved using the method of primal assembly. This involves creating a connectivity matrix between the equations of motion of the plate and those of the mass-spring system to be added. The two are linked together with a set of constraints. (For a more detailed derivation see [13] ).

The resulting response of the plate was then simulated for various test cases. The plate's response was favored over the component's as the main interest was to analyze how the addition of component to the plate would affect the plate's response, and hence the component's base excited shock environment specification. (While the SRS on the component is not shown here, in general it was observed that the component's SRS closely matched that of the plate including a similar or identical knee but at a higher amplitude. The component's SRS was generally no more than 5g's higher than that of the plate.) It was hypothesized that the component would have the greatest effect on the plate's response when its own resonant frequency matched that of the plate and when it's mass was the highest. Since the exact mass and spring constant values of the component will vary widely from case to case, the plate's response was simulated as the parameters of the component were varied over a wide range, as shown in Table 1.

**Table 1** The mass and spring constant of the component attached to the plate were varied to maintain a given resonant frequency relative to that of the first bubble mode of the plate. The mass was set to be a fraction of the bare plate's 35.4 kg mass. The knee frequencies listed are those observed in Fig. 8.

| Component Resonant Freq (Hz) | Mass (kg) | Spring Constant (N/m) | Observed Knee Freq. (Hz) |
|---|---|---|---|
| | $35.4 \times 10^{-1}$ | $1.80 \times 10^{6}$ | 1140.35 |
| 112.53 | $35.4 \times 10^{-2}$ | $1.80 \times 10^{5}$ | 1140.35 |
| | $35.4 \times 10^{-3}$ | $1.80 \times 10^{4}$ | 1140.35 |
| | $35.4 \times 10^{-1}$ | $1.80 \times 10^{8}$ | 1436.75 |
| 1125.3 | $35.4 \times 10^{-2}$ | $1.80 \times 10^{7}$ | 1140.35 |
| | $35.4 \times 10^{-3}$ | $1.80 \times 10^{6}$ | 1140.35 |
| | $35.4 \times 10^{-1}$ | $1.80 \times 10^{10}$ | 1015.94 |
| 11253 | $35.4 \times 10^{-2}$ | $1.80 \times 10^{9}$ | 1140.35 |
| | $35.4 \times 10^{-3}$ | $1.80 \times 10^{8}$ | 1140.35 |
| N/A | 0 | 0 | 1140.35 |

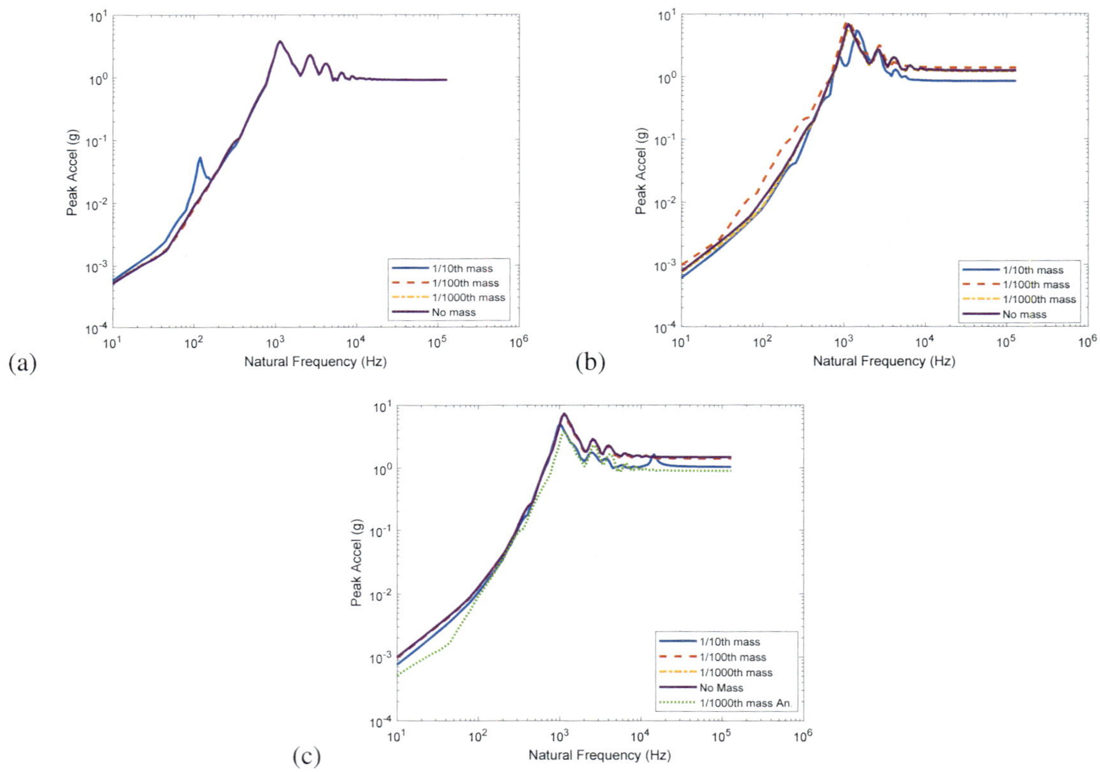


**Fig. 8** The SRS of the plate with a single component attached. The parameters of the component are defined in Table 1. The component's mass was a set fraction of the plate's mass, as listed in the legend. The fixed-base resonant frequency of the component was (a) 1/10th, (b) equal, (c) 10 times that of the 1125.4 Hz first bubble mode of the plate.

The resonant frequency of the device is specified by its fixed base natural frequency $f_R = \sqrt{k/m}/(2\pi)$, or the frequency that the component would have if its base was fixed. Note that, to simplify the implementation with the tools available, the component was actually modeled as a 2DOF system consisting of two masses connected by a spring of stiffness $k$. The mass at the base was given a value of $m/100$ so that the component's mass $m$ is the only important factor. For a particular resonant frequency the component's mass was then varied between 1/10th, 1/100th, and 1/1000th of the plate's base mass. The spring constant was adjusted accordingly to maintain the desired resonance of the device (see Table 1). The resonant frequency of the device was set at three values: 1/10th the resonant frequency of the plate, at resonance and at 10 times the resonant frequency. The simulations were run with a $\omega_p = 1000$ Hz sine pulse input and are shown in Fig. 8. Note, however, that the cases in Fig. 8(c) were computed with MATLAB's ode45 function. This was necessary because the natural frequency of the plate was nearly 1000 Hz in the case where the component mass was 1/10th of the plate mass, so the analytical formula gave an inaccurate result. The three curves shown were computed with ode45, and the analytical solution for the 1/100th mass case is also shown, confirming that the ode45 settings (RelTol=$1e-9$, AbsTol=$1e-12$) were adequate to produce a solution that agrees reasonably well with the analytical one.

The behavior observed in Fig. 8 is mostly as one might expect. When the component has a low natural frequency, as in Fig. 8(a), the assembly basically has the modes of the plate plus one mode where the component bounces on the approximately rigid plate. That mode occurs at 119 Hz for the case where the component has 1/10th of the plate mass, and it causes a peak that, while visible in the SRS, is not particularly large. For the other cases that mode occurs at 114.0 and 112.5 Hz, but the mass is small enough that it does not affect the plate response noticeably. Note, however, that one would see a change in the SRS on the component.

The largest changes occur when the component's natural frequency is equal to the plate's bubble mode frequency, changing the shape of the SRS significantly, especially when the mass is large. When the component's natural frequency is high relative to the bubble mode of the plate, the component basically acts as a rigid mass that lowers the frequency of the bubble mode, shifting the SRS to the left. This shift is noticeable for the 1/10th mass case, but imperceptible for the other cases.

**Table 2** The differences between the SRSs in Fig. 8 are quantified in terms of the maximum and RMS differences between each and the reference case where no component is on the plate.

| Component Resonant Freq (Hz) | Mass (kg) | max. difference (g) | Root Mean Square (g) |
|---|---|---|---|
| | $35.4 \times 10^{-1}$ | 3.059 | 0.533 |
| 112.53 | $35.4 \times 10^{-2}$ | 3.062 | .532 |
| | $35.4 \times 10^{-3}$ | 3.063 | 0.532 |
| | $35.4 \times 10^{-1}$ | 5.237 | 0.786 |
| 1125.3 | $35.4 \times 10^{-2}$ | 1.969 | 0.324 |
| | $35.4 \times 10^{-3}$ | 1.302 | 0.156 |
| | $35.4 \times 10^{-1}$ | 3.4856 | 0.664 |
| 11253 | $35.4 \times 10^{-2}$ | 0.5556 | 0.916 |
| | $35.4 \times 10^{-3}$ | 0.0544 | 0.0088 |
| N/A | 0 | 0 | 0 |

## Conclusion

These results offer some insight into the potential future use of substructuring techniques to simulate resonant plate tests. The simulations show that the force pulse width must be chosen in conjunction with the resonant plate's bubble mode frequency to obtain the typical SRS with a knee at 1kHz. (The force pulse width is set by the programming material on the back of the plate.) When this is done, the SRS is dominated by a single mode - the bubble mode, which was at 1125 Hz for the plate studied here. On the other hand, if the force pulse bandwidth extends to 5kHz, then a higher order bubble mode can respond and the SRS changes considerably. In that case one must include up to 19 modes at least; convergence was obtained with 50 modes in the simulations shown here. Modes beyond the knee frequency have a minor effect on the SRS.

When a component was added to the resonant plate, it changed the SRS in ways that were fairly straightforward to predict, as discussed in the prior section. The changes were also relatively small, mostly because the component mass was always small relative to the resonant plate. If the fixed-base natural frequencies and mass of the component are known, one can presumably use these results to predict the types of changes that might occur in the SRS.

Future work will build on these results, addressing cases where the component has both mass and inertia and interfaces with the resonant plate over a non-negligible area. (In the simulations presented here, the component was modeled as a point mass connected at the exact center of the plate.)

## References


1. Davie, N.T. and Bateman, V.I. "Recent developments in phyroshock simulation using fixtures with tunable resonant frequencies". Technical Report SAND-94-0350C, Sandia National Labs (1994)
2. Soine, D.E., Schoenherr, T.F., and Bouma, A.J. "Resonant bar shock test equipment under offset loading". In Schoenherr, T., Karlicek, A., and Beale, D., editors, *Dynamic Environments Testing, Vol. 7: Proceedings of the 42nd IMAC, A Conference and Exposition on Structural Dynamics 2024*, pages 79–84, Cham (2024) Springer Nature Switzerland.
3. Bouma, A.J., Schoenherr, T.F., and Soine, D.E. "Fixture design and analysis for multi-axis mechanical shock testing". In *Society for Experimental Mechanics Annual Conference and Exposition*, pages 121–124. Springer (2023)
4. Soine, D.E., Schoenherr, T.F., and Heister, J.D. "Evaluation of a multiaxis shock fixture concept". In *Society for Experimental Mechanics Annual Conference and Exposition*, pages 61–67. Springer (2023)
5. Jacobson, E.M., Blough, J.R., DeClerck, J.P., Van Karsen, C.D., and Soine, D. "Frequency based substructuring on resonant plate". In *Special Topics in Structural Dynamics & Experimental Techniques, Volume 5: Proceedings of the 38th IMAC, A Conference and Exposition on Structural Dynamics 2020*, pages 327–333. Springer (2021)
6. Qiu, X., Deshpande, V., and Fleck, N. "Finite element analysis of the dynamic response of clamped sandwich beams subject to shock loading". *European Journal of Mechanics - A/Solids*, 22(6):801–814 (2003)
7. García-Pérez, A., Sorribes-Palmer, F., Alonso, G., and Ravanbakhsh, A. "Overview and application of fem methods for shock analysis in space instruments". *Aerospace Science and Technology*, 80:572–586 (2018)
8. Fraser, J.S. "Abaqus implementation of a hyperelastic damage model for glass-reinforced polymers under shock and impact loading". Master's thesis, The University of Mississippi (2022)
9. Bouma, A.J., Schoenherr, T.F., and Soine, D.E. "Design and analysis of resonant bar fixtures for multi-axis shock response testing". In *Society for Experimental Mechanics Annual Conference and Exposition*. Springer (2024)
10. Schoenherr, T.F., Soine, D.E., and Witt, B.L. "Experimental modal analysis of a resonant plate during a mid-field pyroshock replication test". In *Sensors and Instrumentation, Aircraft/Aerospace and Dynamic Environments Testing, Volume 7: Proceedings of the 40th IMAC, A Conference and Exposition on Structural Dynamics 2022*, pages 41–52. Springer (2022)
11. L. Viale, A.F.L.G., A.P. Daga. "Optimization of irregular-shaped resonant plates for pyroshock testing in the aerospace industry". In *Proceedings of ISMA2024, Leuven, Belgium* (2024)
12. De Klerk, D. *Dynamic Response Characterization of Complex Systems Through Operational Identification and Dynamic Substructuring*. PhD Thesis, Delft University of Technology (2009)
13. Allen, M.S., Rixen, D., Van Der Seijs, M., Tiso, P., Abrahamsson, T., and Mayes, R.L. *Substructuring in Engineering Dynamics: Emerging Numerical and Experimental Techniques*. Springer, CISM International Centre for Mechanical Sciences (2020).
14. Sisemore, C. and Babuška, V. *The science and engineering of mechanical shock*. Springer (2020)
15. Ginsberg, J.H. *Mechanical and structural vibrations: theory and applications*. John Wiley & Sons, New York, 1st edition (2001)
16. Biot, M.A. *Transient oscillations in elastic systems*. PhD thesis, California Institute of Technology (1932)
17. Smallwood, D.O. et al. "An improved recursive formula for calculating shock response spectra". *Shock and vibration bulletin*, 51(2):211–217 (1981)
18. Ahlin, K. "Shock response spectrum calculation–an improvement of the smallwood algorithm". In *70th Shock and Vibration Symposium*. SAVIAC (1999)
19. Irvine, T. "Derivation of the filter coefficients for the ramp invariant method as applied to base excitation of a single-degree-of-freedom system" (2013) http://www.vibrationdata.com/ramp_invariant/ramp_invariant_base.pdf [Accessed 2 Oct 2024].

# Chapter 6
# Condition Monitoring for the Mounts of a Simple Supported Beam Using Transfer Path Analysis

**Michael Kreutz and Daniel J. Rixen**

**Abstract** Large machines like engines or turbines are subject to operational forces that may be hard to model or measure directly. Large machines are often connected to the environment via soft supports, which can degrade over time. Condition monitoring should detect and quantify changes in the dynamics of the joints to ensure the machine's performance. Transfer path analysis (TPA) can be used to characterize the unknown operational forces. Unlike in typical TPA, the characterization of the operational forces is not used to predict responses on the receiver system, but it is used together with measured operational responses to estimate frequency response functions (FRFs) of the system. These FRFs can be exploited in a frequency-based substructuring analysis to isolate and identify the joint dynamics. A numerical case study of a beam connected to the environments via discrete joints is used to validate this approach.

**Keywords** transfer path analysis · frequency-based substructuring · monitoring · joint identification

## Introduction

Large machines, such as engines, turbines, and similar equipment, are often exposed to operational forces that are not fully understood. These machines are typically connected to their environment through multiple soft mounts (joints), which play a crucial role in vibration isolation and ensure the machine's proper function. However, the dynamics of these joints can change over time due to aging. In the context of condition monitoring, it is crucial to be able to detect and identify these changes in the joint dynamics during operation.

Transfer path analysis (TPA) is a method that considers the assembly of a source with an unknown excitation and a receiver structure. Typically, the goal of TPA is to characterize the excitation of the source and predict responses on the receiver. However, in the context of condition monitoring, the goal is different. From an equivalent force and a measured response, the dynamics (i.e., transfer functions, frequency response functions FRFs) of the system should be calculated. Using the calculated transfer functions on the machine, the joint dynamics can be extracted by using frequency-based substructuring.

This contribution presents a workflow to use TPA and frequency-based substructuring to identify in operation the mounts (called joints in this contribution) connecting a machine to a rigid environment. A numerical example of a beam with an operational excitation that is connected to the environments via spring-damper elements is used to validate the workflow. In a TPA step, the joints are considered as part of the receiver system and the beam is considered as source system. An equivalent force at the interface that represents the operational excitation is determined. Measurements of the response on the interface are used to obtain some transfer functions of the system at the interface. These transfer functions are then exploited in a frequency-based substructuring analysis to isolate the dynamics of the mounts.

Section 2 outlines the methodology. Section 3 presents a numerical case study of a beam in soft mounts to validate the method. Finally, a conclusion of the work and outlook for future studies is given.

## Method

This section shortly introduces the TPA and the concept of equivalent forces, which is used to identify transfer functions of assembled systems. The identified transfer functions can be used to estimate the dynamics of a joint for a simplified system.

Michael Kreutz · Daniel J. Rixen
Chair of Applied Mechanics, TUM School of Engineering and Design, Technical University of Munich, Boltzmannstr. 15, 85748 Garching, Germany
e-mail: m.kreutz@tum.de; rixen@tum.de



Walter D'Ambrogio, et al. (eds.), *Dynamic Substructuring & Transfer Path Analysis, Vol. 4*, Conference Proceedings of the Society for Experimental Mechanics Series, https://doi.org/10.13052/9788743801498-6

## *Transfer Path Analysis*

An overview on transfer path analysis (TPA) can be found in [1, 2]. This section repeats the most important points of TPA.

Figure 1a present the setup of a TPA problem. The assembled system $SR$ consists of an active source structure $S$ with an unknown excitation $\boldsymbol{f}_1$ in degrees of freedom (dofs) 1 and a passive receiver structure $R$, where the responses $\boldsymbol{u}_3$ in dofs 3 should be predicted. The structures are connected in the interface with dofs 2. The excitation $\boldsymbol{f}_1$ is hard to be measured or modeled directly and is thus unknown. An example are the forces of a combustion engine that influence the noise-vibration harshness in a passenger car, cf. [3]. TPA aims at characterizing the excitation, identifying critical transfer paths and predicting the response on the receiver structure.

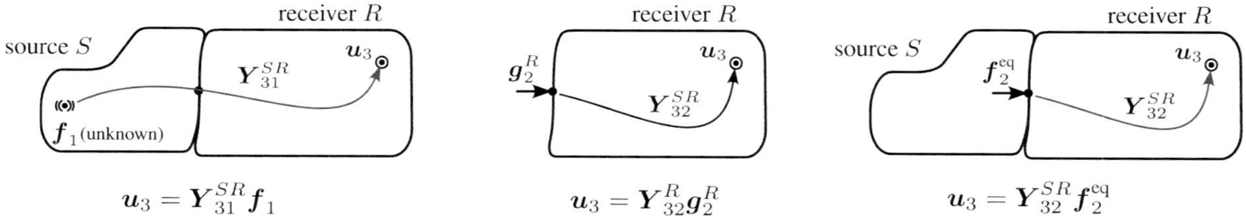


(a) Problem of TPA. (b) Classical TPA. (c) Component-based TPA.

**Fig. 1** Overview of transfer path analysis, [1].

The *classical TPA* approach, see fig. 1b, tries to measure or estimate the internal forces in the interface 2 between the substructures. A disadvantage of the method is that those internal forces depend on the assembly and can thus not be reused if the receiver component is changed.

The *component-based TPA*, see fig. 1c, measures or estimates an equivalent force $\boldsymbol{f}^{\text{eq}}_2$ at the interface to represent the effect of the unknown excitation $\boldsymbol{f}_1$ on the responses $\boldsymbol{u}_3$ of the receiver:

$$\boldsymbol{u}_3 = \boldsymbol{Y}^{SR}_{32}\boldsymbol{f}^{\text{eq}}_2 \tag{1}$$

The equivalent force can be measured on the source structure alone. The equivalent force is the opposite of the force at the interface, if the interface dofs are set to zero $\boldsymbol{u}_2 = 0$ (blocked). The equivalent force can also be found in-situ, by using response measurements and measured frequency response functions (FRFs) on the assembly, see [4]. Other methods to determine the equivalent force can be found in literature, e.g. in [1].

Applying the equivalent force $\boldsymbol{f}^{\text{eq}}_2$ on the interface of the assembled system has the same effect as the original excitation force $\boldsymbol{f}_1$. The equivalent force is a property of the source only and thus it is still valid even if the receiver structure is changed.

The workflow to apply component-based TPA is the following. The equivalent force $\boldsymbol{f}^{\text{eq}}_2$ is measured or estimated, for example as a blocked force of the source structure alone. It is also necessary to measure the FRFs $\boldsymbol{Y}^{SR}_{32}$ of the assembly $SR$ from the interface dofs 2 to the response dofs 3. With this, we unknown responses $\boldsymbol{u}_3$ of the receiver can be estimated according to eq. (1).

Transmissibilty-based TPA is not presented here, but can be found in the literature (see e.g. [1, 2] for an overview).

## *Equivalent force for transfer function estimation*

The idea of an equivalent force from component-based TPA is now used to estimate the FRFs of an assembled system.

Figure 2a shows the presented component-based TPA problem, where the goal is to estimate the response $\boldsymbol{u}_3$ using the equivalent force. Now, the problem is interpreted differently, see fig. 2b, in case it is possible to measure the response $\boldsymbol{u}_2$ or $\boldsymbol{u}_3$ in operation. Here, the goal is to estimate FRFs of the assembly. The previously measured or estimated equivalent force $\boldsymbol{f}^{\text{eq}}_2$ and the response $\boldsymbol{u}_2$ or $\boldsymbol{u}_3$ that is measured in operation are used in an FRF-estimator to obtain the FRFs $\boldsymbol{Y}^{SR}_{22}$ or $\boldsymbol{Y}^{SR}_{32}$ respectively. This idea has been used in [5].

To properly estimate the FRFs in a wide frequency range the equivalent forces must be broadband. This is not always the case, as $\boldsymbol{f}^{\text{eq}}_2$ is a consequence of the unknown operational forces $\boldsymbol{f}_1$, which may only excite some frequencies in practice.

If the interface is not described by a single dof but multiple dofs, a multiple-input multiple-output (MIMO) FRF-estimation has to be used. This presents the further requirements that the equivalent forces $\boldsymbol{f}^{\text{eq}}_2$ must be uncorrelated, cf. [6, p. 372]. As the equivalent forces depend on the dynamics of the source structure, some correlation cannot be avoided.

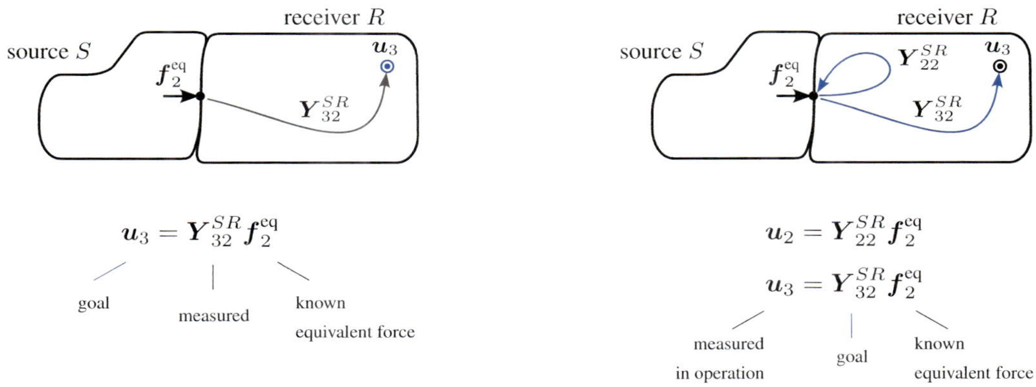


(a) Typical problem of component-based TPA, [1]. (b) For estimating transfer functions.

**Fig. 2** Comparison of component-based TPA with FRF estimation using equivalent forces.

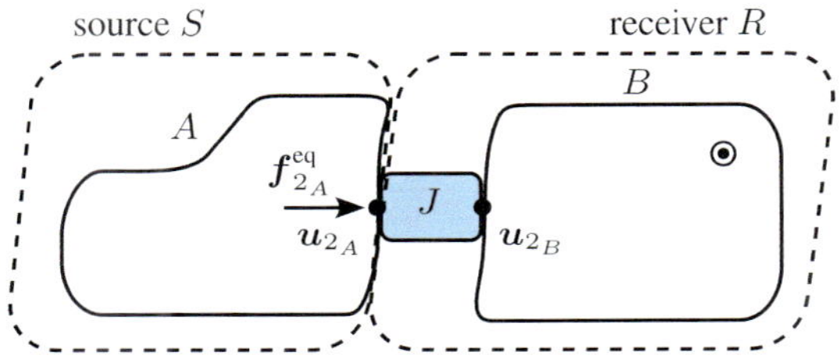


**Fig. 3** Assembly $AJB$ connected by a joint $J$ and divided into a source $S$ and a receiver $R$.

### *Towards condition monitoring of joints*

With the presented method, the FRFs $\boldsymbol{Y}_{32}^{SR}$ and $\boldsymbol{Y}_{22}^{SR}$ can be measured in operation. As long as the source structure and the operational forcing $\boldsymbol{f}_1$ does not change, the equivalent force does not change, and the estimated FRFs are valid. This is even true, if the receiver system changes, as this does not influence the equivalent force $\boldsymbol{f}_2^{\text{eq}}$. With this, it is possible to observe changes in the estimated FRFs. The FRFs would change, if the dynamics of the receiver structure change. The current FRF that is measured during operation can be used for condition monitoring.

Figure 3 shows two substructures $A$ and $B$ that are connected via a joint $J$ that can change and should be monitored. The structure $A$ is considered the source structure $S$, while the part $JB$ of the assembly is considered the receiver structure $R$. In a preliminary step, the equivalent force $\boldsymbol{f}_{2_A}^{\text{eq}}$ is measured or estimated. By measuring or estimating $\boldsymbol{u}_{2_A}$ and $\boldsymbol{u}_{2_B}$ during operation, a part of the FRF matrix of the assembly at the interface dofs can be estimated:

$$\begin{pmatrix} \boldsymbol{u}_{2_A} \\ \boldsymbol{u}_{2_B} \end{pmatrix} = \begin{pmatrix} \boldsymbol{Y}_{2_A,2_A}^{AJB} \\ \boldsymbol{Y}_{2_B,2_A}^{AJB} \end{pmatrix} \boldsymbol{f}_{2_A}^{\text{eq}} \tag{2}$$

Now this matrix is not the full interface FRF-matrix of the assembly, but only the columns corresponding to $2_A$. So, it is for this general case not possible to directly apply the frequency-based substructuring techniques (that were used in [7]). For example, applying primal decoupling, would be not possible here, because an inversion of the submatrix $\boldsymbol{Y}_{2_A,2_A}^{AJB}$ corresponds to the condensation of the dynamics of $AJB$ on point $2_A$. It would not be possible to remove/decouple the dynamics of dynamic substructure $B$ from this.

However, in the special case that the joint is quasi-static, the equations of dual coupling with a quasi-static joint (via a pseudo-inverse) can be rearranged to obtain the quasi-static joint. This idea was presented in [8], where FRF-measurements of internal points (away from the interface) were used. It is also possible to use only interface dofs, which can be seen in [9] and is given here shortly in the notation used in this paper. The coupling equation for a quasi-static joint to obtain the assembly FRF at dofs $2_A$ is given as:

$$\boldsymbol{Y}_{2_A,2_A}^{AJB} = \boldsymbol{Y}_{2_A,2_A}^{A} - \boldsymbol{Y}_{2_A,2_A}^{A}\left(\boldsymbol{Y}_{2_A,2_A}^{A} + \boldsymbol{Y}_{2_B,2_B}^{B} + \tilde{\boldsymbol{Y}}^{J}\right)^{-1}\boldsymbol{Y}_{2_A,2_A}^{A}, \quad \text{with} \quad \boldsymbol{u}_{2_B} - \boldsymbol{u}_{2_A} = \tilde{\boldsymbol{Y}}^{J}\boldsymbol{\lambda} \tag{3}$$

This equation can be rearranged to obtain the joint dynamics

$$\tilde{\boldsymbol{Y}}^{J} = \boldsymbol{Y}^{A}_{2_A,2_A}\left(\boldsymbol{Y}^{A}_{2_A,2_A} - \boldsymbol{Y}^{AJB}_{2_A,2_A}\right)^{-1}\boldsymbol{Y}^{A}_{2_A,2_A} - \boldsymbol{Y}^{A}_{2_A,2_A} - \boldsymbol{Y}^{B}_{2_B,2_B} \quad (4)$$

This requires additional measurements on the individual substructures $A$ and $B$, but enables isolation of a quasi-static joint.

### *Estimation of joint dynamics connected to a rigid wall*

Now, a special case is considered, where the source structure is connected to the environment via a joint, see fig. 4. This can be seen as replacing substructure $B$ with a rigid wall. So, for this system, the full interface FRF matrix can be estimated,

$$\boldsymbol{u}_{2_A} = \boldsymbol{Y}^{AJB}_{2_A,2_A}\boldsymbol{f}^{\text{eq}}_{2_A}, \quad (5)$$

as for this case $\boldsymbol{u}_{2_B} = \boldsymbol{0}$ and thus $\boldsymbol{Y}^{AJB}_{2_A,2_B} = \boldsymbol{Y}^{AJB}_{2_B,2_A} = \boldsymbol{Y}^{AJB}_{2_B,2_B} = \boldsymbol{0}$. So, for this case a dynamic decoupling (frequency-based substucturing, see [2]) procedure can be applied that does not require the joint to be quasi-static.

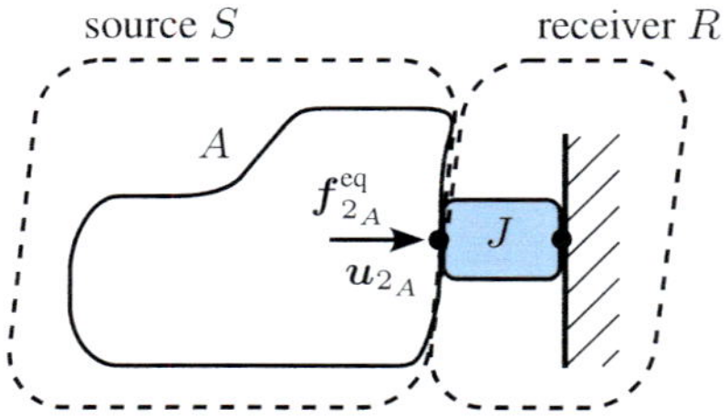


**Fig. 4** Substructure $A$ that is connected via a joint $J$ to the rigid environment.

Here, a primal decoupling is applied to estimate the dynamic stiffness of the joint, see [2, 7],

$$\boldsymbol{Z}^{J}_{2_A,2_A} = \boldsymbol{Z}^{AJB}_{2_A,2_A} - \boldsymbol{Z}^{A}_{2_A,2_A}, \quad \text{with} \quad \boldsymbol{Z}^{AJB}_{2_A,2_A} = \left(\boldsymbol{Y}^{AJB}_{2_A,2_A}\right)^{-1} \quad \text{and} \quad \boldsymbol{Z}^{A}_{2_A,2_A} = \left(\boldsymbol{Y}^{A}_{2_A,2_A}\right)^{-1}. \quad (6)$$

It is necessary to measure the FRF $\boldsymbol{Y}^{A}_{2_A,2_A}$ of substructure $A$, which has to be done on the disassembled structure, for example before the operation, when determining the equivalent forces. The resulting $\boldsymbol{Z}^{J}_{2_A,2_A}$ can be used to parameterize a simple physical model of the joint, e.g. as springs and dampers, cf. [7].

## Numerical case study

The procedure presented in section 2.4 for a structure that is connected to the rigid environment via a joint is applied to a simulated example. First, the simulation setup is presented. Then the results for the estimation of the joint dynamics are given.

### *Simulation setup*

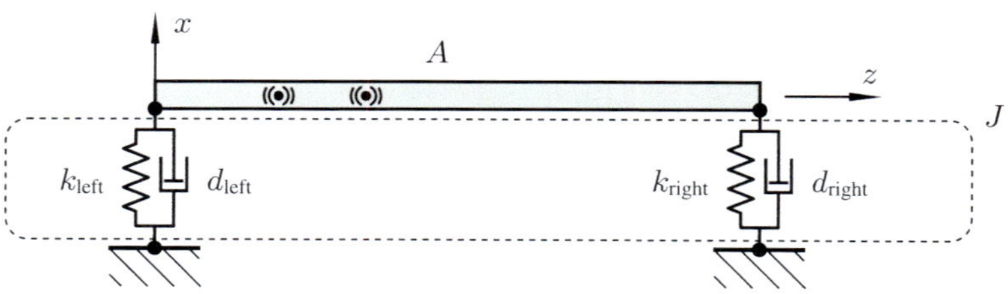


**Fig. 5** Simulation setup.

The system consists of a 2D finite element TIMOSHENKO-beam which is connected to the environment via a spring-damper system on its left and right end, see fig. 5. The Young's modulus $E$, the density $\rho$, the Poisson-ratio $\nu$, the length $L$, the diameter of the round cross-section $d$ and the mesh size $\Delta z$ are given as

$$E = 211 \cdot 10^9\,\mathrm{N\,m^{-2}}, \quad \rho = 7860\,\mathrm{kg\,m^{-3}}, \quad \nu = 0.3, \quad L = 0.5\,\mathrm{m}, \quad d = 0.01\,\mathrm{m}, \quad \Delta z = 0.01\,\mathrm{m}. \tag{7}$$

The parameters of the spring-damper elements are

$$k_{\text{left}} = 1 \cdot 10^6\,\mathrm{N/m}, \quad d_{\text{left}} = 100\,\mathrm{N\,s\,m^{-1}}, \quad k_{\text{right}} = 2 \cdot 10^5\,\mathrm{N/m}, \quad d_{\text{right}} = 200\,\mathrm{N\,s\,m^{-1}}. \tag{8}$$

The operational forces $\boldsymbol{f}_1$ are applied in $x$-direction at positions $z = \begin{pmatrix}0.05, & 0.2\end{pmatrix}\,\mathrm{m}$ as constant spectrum with random phase in the frequency domain until $2000\,\mathrm{Hz}$.

In the simulation model, the true equivalent force $\boldsymbol{f}_{2_A}^{\text{eq}}$ and the operational response $\boldsymbol{u}_{2_A}^{AJB}$ are computed in the frequency domain. They are then transformed in the time domain via an inverse fast Fourier transformation, where noise is added to the time signals. The transformation to the time domain is used, because in a real experimental application only time signals are measured directly. From the noisy time signals, the auto- and cross-power spectral densities are calculated using Welch's method and are given to a $H_1$-FRF-estimator. For these operations the free, open *Matlab*-toolbox *Abravibe* [10, 11] is used.

In detail, the steps for the simulated experiment are:

1. Obtain the equivalent force $\boldsymbol{f}_{2_A}$ from the source structure:

$$\boldsymbol{f}_{2_A}^{\text{eq}} = \left(\boldsymbol{Y}_{2_A,2_A}^{A}\right)^{-1} \boldsymbol{Y}_{2_A,1}^{A}\,\boldsymbol{f}_1 \tag{9}$$

   In an experimental application, the equivalent force would be measured, for example as blocked forces or in-situ.
2. Obtain the operational displacement $\boldsymbol{u}_{2_A}$ in the assembly:

$$\boldsymbol{u}_{2_A} = \boldsymbol{Y}_{2_A,1}^{AJB}\,\boldsymbol{f}_1 \tag{10}$$

   In an experimental application, this would be measured or estimated during the operation.
3. Transform $\boldsymbol{f}_{2_A}^{\text{eq}}$ and $\boldsymbol{u}_{2_A}$ in the time domain via an inverse FFT:

$$\tilde{\boldsymbol{f}}_{2_A}^{\text{eq}}(t) = \mathrm{IFFT}\left(\boldsymbol{f}_{2_A}^{\text{eq}}(\omega)\right), \qquad \tilde{\boldsymbol{u}}_{2_A}(t) = \mathrm{IFFT}\left(\boldsymbol{u}_{2_A}(\omega)\right) \tag{11}$$

   The time signals are periodically repeated and noise is added to simulate measurement noise.
4. Apply Welch's method to obtain auto- and cross power spectral densities and estimate $\boldsymbol{Y}_{2_A,2_A,\text{estim}}^{AJB}$ using an $H_1$-FRF-estimator, [6]:

$$\boldsymbol{Y} = \boldsymbol{G}_{\boldsymbol{yx}}\boldsymbol{G}_{\boldsymbol{xx}}^{-1}, \qquad \text{here, specifically:} \quad \boldsymbol{Y}_{2_A,2_A,\text{estim}}^{AJB} = \boldsymbol{G}_{\boldsymbol{u}_{2_A}\boldsymbol{f}_{2_A}^{\text{eq}}}\boldsymbol{G}_{\boldsymbol{f}_{2_A}^{\text{eq}}\boldsymbol{f}_{2_A}^{\text{eq}}}^{-1}, \tag{12}$$

   where $\boldsymbol{G}$ denotes an averaged auto or cross power spectral density. It can be seen that the input signal $\boldsymbol{x} = \boldsymbol{f}_{2_A}^{\text{eq}}$ should be uncorrelated, so that the inversion is well-conditioned.
   The same procedure would be applied in an experimental application.
5. Estimate the joint stiffness from $\boldsymbol{Y}_{2_A,2_A,\text{estim}}^{AJB}$

$$\boldsymbol{Z}_{2_A,2_A,\text{estim}}^{J} = \left(\boldsymbol{Y}_{2_A,2_A,\text{estim}}^{AJB}\right)^{-1} - \left(\boldsymbol{Y}_{2_A,2_A}^{A}\right)^{-1}. \tag{13}$$

   Select a clean frequency range in $\boldsymbol{Z}_{2_A,2_A,\text{estim}}^{J}$ and fit stiffness and damping in the real and imaginary part of $\boldsymbol{Z}_{2_A,2_A,\text{estim}}^{J}$ respectively for $\boldsymbol{Z}_{2_A,2_A,\text{identified}}^{J} = \mathrm{diag}\left(k_i + j\omega d_i\right)$.

## Results

The correlation between two input signals $x_n$ and $x_m$ can be computed by

$$\gamma_{x_n,x_m}^2 = \frac{\left|G_{x_n,x_m}\right|^2}{G_{x_n,x_n}G_{x_m,x_m}}, \tag{14}$$

where $G_{x_n,x_n}$ and $G_{x_m,x_m}$ are the auto-power spectral densities of $x_n$ and $x_m$ respectively and $G_{x_n,x_m}$ is the cross-power spectral density between $x_n$ and $x_m$. Figure 6 gives the correlation between the two entries of the blocked force $\boldsymbol{f}_{2_A}^{\text{eq}}$.

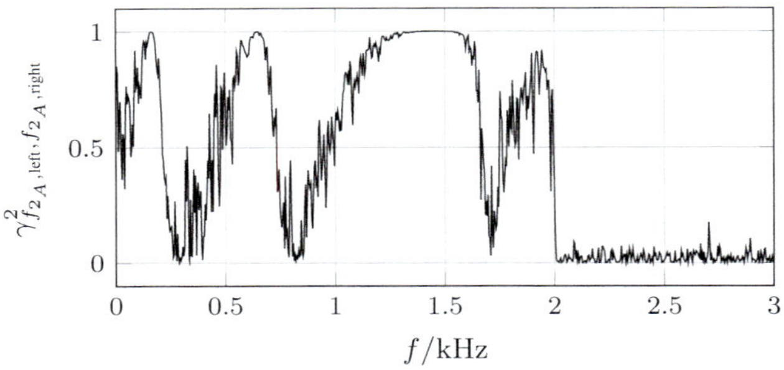


**Fig. 6** Correlation $\gamma^2_{f_{2_A,\text{left}},f_{2_A,\text{right}}}$ between entries of equivalent force $\boldsymbol{f}^{\text{eq}}_{2_A}$.

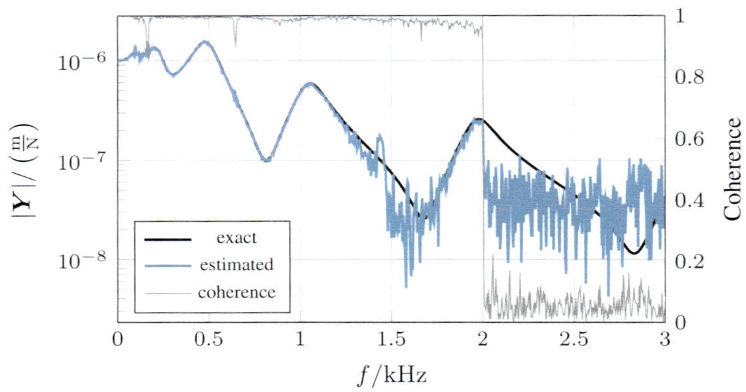


**Fig. 7** Magnitude of the assembly FRF $|\boldsymbol{Y}^{AJB}_{2_A,2_A}|$ from estimation and exact values. Exemplary entry for excitation and response on the left side.

The correlation of the operational forces is not shown here, because it is zero for the entire frequency range meaning it is fully uncorrelated. This is the case, because we chose $\boldsymbol{f}_1$ to have a random phase in the frequency domain. Naturally, the correlation between the equivalent forces is not zero, because it depends on the dynamics of substructure $A$, see eq. (9). The FRF-estimation is ill-conditioned in regions with high correlation, because it involves an inversion of the auto power spectral density of the input, see eq. (12).

Figure 7 shows the estimated FRF $|\boldsymbol{Y}^{AJB}_{2_A,2_A,\text{estim}}|$ for the first diagonal entry and compares it to the exact values. Also, the multiple coherence for this FRF is given. It can be seen that the results agree well and that the coherence is high in the frequency range of excitation (until $2000\,\text{Hz}$). The effect of adding noise can be seen especially well in the region with low response amplitude. It should be noted that the effect of a large correlation between the equivalent forces seems to only have a negative effect on the coherence, where the correlation is almost one. These are the frequencies, where the coherence drops sharply in fig. 7.

Figure 8 shows the diagonal entries of the dynamic joint stiffness $\boldsymbol{Z}^J_{2_A,2_A}$ in real and imaginary parts. It gives the estimated and the exact values of $\boldsymbol{Z}^J_{2_A,2_A}$ and the fitted values in the marked frequency range. Additionally, the correlation between the input signal from fig. 6 is also plotted in the figure. The fitted stiffness and damping values and its deviation to the exact values is given in table 1. It can be seen that the identified stiffness and damping values agree very well for this example. It is very interesting that the values of $\boldsymbol{Z}^J_{2_A,2_A,\text{estim}}$ appear noisy or give bad results when the correlation between the input signals is bad. This is not obvious from looking at the coherence in fig. 7 which was good in the whole frequency range. So, to assess the quality of the identified stiffness, it is necessary to not only check the coherence of the estimated FRFs but also the correlation between the time signals. The correlation does not have to be exactly zero, as it can be seen that the estimated joint stiffness is also relatively clean in regions with some correlation between the input signals.

In appendix A, a case is considered, where only one force $f_1$ is applied on the beam at position $z = 0.05\,\text{m}$. This causes a very high correlation of the equivalent forces $\boldsymbol{f}^{\text{eq}}_{2_A}$. This causes a bad estimation of the assembly FRFs $\boldsymbol{Y}^{AJB}_{2_A,2_A,\text{estim}}$ and thus also of the estimated dynamic joint stiffness $\boldsymbol{Z}^J_{2_A,2_A,\text{estim}}$, which is why the joint identification fails for this case. This effect cannot be detected by the coherence of the FRF, but only through considering the correlation of the input signals.

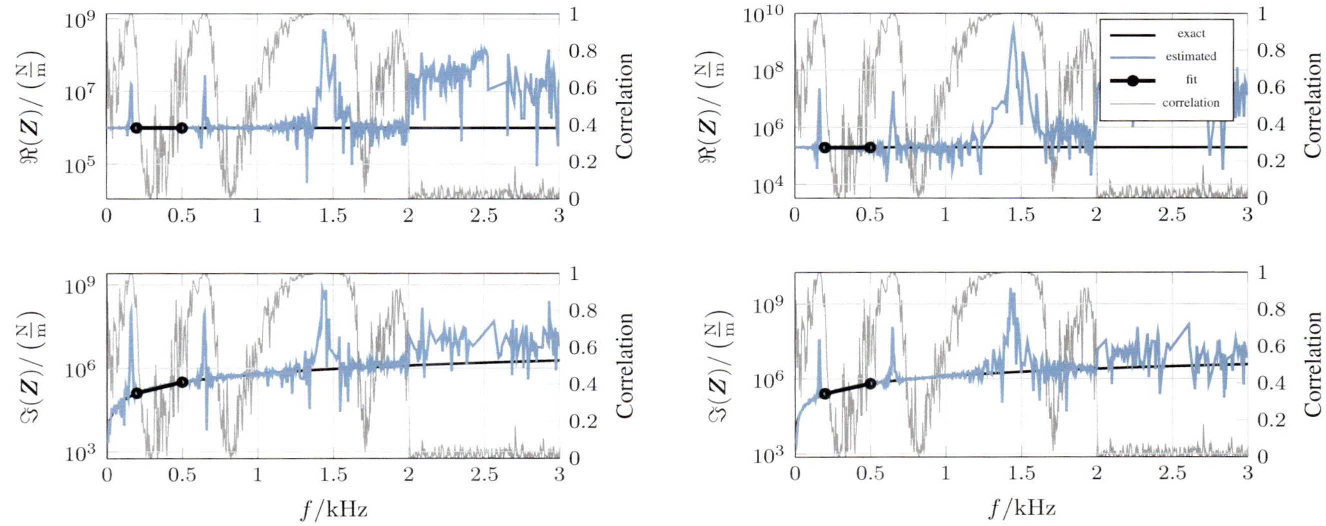


(a) Left joint: $Z^J_{2_{A,\mathrm{left}},2_{A,\mathrm{left}}}$.

(b) Right joint: $Z^J_{2_{A,\mathrm{right}},2_{A,\mathrm{right}}}$.

**Fig. 8** Dynamic joint stiffness $Z^J_{2_A,2_A}$.

**Table 1** Identified stiffness and damping.

| | exact value | estimated value | relative error |
|---|---|---|---|
| $k_{\mathrm{left}}$ | $1 \cdot 10^6\,\mathrm{N\,m^{-1}}$ | $9.999 \cdot 10^5\,\mathrm{N\,m^{-1}}$ | $-0.01\,\%$ |
| $d_{\mathrm{left}}$ | $100\,\mathrm{N\,s\,m^{-1}}$ | $101\,\mathrm{N\,s\,m^{-1}}$ | $1.36\,\%$ |
| $k_{\mathrm{right}}$ | $2 \cdot 10^5\,\mathrm{N\,m^{-1}}$ | $1.951 \cdot 10^5\,\mathrm{N\,m^{-1}}$ | $-2.45\,\%$ |
| $d_{\mathrm{right}}$ | $200\,\mathrm{N\,s\,m^{-1}}$ | $203\,\mathrm{N\,s\,m^{-1}}$ | $1.37\,\%$ |

The presented difficulties due to a possible correlation naturally only occurs, when a MIMO-FRF-estimation is performed, i.e. if the joint has more than one dof. In the case, where only a single one-dof joint should be identified, the FRF estimation does not have these problems.

## Conclusion and Outlook

This contribution presented a method to utilize the concept of equivalent forces, from component-based TPA, to estimate the FRFs of an assembly with a joint. In a simulation example of a simple supported beam, it was shown how this method can be used to estimate the dynamics of the mounts/joints from operational responses. This can be used in a condition monitoring to detect and quantify changes in the dynamics of the joints.

It was shown that the methods requires equivalent forces that are broadband, i.e. that excite a larger frequency range and it requires that the equivalent forces are sufficiently uncorrelated. The correlation is influenced by the correlation of the unknown operational forces and the dynamics of the active source structure. If the model of the joint is known, e.g. just a direct stiffness and damping, it may be enough if the correlation of the equivalent forces is low at some frequencies, but this was not investigated in detail here.

In a next step, the method should be applied on a real experimental example. It can be investigated how realistic operational forces lead to a correlation of the equivalent forces and thus the degradation of the joint identification. It can be discussed if the method can detect and quantify realistic changes in the joint parameters.

Another point that was not fully discussed in this contribution is how to estimate the joint dynamics for more complex assemblies, where the joint connects two flexible substructures.

## Appendix: Results for high correlation of equivalent forces

In this section, the results for the presented simulation model in section 3 are given, when only one operational force $\boldsymbol{f}_1$ is applied. It can be seen that this leads to a high correlation of the equivalent forces, see fig. 10. In this case the FRF-estimation (fig. 9) and the decoupling of the joint stiffness (fig. 10) give bad results.

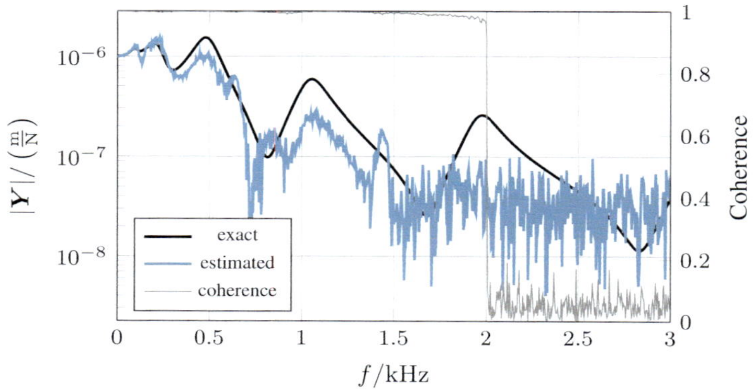


**Fig. 9** Magnitude of the assembly FRF $|\boldsymbol{Y}^{AJB}_{2_A,2_A}|$ from estimation and exact values in the case of only one operational force $f_1$. Exemplary entry for excitation and response on the left side.

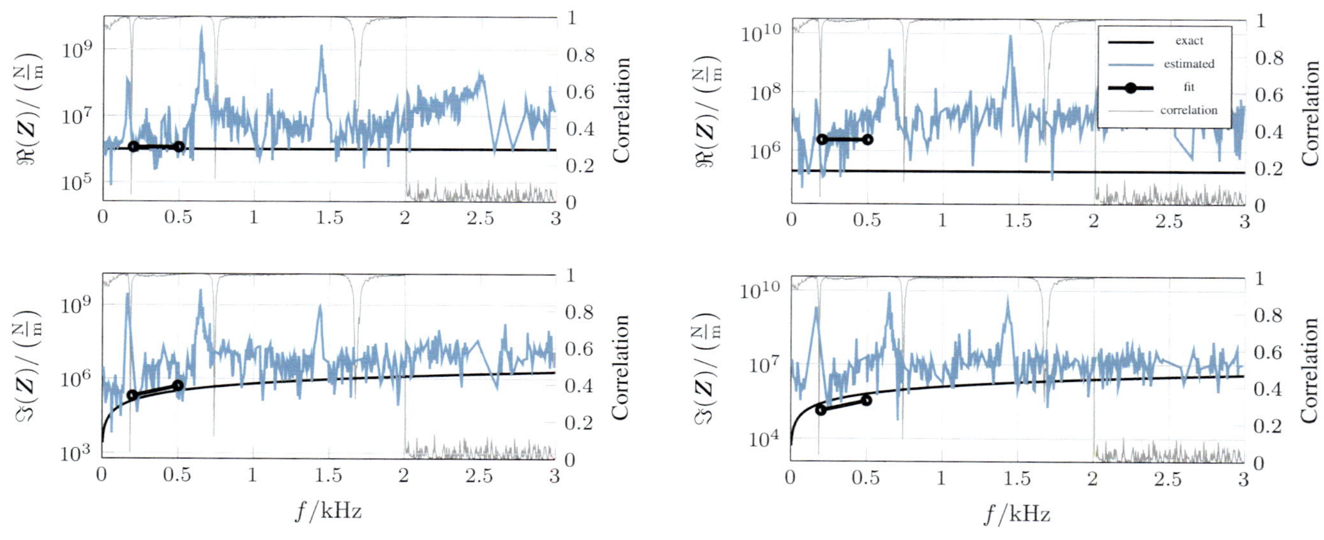


(a) Left joint: $\boldsymbol{Z}^J_{2_{A,\mathrm{left}},2_{A,\mathrm{left}}}$.

(b) Right joint: $\boldsymbol{Z}^J_{2_{A,\mathrm{right}},2_{A,\mathrm{right}}}$.

**Fig. 10** Dynamic joint stiffness $\boldsymbol{Z}^J_{2_A,2_A}$ in the case of only one operational force $f_1$.

This example highlights the importance of checking the correlation of the input signals for the FRF-estimation. In the presented method, these are the equivalent forces $\boldsymbol{f}^{\mathrm{eq}}_{2_A}$, see eq. (12).

## References

[1] Seijs, M. V. van der, Klerk, D. de, and Rixen, D. J. "General framework for transfer path analysis: History, theory and classification of techniques". In: *Mechanical Systems and Signal Processing* 68-69 (2016), pp. 217–244. DOI: 10.1016/j.ymssp.2015.08.004.

[2] Allen, M. S., Rixen, D., Seijs, M. van der, Tiso, P., Abrahamsson, T., and Mayes, R. L. *Substructuring in Engineering Dynamics*. CISM International Centre for Mechanical Sciences. Cham: Springer International Publishing, 2020. DOI: 10.1007/978-3-030-25532-9.

[3] El Mahmoudi, A., Trainotti, F., Park, K., and Rixen, D. J. "In-Situ Source Characterization for NVH Analysis of the Engine-Transmission Unit". In: *Conference Proceedings of the Society for Experimental Mechanics Series*. Vol. 4. 2021, pp. 79–91. DOI: 10.1007/978-3-030-47630-4_7.

[4] Meggitt, J. and Moorhouse, A. "On the completeness of interface descriptions and the consistency of blocked forces obtained in situ". In: *Mechanical Systems and Signal Processing* 145 (Nov. 2020), p. 106850. DOI: 10.1016/j.ymssp.2020.106850.

[5] Ocepek, D., Vrtač, T., Čepon, G., and Boltežar, M. "Estimation of the frequency response functions for operational assemblies using independent source characterization". In: *Mechanical Systems and Signal Processing* 182.July 2022 (2023). DOI: 10.1016/j.ymssp.2022.109542.

[6] Brandt, A. *Noise and Vibration Analysis*. Second. Wiley, Oct. 2023. DOI: 10.1002/9781118962176.

[7] Kreutz, M., Trainotti, F., Gimpl, V., and Rixen, D. J. "On the robust experimental multi-degree-of-freedom identification of bolted joints using frequency-based substructuring". In: *Mechanical Systems and Signal Processing* 203 (Nov. 2023), p. 110626. DOI: 10.1016/j.ymssp.2023.110626.

[8] Özşahin, O., Ertürk, A., Özgüven, H. N., and Budak, E. "A closed-form approach for identification of dynamical contact parameters in spindle-holder-tool assemblies". In: *International Journal of Machine Tools and Manufacture* 49.1 (2009), pp. 25–35. DOI: 10.1016/j.ijmachtools.2008.08.007.

[9] Trainotti, F., Brøns, M., Klaassen, S., and Rixen, D. J. "Revisiting the Dual Admittance-Based Quasi-Static Formulation for the Identification of Linear Joint Dynamics with Dynamic Substructuring". In: *IMAC XLIII*. Orlando, FL, USA: Society for Experimental Mechanics, 2025.

[10] Brandt, A. "The ABRAVIBE toolbox for teaching vibration analysis and structural dynamics". In: *Conference Proceedings of the Society for Experimental Mechanics Series* 6 (2013), pp. 131–141. DOI: 10.1007/978-1-4614-6546-1_13.

[11] Brandt, A. *ABRAVIBE - A MATLAB toolbox for noise and vibration analysis*. www.abravibe.com. 2024.

# Chapter 7
# Frequency Response Function Expansion using a Symmetry Preserving Gysin Expansion Technique

**Christopher Page and Peter Avitabile**

**Abstract** The theoretical basis for frequency based substructuring (FBS) has been established for decades. Despite this, using FBS with measured data is extremely challenging due to measurement errors and practical limitations related to measuring rotational degrees-of-freedom (r-DoF). This paper revisits a previously proposed approach, known as Gysin expansion, that models the response using columns of an impedance matrix from a finite element model (FEM), expanding measured FRFs in a least squares sense. The original technique is shown to be robust with respect to FEM errors, but the expanded FRFs lack reciprocity. FRF reciprocity implies a passive structural system, which is an important characteristic for FBS applications. In this work, a novel extension of Gysin expansion is presented that ensures the expanded FRFs are reciprocal (symmetric) and thus describe a passive structural system. The extended method is also shown to be robust with respect to modeling errors and can be used to estimate unmeasured r-DoF. Moreover, the technique can reconstruct an unmeasured drive-point partition of an FRF matrix using measured cross-FRFs only. Because experimental techniques used to acquire drive-point measurements often introduce bias errors that render FBS predictions functionally useless, this new approach to FRF expansion helps mitigate the effects of some measurement errors on FBS. An Example using numerically derived data is presented. Similarities to other techniques, such as System Equivalent Modeling Mixing (SEMM) are highlighted.

**Keywords** Gysin expansion · Reciprocity · Expansion techniques · Rotational DoF · Frequency based substructuring

## Introduction

Frequency based substructuring requires frequency response functions (FRFs) at substructure interfaces that are often inaccessible or unmeasurable using conventional testing methodologies. Techniques for FRF expansion include the dynamic expansion [1], the virtual point transformation (VPT) [2] and system equivalent mixing model (SEMM) [3, 4] amongst others. However, these expansion techniques result in expanded FRFs which violate important inductive biases from modal theory, e.g., reciprocity, passivity or minimality (i.e., rank-1 modal residues). This paper presents an enhanced FRF expansion methodology which produces reciprocal FRFs [5]. The methodology is applied to FRFs expanded using a formulation of dynamic expansion known as Gysin expansion [6] which is more robust to modeling errors between the FEM derived basis functions and experimental data. The paper highlights that the long established Gysin expansion technique is identical the more recently proposed SEMM.

## FRF Expansion Methodologies

Dynamic expansion is an expansion technique that relies on basis functions derived from a FEM's dynamic stiffness matrix. The expansion may be expressed in an impedance or admittance form. The impedance form is derived by partitioning the

Christopher Page
Noise Control Engineering, 85 Rangeway Road, Billerica, Massachusetts 01862
e-mail: c.page@noise-control.com

Peter Avitabile
Structural Dynamics and Acoustic Systems Laboratory, University of Massachusetts - Lowell, One University Drive, Lowell, Massachusetts 01854
e-mail: Peter_Avitabile@uml.edu



Walter D'Ambrogio, et al. (eds.), *Dynamic Substructuring & Transfer Path Analysis, Vol. 4*, Conference Proceedings of the Society for Experimental Mechanics Series, https://doi.org/10.13052/9788743801498-7

FEM's dynamic stiffness matrix along a set of, i, interior DoF and b, boundary/measured DoF as

$$\begin{bmatrix} \mathbf{Z}_{ii} & \mathbf{Z}_{ib} \\ \mathbf{Z}_{bi} & \mathbf{Z}_{bb} \end{bmatrix} \begin{bmatrix} u_i \\ u_b \end{bmatrix} = \begin{bmatrix} f_i \\ f_b \end{bmatrix} \tag{1}$$

where $u_i$ represents the displacements at the interior DoF, $u_b$ the displacements at the boundary DoF, $f_i$ are the external forces on the interior DoF and $f_b$ are the external forces acting on the boundary DoF. The impedance form of dynamic expansion expresses the displacement at the interior DoF in terms of the external forces on the interior DoF and the measured/boundary DoF according to

$$u_i = \mathbf{Z}_{ii}^{-1} f_i - \mathbf{Z}_{ii}^{-1} \mathbf{Z}_{ib} u_b \tag{2}$$

By leveraging block matrix inversion the admittance form of dynamic expansion can be shown to be

$$u_i = \mathbf{Y}_{ii} - \mathbf{Y}_{ib} \mathbf{Y}_{bb}^{-1} \left( \mathbf{Y}_{bb} - \hat{\mathbf{Y}}_{bb} \right) \mathbf{Y}_{bb}^{-1} \mathbf{Y}_{bi} f_i + \mathbf{Y}_{ib} \mathbf{Y}_{bb}^{-1} \hat{\mathbf{Y}}_{bb} f_b \tag{3}$$

and by extension the entire expanded FRF matrix in block format becomes

$$\mathbf{Y}^{(\text{expanded})} = \begin{bmatrix} \mathbf{Y}_{ii} - \mathbf{Y}_{ib} \mathbf{Y}_{bb}^{-1} \left( \mathbf{Y}_{bb} - \hat{\mathbf{Y}}_{bb} \right) \mathbf{Y}_{bb}^{-1} \mathbf{Y}_{bi} & \mathbf{Y}_{ib} \mathbf{Y}_{bb}^{-1} \hat{\mathbf{Y}}_{bb} \\ \hat{\mathbf{Y}}_{bb} \mathbf{Y}_{bb}^{-1} \mathbf{Y}_{bi} & \hat{\mathbf{Y}}_{bb} \end{bmatrix} \tag{4}$$

where in (3) and (4), $\hat{\mathbf{Y}}_{bb}$ indicates admittance functions that come from measurements with all other admittance functions synthesized from the FEM. The admittance form of dynamic expansion is equivalent to SEMM with a conforming interface [3].

Equation (4) illustrates a particular defect with dynamic expansion. Namely when there are inconsistencies between the measured admittance functions and the FEM synthesized FRFs at the boundary DoF, the boundary partition admittance functions do not cancel each other out, producing spurious peaks in the FRFs at resonant frequencies of the FEM constrained at the boundary DoF. In the cross-FRFs imperfect pole cancellation in the term $\hat{\mathbf{Y}}_{bb} \mathbf{Y}_{bb}^{-1}$ has the same result. When $\hat{\mathbf{Y}}_{bb} = \mathbf{Y}_{bb}$ the dynamic expansion process is exact. The defect will be illustrated using an analytical case study later in the paper.

## Gysin Expansion

Partitioning the columns of the finite element model dynamic stiffness matrix along interior and boundary DoF as

$$\begin{bmatrix} \mathbf{Z}_{ni} & \mathbf{Z}_{nb} \end{bmatrix} \begin{bmatrix} u_i \\ u_b \end{bmatrix} = \begin{bmatrix} f_i \\ f_b \end{bmatrix} \tag{5}$$

Gysin expansion estimates the response at the interior DoF by

$$u_i = \mathbf{Z}_{ni}^{\dagger} \begin{bmatrix} f_i \\ f_b \end{bmatrix} - \mathbf{Z}_{ni}^{\dagger} \mathbf{Z}_{nb} u_b \tag{6}$$

where $\dagger$ denotes the generalized inverse. Functionally, Gysin expansion is similar to dynamic expansion. However, all $n$ finite element DoF are used in the expansion, therefore any error between the tested structure and finite element model is distributed across all FEM DoF. In this way, Gysin expansion avoids the defects of dynamic expansion when the finite element model and measured data are inconsistent. Gysin expansion is the impedance equivalent to SEMM over an extended interface [4].

## Symmetry Constrained Gysin Expansion

A formulation of Gysin expansion which respects reciprocity is obtained by recognizing the expansion process as a linear regression problem in the columns of the FEM's dynamic stiffness matrix associated with the interior DoF. Modifying (5) such that the interior DoF are replaced by the FRFs between interior DoF, the forces on the interior DoF are replaced by an identity matrix, the forces on the boundary DoF by a null matrix and the boundary DoF response by the cross-FRFs between the interior and boundary DoF gives

$$\mathbf{Z}_{ni} \mathbf{Y}_{ii} = \begin{bmatrix} \mathbf{I}_i \\ 0 \end{bmatrix} - \mathbf{Z}_{nb} \hat{\mathbf{Y}}_{bi} \tag{7}$$

Introducing a symmetry constraint for the form

$$\mathbf{B}\text{vec}\left(\mathbf{Y}_{ii}\right) = 0 \tag{8}$$

where $\mathbf{B}$ is a signed Boolean matrix that "picks-out" those entries in $\mathbf{Y}_{ii}$ that are related by symmetry, i.e. $\mathbf{Y}_{kj} - \mathbf{Y}_{jk} = 0$, q and combining the constraint with a vectorized form of (7) produces a constrained linear regression problem of the form

$$\begin{bmatrix} \left(\mathbf{I}_i \otimes \mathbf{Z}_{ni}^T \mathbf{Z}_{ni}\right) & \mathbf{B}^T \\ \mathbf{B} & 0 \end{bmatrix} \begin{bmatrix} \text{vec}\left(\mathbf{Y}_{ii}\right) \\ \lambda \end{bmatrix} = \begin{bmatrix} \text{vec}\left(\mathbf{Z}_{ni}^T \begin{bmatrix} \mathbf{I}_i \\ 0 \end{bmatrix} - \mathbf{Z}_{ni}^T \mathbf{Z}_{nb} \hat{\mathbf{Y}}_{bi}\right) \\ 0 \end{bmatrix} \tag{9}$$

In (9), $\lambda$ is a vector of Lagrange multipliers associated with the symmetry constraint. Solving (9) for the Lagrange multipliers, the FRFs expanded to the interior DoF can be shown to be

$$\text{vec}\left(\mathbf{Y}_{ii}\right) = \text{vec}\left(\mathbf{Y}_{ii}^0\right) - \left(\mathbf{I}_i \otimes \left(\mathbf{Z}_{ni}^T \mathbf{Z}_{ni}\right)^{-1}\right) \mathbf{B}^T \left[\mathbf{B}\left(\mathbf{I}_i \otimes \left(\mathbf{Z}_{ni}^T \mathbf{Z}_{ni}\right)^{-1}\right) \mathbf{B}^T\right]^{-1} \mathbf{B}\text{vec}\left(\mathbf{Y}_{ii}^0\right) \tag{10}$$

where $\mathbf{Y}_{ii}^0$ represents the Gysin expanded FRFs with no reciprocity constraint.

## Numerical Case Study

In this section the defects of conventional dynamic expansion are demonstrated using data synthetically generated from two FEMs. Standard Gysin expansion is shown to eliminate the defects of dynamic expansion but produce expanded FRFs that are not reciprocal. Finally the symmetry constrained Gysin expansion process is applied resulting in reciprocal FRFs.

Figure (1) presents the finite element models used to generated data for this case study. The model on the left was used to generate the measured data. The model on the right was used for dynamic expansion. Modeling errors were introduced by using two different physics models for beam bending, perturbing material properties and using different boundary conditions for the 'test' finite element model and 'expansion' model. These differences are summarized in the table in Figure (1).

**Table:** Example of Errors Between Tested and Modeled Structure

| Model | Physics | Length [in.] | Width [in.] | Thickness [in.] | Elastic Modulus [psi] | Density [lb/in³] | Boundary Rotary Stiffness [lbf-in/rad] | Boundary Rotary Stiffness [lbf/in] |
|---|---|---|---|---|---|---|---|---|
| Test | Timoshenko | 35.8125 | 2.85 | 1.02 | 11E6 | 0.103 | 1E4 (left) /9.87E3 (right) | 1.17E3 (left) /9.3E3 (right) |
| FEM | Euler-Bernoulli | 36 | 3 | 1 | 10E6 | 0.1 | 0 (left) / 0 (right) | 1E3 (left) / 1E3 (right) |

**Fig. 1** Beam Model for Gysin Expansion Study

The test finite element model is used to simulated FRFs at four response points due to inputs at three reference DoF. The response measurements are indicated by the green squares in Figure (1) and the input points are marked using the impact hammer. No drive-point FRFs are acquired. Dynamic expansion is be used to estimate the drive-point partition at the reference DoF. Figure (2) compares the dynamically expanded FRFs at the reference DoF (dashed orange) with the ground truth FRFs obtained from the test model. Despite the modeling error, the dynamically expanded FRFs are accurate up to 200 Hz. However, two prominent spurious resonances at 591 and 705 Hz affect the accuracy of the expansion from 200-1000 Hz for most FRFs. Each of these spurious resonances are associated with natural frequencies of the 'FEM' model constrained at those DoF co-located with the response measurements. Figure (3) shows the same FRFs expanded using Gysin expansion. By using Gysin expansion, the spurious resonances have been eliminated completely and the expanded FRFs have excellent agreement with the ground truth FRFs. However, Figure (4) compares expanded cross-FRFs illustrating that these FRFs are not reciprocal.

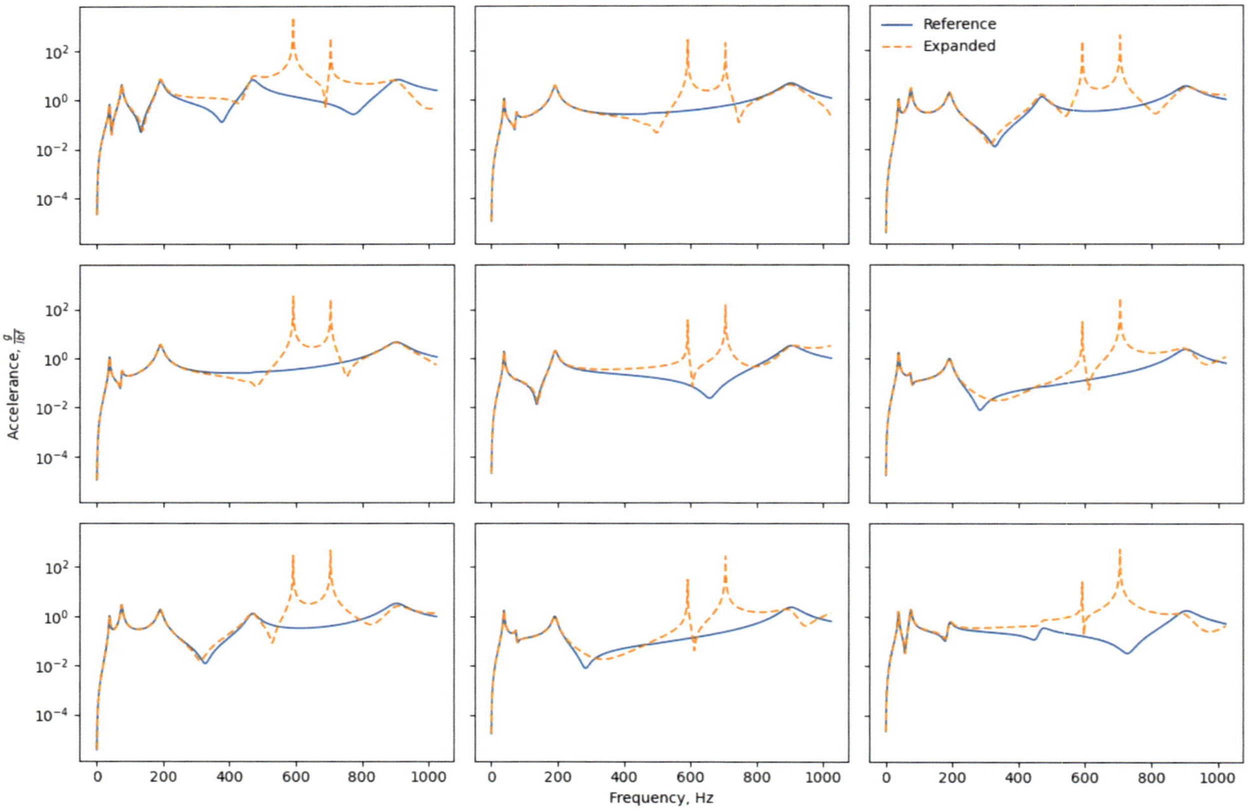


**Fig. 2** Defects of Dynamic Expansion

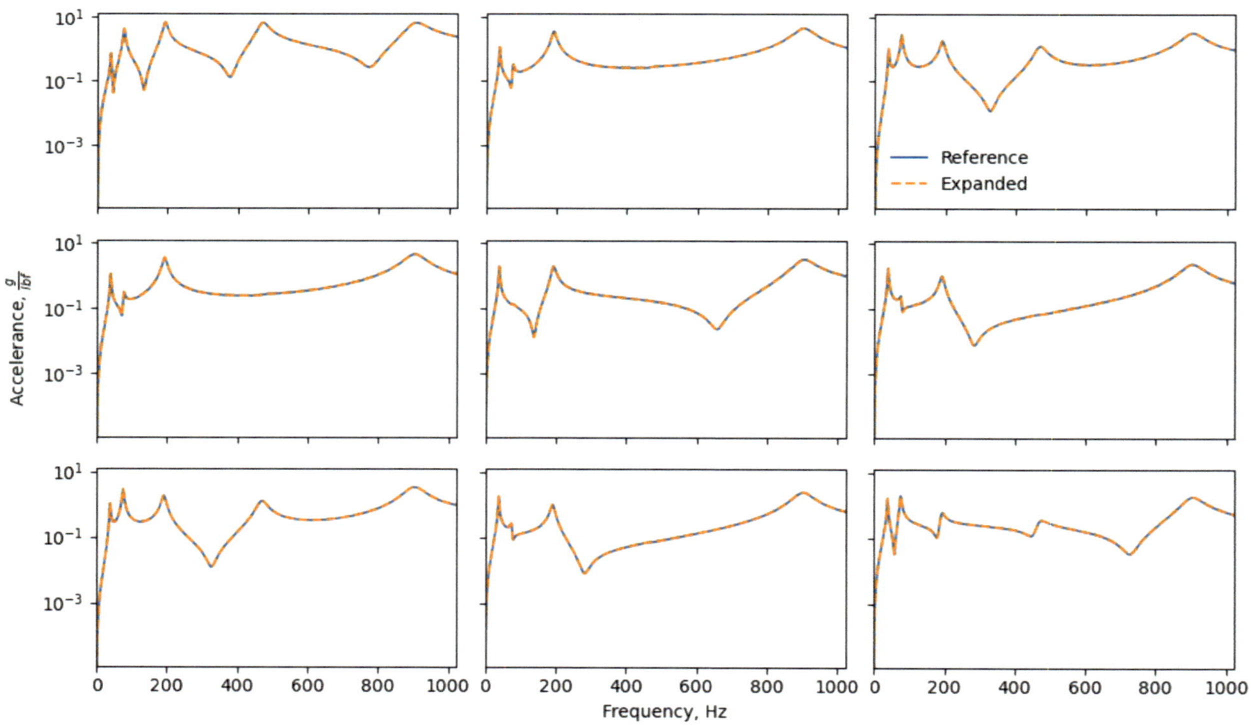


**Fig. 3** Gysin Expanded FRFs

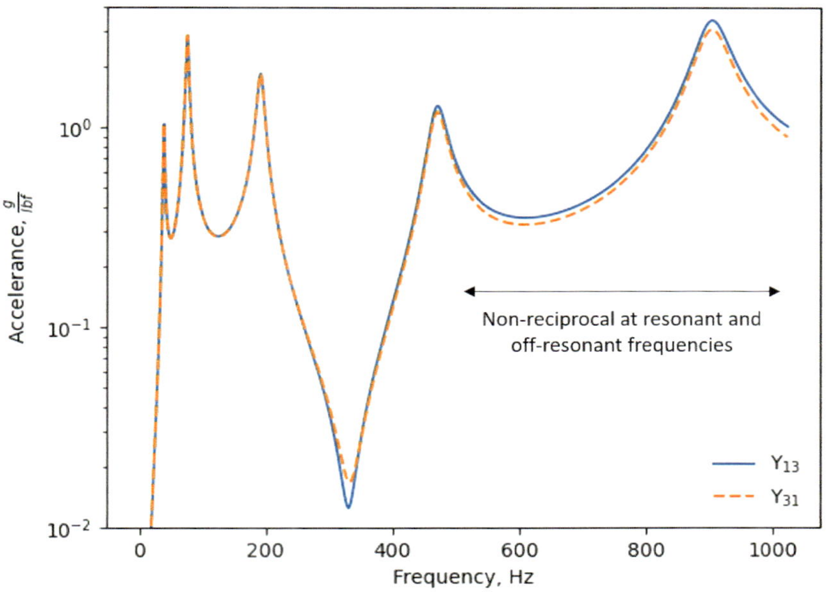


**Fig. 4** Lack of Reciprocity in Gysin Expanded FRFs

The expanded drive point partition is presented in Figure (5) and Figure (6) shows two reciprocal cross-FRFs from expanded drive point partition illustrating that the symmetry constrained was observed through the expansion process.

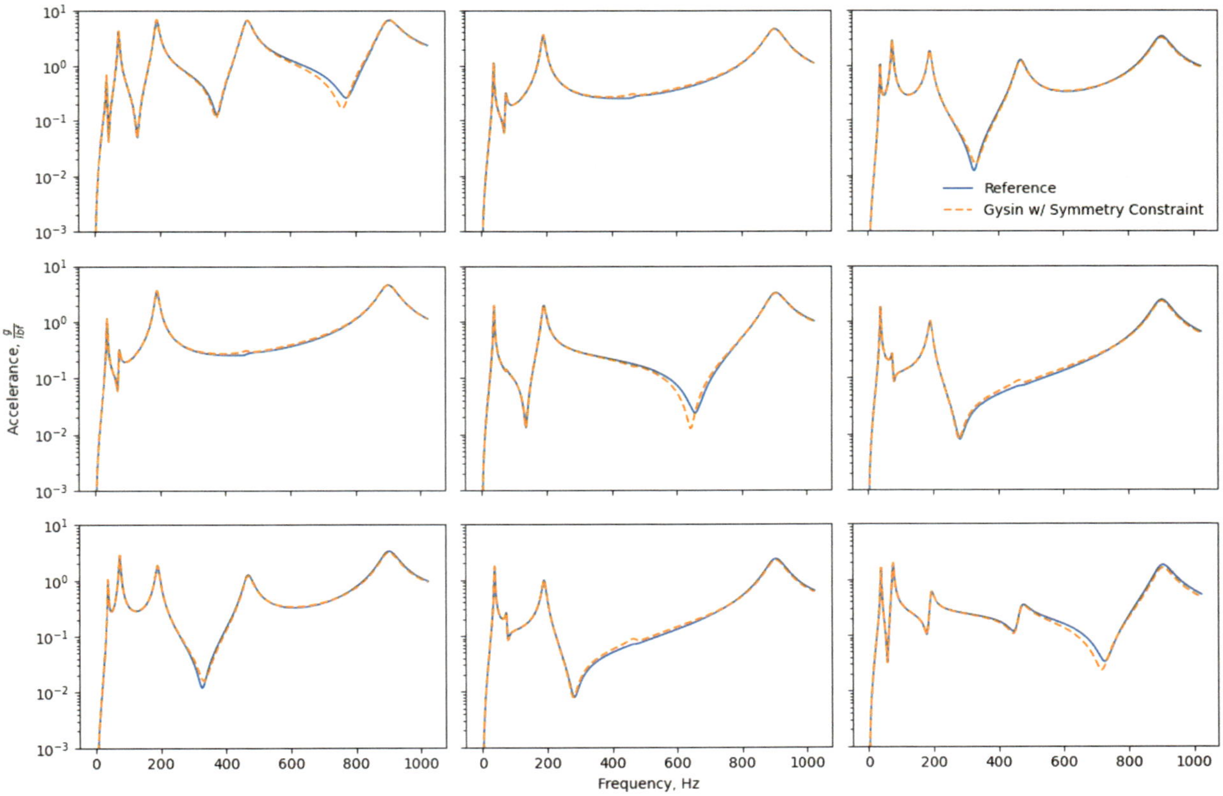


**Fig. 5** Gysin Expanded FRFs with Reciprocity Constraint

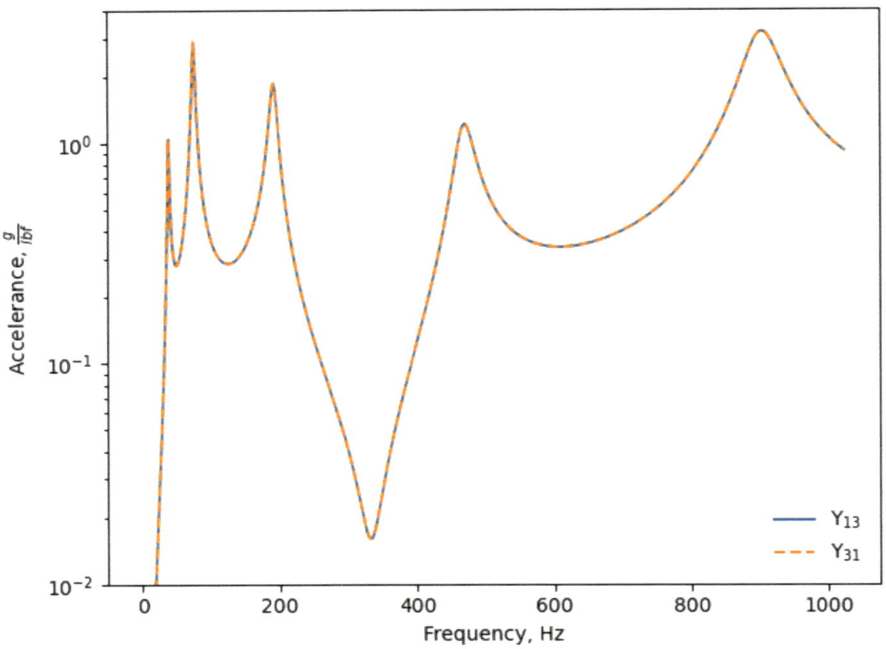


**Fig. 6** Gysin Expanded FRFs with Reciprocity Constraint; Comparison of Reciprocal Measurements

## Conclusion

In this paper, Gysin expansion was shown to eliminate defects with dynamic expansion when there are inconsistencies between the FEM basis functions and measured data. However, the Gysin expanded FRFs are not reciprocal. The Gysin expansion process was re-cast as a linear regression problem and a symmetry constraint was introduced to ensure expanded FRFs maintain reciprocity. The methodology was demonstrated using data synthetically generated from finite element models.

## References


1. Paz, M. "Dynamic condensation". *AIAA Journal*, 23(5) (1985)
2. Van der Seijs, M., van den Bosch, D., Rixen, D., and Klerk, D. "An improved methodology for the virtual point transformation of measured frequency response functions in dynamic substructuring". In *Proceedings of COMPDYN 2013*, pages 4334–4347 (2013)
3. Klaassen, S.W.B. and Seijs, M.V.v.d. "Introducing SEMM: A novel method for hybrid modelling". In Linderholt, A., Allen, M.S., Mayes, R.L., and Rixen, D., editors, *Dynamics of Coupled Structures, Volume 4*, Conference Proceedings of the Society for Experimental Mechanics Series, pages 117–125. Springer International Publishing (2018)
4. Klaassen, S.W.B., van der Seijs, M.V., and de Klerk, D. "System equivalent model mixing". *Mechanical Systems and Signal Processing*, 105:90–112 (2018)
5. Page, C. *A Unified Substructuring Approach to Fixture Neutralization*. Phd thesis, University of Massachusetts - Lowell, Lowell, CA (2020)
6. Gysin, H. "Comparison of expansion method for FE modeling error localization". In *Proceedings of the 8th International Modal Analysis Conference* (1990)

# Chapter 8
# Cutting Force Estimation in Sensor-Equipped Metal Cutting Tools Using Strain-Force Transfer Function

**Wu Peng, Anders Liljerehn, Martin Magnevall, and Dan Östling**

**Abstract** Sensor-equipped cutting tools are becoming more common in the metal cutting industry as the demand for process monitoring and automation grows. These products are used for monitoring cutting forces and vibrations during machining with the objective of providing valuable insight into tool deflection, tool wear, surface finish, and process stability. The number of sensors and their position in the cutting tool, are often restricted by price and the thermo-mechanical loads acting in the cutting zone. Obtaining accurate measurements is a challenge due to mechanical filtering between the sensor positions and the tool tip as well as due to measurement noise from the sensors and electronics. This paper presents a method to estimate dynamic loads and tool tip vibrations using recorded data from strain sensors in a cutting tool. This is achieved through output-only calibration of an analytical model of the cutting tool, which defines the strain-force relationship based solely on the measured strain response data. This approach provides key parameters, including the first resonance frequency and the damping ratio of the cutting tool. The model is then applied to eliminate unwanted dynamic effects from the measured strain data, enhancing the accuracy of predictions of forces and tool tip displacement.

**Keywords** Cutting force estimation · Deflection prediction · Metal cutting dynamics · Sensor-equipped cutting tools · Strain-force transfer function

## Introduction

As the demand for process monitoring and automation grows, sensor-equipped cutting tools are becoming increasingly common in the metal cutting industry [1]. This concept concerns expansion of existing products by adding sensors and communication devices on the tool body. Typically, two types of sensors are integrated: accelerometers, which measure vibration levels during machining, and strain sensors, which capture strain data used to estimate the deflection at the cutting tool tip. Measured data through these sensors provides the important insight into monitoring in-process conditions in terms of tool wear, surface finish and cutting process stability [2].

The machining process creates intense pressure and temperatures in the cutting zone, requiring sensors to be placed at a distance from the tool tip [3], as shown in Figure 1. This presents a challenge for accelerometers, which are ideally positioned as close to the tool tip as possible, where vibration amplitude is greatest. If not, accurately monitoring the vibration at the tool tip becomes difficult [4]. Another issue in these applications is signal distortion, caused by mechanical filtering through the transfer path [5, 6]. This means that the vibration transmitted from the tool tip to the sensor is affected by the dynamic properties of the tool body [7]. The conventional static calibration of strain sensor is hence insufficient to predict the dynamic loads acting at the tool tip. This is exemplified in Figure 2, where a statically calibrated strain sensor is used to predict the dynamic forces acting on the tool tip generated by a shaker and compared to the measurements from a laboratory-grade force sensor. The prediction of dynamic force at very low frequencies is accurate, but amplitude and phase deviations occur

Wu Peng · Anders Liljerehn
Department of Cutting Tool Research, AB Sandvik Coromant, Sandviken, SE 81181
e-mail: wu.peng@sandvik.com; anders.c.liljerehn@sandvik.com

Wu Peng · Martin Magnevall
Department of Mechanical Engineering, Blekinge Institute of Technology, Karlskrona, SE 37179
e-mail: wu.peng@sandvik.com; martin.magnevall@bth.se

Dan Östling
Department of Damped Solutions, Sandvik Teeness AS, Trondheim, NO 7053
e-mail: dan.ostling@sandvik.com



Walter D'Ambrogio, et al. (eds.), *Dynamic Substructuring & Transfer Path Analysis, Vol. 4*, Conference Proceedings of the Society for Experimental Mechanics Series, https://doi.org/10.13052/9788743801498-8

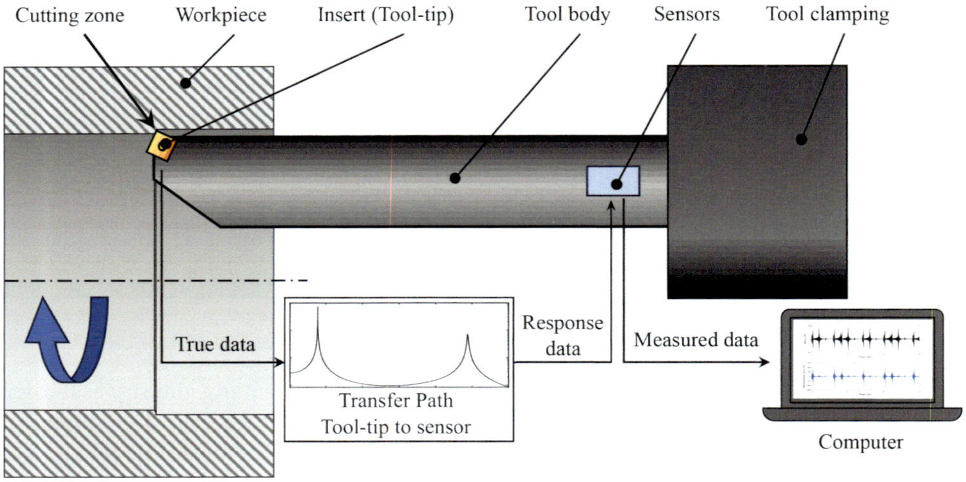


**Fig. 1** An illustration of the information gets distorted between tool tip and sensor location on a sensor-equipped cutting tool in a cutting process.

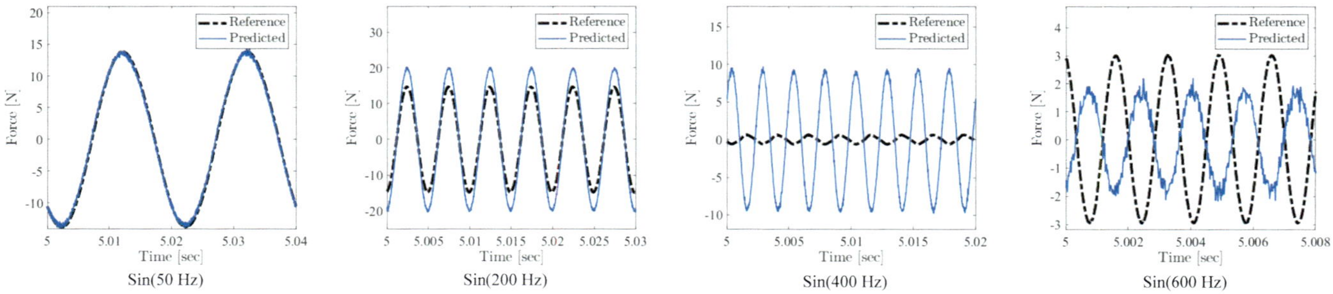


**Fig. 2** Dynamic loads acting on the tool tip predictions using statically calibrated strain sensor on the cutting tool.

at higher frequencies. This issue presents a challenge in estimating not only the force but also the deflection at the tool tip using data from the strain sensor.

Several studies has been made on strain gauge-based dynamometer design for cutting force measurements. For example, Süleyman and Faruk [8] used octagonal rings to develop strain gauge based dynamometer for turning tool, maximizing sensitivity and minimizing cross-sensitivity to reduce the effect of the mechanical transfer path. This method has been proven to work under certain condition but there still remains challenges in measuring cutting forces with high accuracy. Östling et al. [9] presented a sensor-equipped boring bar to detect vibration level and cutting forces in turning process. This sensor-equipped tool uses strain gauges to measure static forces but does not take into account the mechanical transfer path necessary to estimate the cutting forces at higher frequencies.

This study evaluates the method presented in [10], where force estimations were performed on a cantilever beam equipped with a strain sensor. In this case, the study is extended to consider a physical cutting tool, assessing the method's ability to predict transient loads and, additionally, to estimate tool deflection. A novel aspect of this research is that model calibration is conducted solely on output data recorded from a strain sensor mounted on the cutting tool. This is accomplished by first creating an analytical model representation of the cutting tool describing the strain-force relationship. The model of strain-force transfer function is used to remove unwanted dynamic effects of the transfer path from the measured strain data to improve the estimation of the forces acting on the cutting tool. The estimated force is used as the input of conventional displacement-force transfer function to estimate the displacement response at the tool tip. The model representation requires calibration to ensure accurate predictions. This is challenging to performer by the end user of the tools where access to modal hammers or shaker equipment is limited. The paper hence evaluates the possibility to calibrate the model using response data only. This strategy would allow for calibration by simply exciting the system without recording the input force.

For reference evaluation, a direct drive-point frequency response function (FRF) is obtained using the measured input force and the accelerometer response at the tool tip. The accelerance frequency response is recalculated to displacement, after which a state-space system identification scheme is applied to the displacement frequency response function, allowing for a direct estimation of the displacement response from the measured force data.

## Method

This study investigates the single-direction vibration of a cutting tool equipped with a strain sensor near the clamping fixture, see Figure 3. The tool's specifications are detailed in Table 1. To estimate the force acting on the tool tip using the strain response from the sensor-equipped cutting tool, an analytical model is used to describe the strain-force relationship [10].

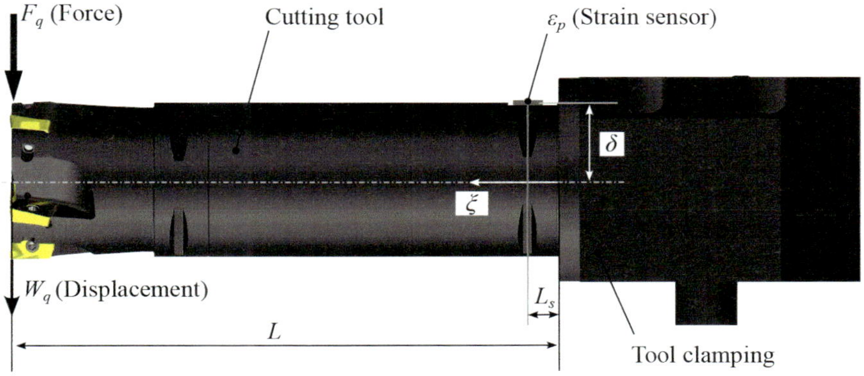


**Fig. 3** An illustration of cutting tool (R390-066C6-18M060) with the extension adapter (C6-391.01-63 140A) and tool clamping (C6-RC2090-42060), equipped with a piezoelectric strain sensor.

**Table 1** Specification of sensor-equipped cutting tool.

| | Parameter | unit |
|---|---|---|
| Mass, $m$ | 5.515 | kg |
| Length, $L$ | 203 | mm |
| Diameter | 63 | mm |
| Distance from neutral axis, $\delta$ | 31 | mm |
| Sensor position, $L_s$ | 14 | mm |
| Density, $\rho$ | 7850 | $[kg/m^3]$ |
| Young's modulus, $E$ | $206 \times 10^9$ | $[Pa]$ |

The proposed methodology begins with an initial calibration stage Figure 4a, where the structure is excited, and the strain response, $\epsilon_p$, is measured. From the measured strain the first resonance frequency, $\omega_{m1}$ and associated damping, $\zeta_1$ are calculated. These extracted parameters are then used to update the analytical model, enabling the derivation of a calibrated representation of the strain-force transfer function, $H^{\epsilon}_{pq}(z)$, the inverse strain-force transfer function, $(H^{\epsilon}_{pq}(z))^{-1}$ and a displacement-force transfer function, $H^{d}_{qq}(z)$.

When the analytical model have been calibrated, the following steps are taken to complete the predictions for force and displacement: First, the strain response from the operational excitation force, $F_o$, is combined with the inverted strain-force transfer function, $(H^{\epsilon}_{pq}(z))^{-1}$ to calculate the predicted excitation force, $F_q$. Subsequently, the predicted displacement, $W_q$, can be derived from the predicted force using the displacement-force transfer function. A schematic diagram illustrating these steps is presented in Figure 4b.

The cutting tool is modeled as a cantilever beam, and the strain-force transfer function is described by its superposition residues, $\mathbb{R}^{\epsilon}_{pqr}$, and poles, $\lambda_{pqr}$, as:

$$H^{\epsilon}_{pq}(s) = \frac{\epsilon_p(s)}{F_q(s)} = \sum_{r=1}^{Nr} \frac{\mathbb{R}^{\epsilon}_{pqr}}{s-\lambda_r} + \frac{\mathbb{R}^{\epsilon *}_{pqr}}{s-\lambda^*_r} \tag{1}$$

where $\epsilon_p$ and $F_q$ refers to the strain and force data with subscripts $p$ and $q$ being the response and excitation points. $Nr$ refers to the total number of modes included, and $s$ is the Laplace variable. Evaluating equation 1 along the $j\omega$ axis, the residues

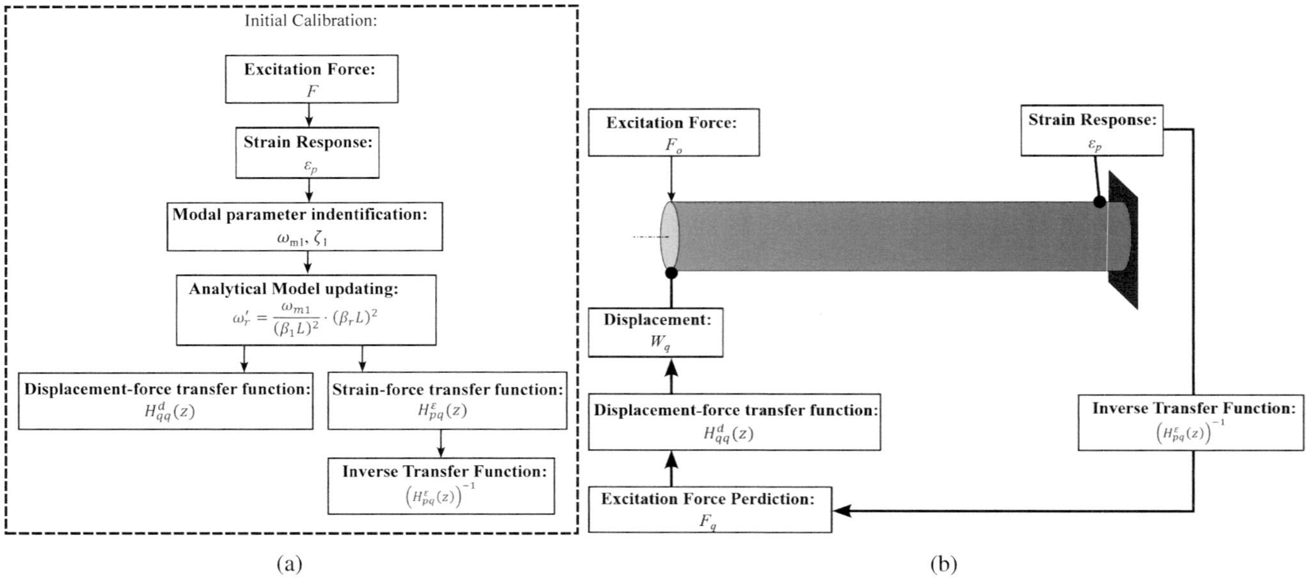


**Fig. 4** (a): initial calibration steps for analytical model. (b):A schematic diagram of the method for estimating the force acting on the tool tip using the strain response of the cutting tool.

and poles can be expressed as:

$$\mathbb{R}_{pqr}^{\epsilon}, \mathbb{R}_{pqr}^{\epsilon *} = \mp j \frac{\phi_{pr}^{\epsilon}(\xi_p) \cdot \phi_{qr}(\xi_q)}{2\omega_{dr} m_r} \tag{2}$$

$$\lambda_{pqr}, \lambda_{pqr}^{*} = -\zeta_r \omega_r \pm j\omega_r \sqrt{1-\zeta_r^2} = -\zeta_r \omega_r \pm j\omega_{dr} \tag{3}$$

$$\omega_r = (\beta_r L)^2 \sqrt{\frac{EI}{mL^3}} \tag{4}$$

$$\phi_{qr}(\xi_q) = C_1 \cos(\beta_r \xi_q) + C_2 \sin(\beta_r \xi_q) + C_3 \cosh(\beta_r \xi_q) + C_4 \sinh(\beta_r \xi_q) \tag{5}$$

$$\phi_{pr}^{\epsilon}(\xi_p) = \delta \beta_r^2 \left[C_1 \cos(\beta_r \xi_p) + C_2 \sin(\beta_r \xi_p) - C_3 \cosh(\beta_r \xi_p) - C_4 \sinh(\beta_r \xi_p)\right] \tag{6}$$

where $\phi_{pr}^{\epsilon}$ and $\phi_{qr}$ refer to the individual mode shapes at the response and excitation points. $\omega_{dr}$ refers to the damped resonance frequency of the mode $r$ obtained from the poles and damping ratio $\zeta_r$. $\omega_r$ refers to the undamped resonance frequency and $\beta_r$ is the roots of the Euler-Bernoulli cantilever beam motion equation [11]. In equation 2, modal mass $m_r$ of the beam is a measure of amount of mass moving in the given mode $r$. The modal mass of the beam can be calculated by equation 7, as presented in [12].

$$m_r = \int_0^L m_\rho \cdot |\phi_r(\xi)|^2 d\xi \tag{7}$$

The strain-force transfer function created in equation 1 need to be calibrated in order to accurately capture the boundary conditions in a real setup. The boundary conditions are typically more flexible than the idealized fixed conditions assumed in the analytical model. As a result, the resonance frequencies and peak values of the transfer function calculated from the model do not match those of the measured transfer function. To improve accuracy, the model is adapted to the cutting tool configuration based on the impulse strain response recorded from the strain sensor. This is achieved by applying a Fourier transform to this strain data allows the first resonance frequency $\omega_{m1}$ of the cutting tool to be obtained. To determine the damping ratio $\zeta_r$ of the cutting tool, logarithmic decrement approach is utilized [13, 14]. Figure 5 illustrates that logarithmic decrement relates the damping ratio to the strain response peaks through equation 8.

$$\zeta_r(\delta_N, N) = \frac{\delta_N/(2\pi N)}{\sqrt{1+(\delta_N/(2\pi N))^2}} \tag{8}$$

where the logarithmic decrement is defined as:

$$\delta_N = \ln\left(\frac{\mu\epsilon_1}{\mu\epsilon_N}\right) \tag{9}$$

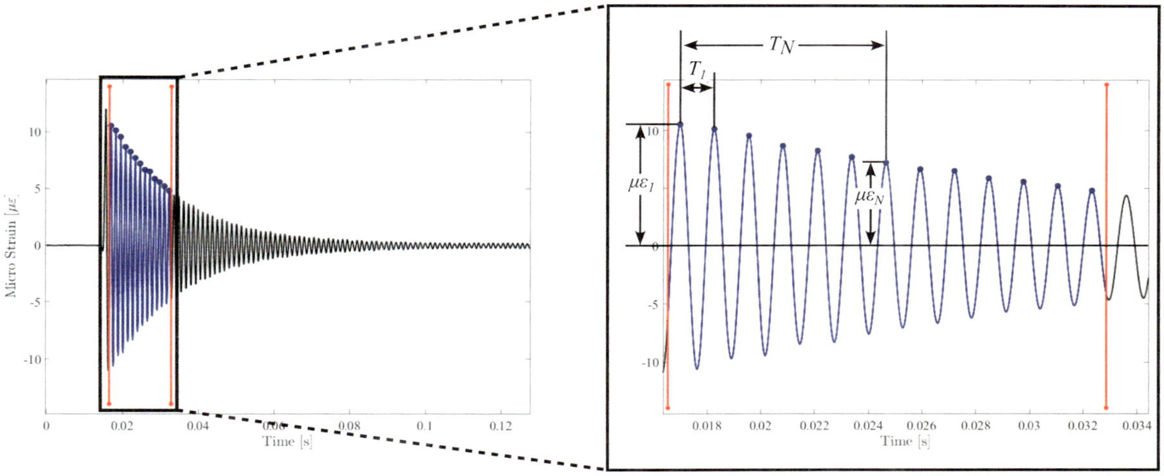


**Fig. 5** An example of the measurement of impulse strain response recorded from sensor-equipped cutting tool taken in a logarithmic decrement analysis. In this case the period $T_N$ is six times the damped period $T_1$, so $N = 6$.

Using the first resonance frequency $\omega_{m1}$ along with the equation 4, the resonance frequency of the first mode can be updated and the rest will follow through equation 10. The peaks values at each resonance frequency of the transfer function can be corrected by determined damping ratio $\zeta_r$.

$$\omega_r^{'} = \frac{\omega_{m1}}{(\beta_1 L)^2} \cdot (\beta_r L)^2 \tag{10}$$

To effectively perform calculations using sampled strain data from the sensor, the transfer function must be converted into the discrete-time domain using the digital filter form of the equation 11 [15].

$$H^{\epsilon}(z) = \frac{B(z)}{A(z)} = \frac{\sum_{k=0}^{n_b} b(k) z^{-k}}{1 + \sum_{k=1}^{n_a} a(k) z^{-k}} = b(0) \frac{\prod_{k=1}^{n_b} (1 - Z_k z^{-1})}{\prod_{k=1}^{n_a} (1 - P_k z^{-1})} \tag{11}$$

where $b(k)$ and $a(k)$ are the coefficients of the numerator and denominator polynomials for the filter, while $n_b$ and $n_a$ denote the total number of coefficients. The roots of the numerator polynomial, $Z_k$, correspond to the zeros of the filter, while the roots of the denominator polynomial, $P_k$, are the poles. The selection of $n_b$ and $n_a$ is performed by visual inspection to ensure a satisfactory fit between the original transfer function and its filter representation.

To estimate forces from strain response data, as described in Figure 4b, equation 11 needs to be inverted. To ensure stability of the inverse filter, which may become unstable due to zeros, $Z_k$, located outside the unit circle, the filter form in equation 11 must be transformed into a minimum-phase representation, denoted by $H^{\epsilon}_{min}$, as outlined in [5]. This minimum-phase filter accurately models the strain-force transfer function of the cutting tool, enabling the estimation of the force acting on the tool tip using strain response data.

Furthermore, to estimate the tool tip displacement using the predicted force, the displacement-force transfer function of the cutting tool is represented by equation 12, as described in [16].

$$H^{d}_{qq}(s) = \frac{W_q(s)}{F_q(s)} = \sum_{r=1}^{N} \frac{\mathbb{R}_{qqr}}{s - \lambda_r} + \frac{\mathbb{R}^{*}_{qqr}}{s - \lambda_r^{*}} \tag{12}$$

$$\mathbb{R}_{qqr}, \mathbb{R}^{*}_{qqr} = \mp j \frac{\phi_{qr}\phi_{qr}}{2\omega_{dr} m_r} \tag{13}$$

where $W_q$ refers to the Laplace transformed transversal displacement at the tool tip, as demonstrated in Figure 3. The individual displacement mode shapes, $\phi_{qr}$, are expressed by equation 5. The other parameters in this model have been obtained in previous steps.

To effectively calculate the displacement from predicted force data, the displacement-force transfer function must also be transformed into digital filter form through equation 11. These considerations and models form the basis of the method to accurately estimate force acting on the cutting tool and tool tip displacement using the measured data solely from the strain sensor.

## Experimental Verification and Results

The experimental test setup, for method validation, is shown in Figure 6. The complete test structure was fixed on a rigid sprung steel module. The cutting tool equipped with a piezoelectric strain sensor used to measure strain response near the tool clamping where the tool experiences the highest strain level. The impulse hammer, equipped with the force sensor, was used to excite the cutting tool and measure the force as a reference. Additionally, an accelerometer was mounted at the tool tip to measure the acceleration data, used to validate the displacement estimation. The equipment used in the test is listed in Table 2.

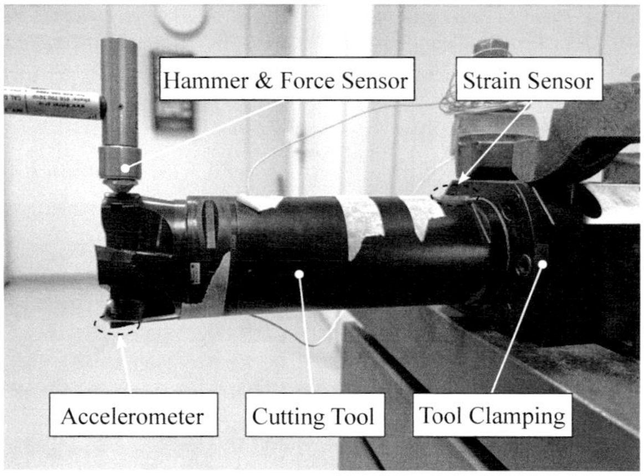


**Fig. 6** An illustration of the experimental setup in the lab.

**Table 2** Experiment equipment.

| Equipment | Model |
|---|---|
| Impulse hammer | Dytran 5800B4 |
| ICP Strain sensor | PCB 740M04 |
| Accelerometer | Dytran 3224A1 |
| Voltage Input Module | NI 9234 |

The analytical model was updated using the extracted parameters in terms of first resonance frequency ($\omega_{m1} = 784$ Hz) and corresponding damping ratio ($\zeta_1 = 0.95$ %) obtained from the measured strain response to improve the model correlation to the strain response. The comparison between the strain FRFs from updated model, $H^{\epsilon}_{ana}$, and the measurement ,$H^{\epsilon}_{mea}$, is shown in Figure 7. Good similarity confirms the updated model accurately captures the real setup configuration. The strain FRF of the minimum-phase filter, $H^{\epsilon}_{min}$, is also similar to the measurement.

To validate accuracy of transformation between the analytical model, $H^{\epsilon}_{ana}$, and its minimum-phase digital filter form, $H^{\epsilon}_{min}$, a combined strain FRF is introduced by equation 14 and shown in Figure 8.

$$H^{\epsilon}_{com} = H^{\epsilon}_{ana} \cdot (H^{\epsilon}_{min})^{-1} \tag{14}$$

The magnitude of the combined strain FRF remains unity confirming good alignment of the transformation from the analytical model to its minimum-phase filter form. Furthermore, to examine the frequency limitation of the model for force estimation, the phase response $\Theta(\omega)$ of the combined strain FRF ,$H^{\epsilon}_{com}$, is presented in Figure 8. The phase remains approximately linear up to 2300 Hz indicating a frequency limitation of the model [10].

To avoid amplitude distortions caused by the nonlinear phase response, a low-pass filter with a cutoff frequency of 2300 Hz needs to be applied to the estimated forces using inverse filtered strain response data. Additionally, the linear phase

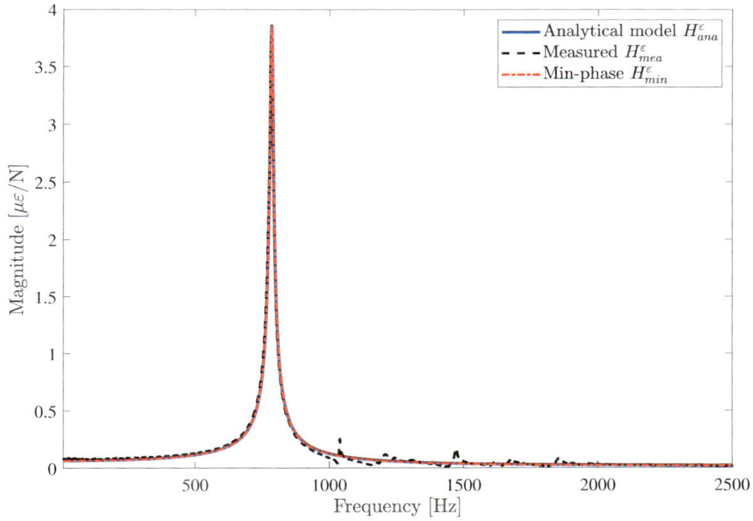


**Fig. 7** comparison of the strain FRFs from the analytical model after model updating $H^{\epsilon}_{ana}$, the measurement $H^{\epsilon}_{mea}$, and the minimum phase transformation $H^{\epsilon}_{min}$.

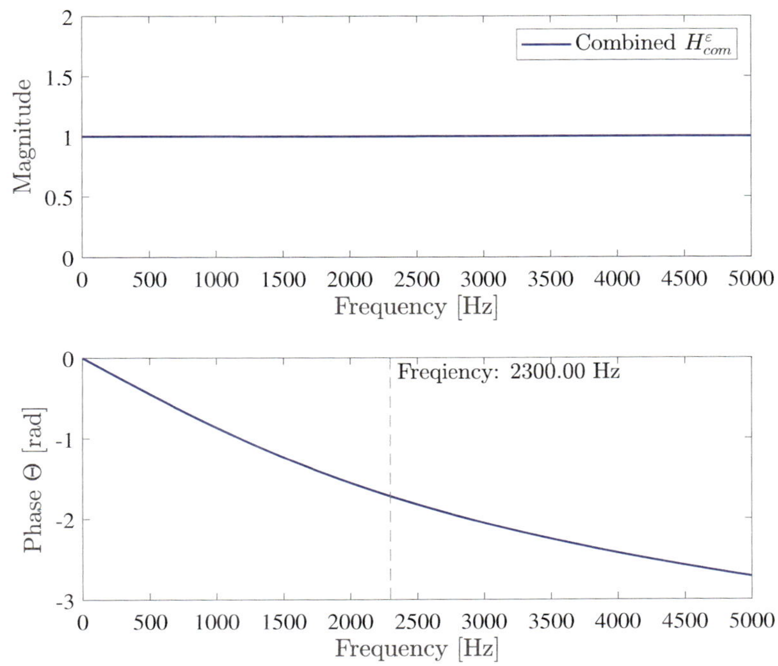


**Fig. 8** Illustration of the combined strain FRF, $H^{\epsilon}_{com}$, by the strain FRF of analytical model, $H^{\epsilon}_{ana}$, multiplying the strain FRF of inverse minimum-phase filter, $(H^{\epsilon}_{min})^{-1}$.

response of the transfer function corresponds to a delay in the time domain. The time delay can be calculated based on the linear phase response by equation 15 [17]. In Figure 8, the time delay is approximately 0.126 ms.

$$D(\omega) \triangleq -\frac{d}{d\omega}\Theta(\omega) \tag{15}$$

The results of estimated and measured impact forces are shown in Figure 9. The test result shows good agreement between the measured and predicted impact forces, highlighting the capability of the proposed method for estimating forces acting on the tool tip based on measured strain response data.

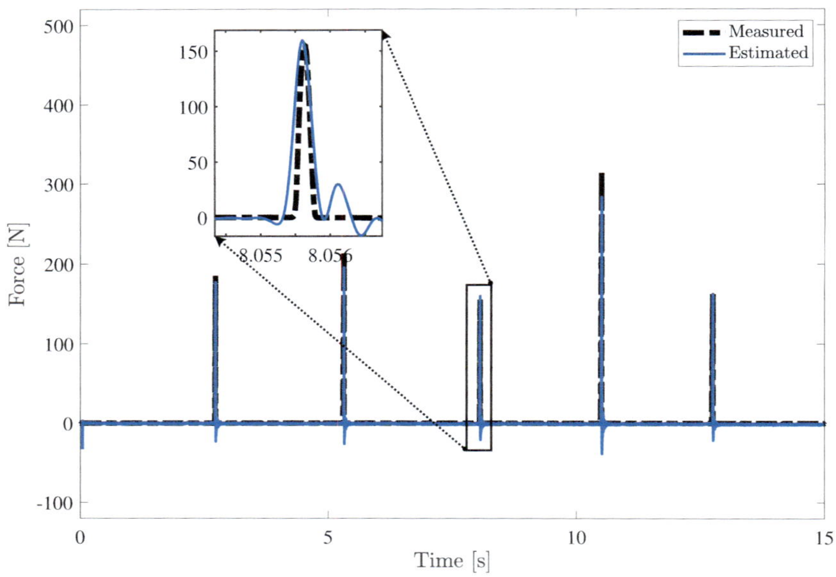


**Fig. 9** Comparison between measured and estimated impact forces.

Using the estimated force as input to the displacement-force model, the tool tip displacement is calculated. To obtain a reliable tool tip displacement estimation as reference for validation, an approach is presented in [18]. In this method, the acceleration FRF of the experimental setup is estimated based on the impulse hammer and accelerometer measurement. The measured acceleration FRF is then converted to displacement. A state-space model is then fitted to the measured displacement FRF using MATLAB's system identification toolbox[19], Figure 10.

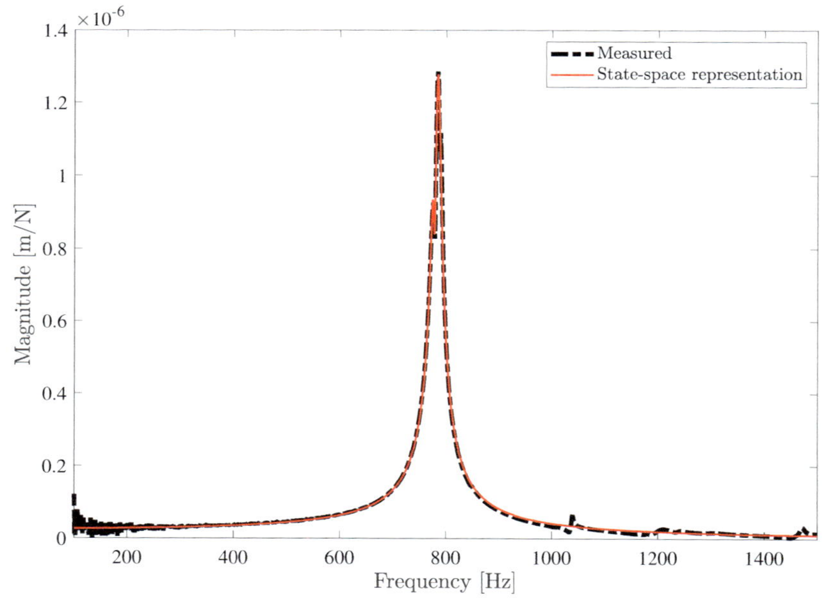


**Fig. 10** Comparison between the measured displacement FRF and the state-space model.

The comparison between the displacement obtained from the proposed method based on strain measurements and inverse filtering and the validation approach is shown in Figure 11. The results shows good similarity, indicating that the tool tip displacement can be accurately estimated based on the measured strain by the proposed method.

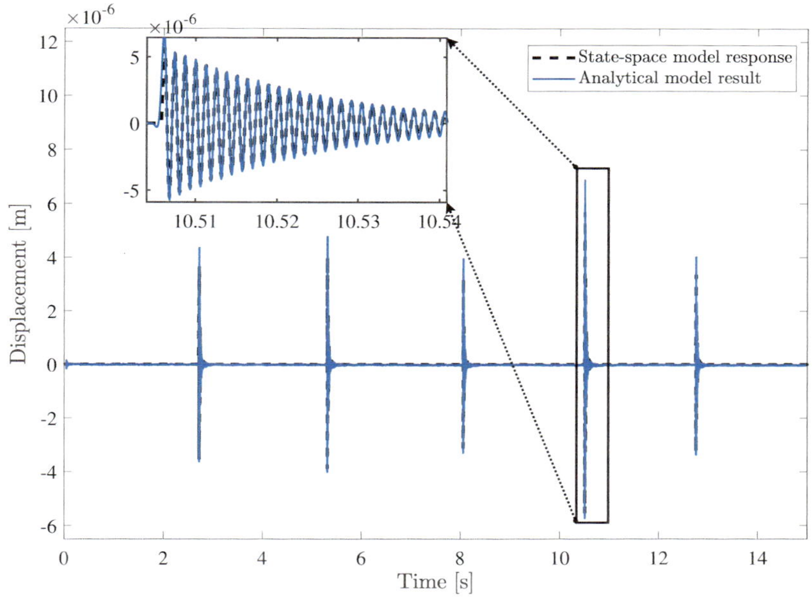


**Fig. 11** Comparison between the results of transient displacement at cutting tool tip from state-space model response and analytical model estimation.

## Conclusion

A method to estimate the force and displacement based on strain measurements has been tested on a physical cutting tool. The method uses inverse filtering of measured strain to estimate the applied force. The estimated force is then used together with a model of the cutting tool to estimate the displacement at the tool-tip. Both these steps relies on having an accurate model representing the cutting tool under the given configuration.

The feasibility of calibrating the model to align with the real cutting tool configurations using only response data have been successfully validated. The limited need for measurement data in combination with fast model updating represents a significant advantage for practical implementation, as it enables possibility for in-field calibration without the need for complex setups involving modal hammers or shakers.

Further, the results indicate that the proposed method is able to accurately estimate the dynamic forces and tool tip displacements. The strong agreement between the estimated and measured results demonstrates the model's effectiveness in eliminating unwanted dynamic effects of the cutting tool across a wide frequency range.

Overall, this study highlights the potential for accurately predicting dynamic forces and tool tip displacements through the use of strain sensors in cutting tools, facilitating improved performance monitoring and optimization in machining operations.

**Acknowledgments** The authors wish to acknowledge the financial support provided by AB Sandvik Coromant.

## References

1. Brinksmeier, E., Heinzel, C., Wilkens, A., Lang, W., and Seedorf, T. "Monitoring of machining processes using sensor equipped tools". *Advanced Engineering Materials*, 12(7):641–645 (2010)
2. Korkmaz, M.E., Gupta, M.K., Li, Z., Krolczyk, G.M., Kuntoğlu, M., Binali, R., Yaşar, N., and Pimenov, D.Y. "Indirect monitoring of machining characteristics via advanced sensor systems: A critical review". *The International Journal of Advanced Manufacturing Technology*, 120(11-12):7043–7078 (2022)
3. Bhuiyan, M. and Choudhury, I. "13.22—review of sensor applications in tool condition monitoring in machining". *Comprehensive Materials Processing*, 13:539–569 (2014)
4. Kammer, D.C. "Estimation of structural response using remote sensor locations". *Journal of guidance, control, and dynamics*, 20(3):501–508 (1997)
5. Magnevall, M., Lundblad, M., Ahlin, K., and Broman, G. "High frequency measurements of cutting forces in milling by inverse filtering". *Machining science and technology*, 16(4):487–500 (2012)

6. Castro, L.R., Viéville, P., and Lipinski, P. "Correction of dynamic effects on force measurements made with piezoelectric dynamometers". *International Journal of Machine Tools and Manufacture*, 46(14):1707–1715 (2006)
7. Tounsi, N. and Otho, A. "Identification of machine–tool–workpiece system dynamics". *International Journal of Machine Tools and Manufacture*, 40(9):1367–1384 (2000)
8. Yaldız, S. and Ünsaçar, F. "Design, development and testing of a turning dynamometer for cutting force measurement". *Materials & design*, 27(10):839–846 (2006)
9. Östling, D., Jensen, T., Tjomsland, M., Standal, O., and Mugaas, T. "Cutting process monitoring with an instrumented boring bar measuring cutting force and vibration". *Procedia CIRP*, 77:235–238 (2018)
10. W. Peng, M. Magnevall, A. Liljerehn and D. Östling. "Cutting force estimation in metal cutting tools: a study on a sensor-equipped cantilever beam". *Presented at ISMA2024 Conference on Noise and Vibration Engineering* (2024)
11. Thomson, W.T. "Theory of vibration with applications". *NASA STI/Recon Technical Report A*, 93:39794 (1993)
12. Aenlle, M., Juul, M., and Brincker, R. "Modal mass and length of mode shapes in structural dynamics". *Shock and Vibration*, 2020:1–16 (2020)
13. Nishiwaki, N., Masuko, M., Ito, Y., and Okumura, I. "A study on damping capacity of a jointed cantilever beam: 1st report; experimental results". *Bulletin of JSME*, 21(153):524–531 (1978)
14. Lamarque, C.H., Pernot, S., and Cuer, A. "Damping identification in multi-degree-of-freedom systems via a wavelet-logarithmic decrement—part 1: theory". *Journal of Sound and Vibration*, 235(3):361–374 (2000)
15. Şafak, E. "Identification of linear structures using discrete-time filters". *Journal of Structural Engineering*, 117(10):3064–3085 (1991)
16. Brandt, A. *Noise and vibration analysis: signal analysis and experimental procedures*. John Wiley & Sons (2023)
17. Smith, J.O. *Introduction to digital filters: with audio applications*, volume 2. Julius Smith (2007)
18. Gindy, M., Vaccaro, R., Nassif, H., and Velde, J. "A state-space approach for deriving bridge displacement from acceleration". *Computer-Aided Civil and Infrastructure Engineering*, 23(4):281–290 (2008)
19. Johansson, R., Robertsson, A., Nilsson, K., and Verhaegen, M. "State-space system identification of robot manipulator dynamics". *Mechatronics*, 10(3):403–418 (2000)

# Chapter 9
# Enhanced Mode Selection in Modal Domain Substructuring on the Round Robin Structure

**Jure Korbar, Miha Pogačar, and Gregor Čepon**

**Abstract** Dynamic substructuring (DS) techniques present an indispensable tool for estimating assembly dynamics based on the individual subcomponents' dynamics. The reliability of DS is heavily subject to the boundary conditions at the interface, the experimentalist's ability to obtain a sufficiently accurate representation of the interface, the domain where dynamic substructuring is performed, and the considerations regarding the weakening of the compatibility and equilibrium constraints. This paper addresses the Thin & Thick Wing challenge on the Round Robin benchmark structure with the objective of estimating the dynamics of the assembled Frame and Thick Wing. DS is performed in the modal domain, which is known to be highly sensitive to the selection of the participating modes in the substructuring methodology, as well as the weakening formulation. In this work, the modal-based substructuring is augmented by a novel methodology, addressing the problem of selecting appropriate modes for the substructuring procedure. This enhanced approach aims to improve the accuracy and robustness of modal-based substructuring, facilitating its utilization due to the reduced amount of guesswork within the existing approaches.

**Keywords** Dynamic substructuring · Modal substructuring · Round robin structure · Virtual point transformation · Mode selection

## Introduction

Dynamic substructuring techniques allow to obtain the dynamic model of a structure by establishing relations between the individual subcomponents with known dynamic models in terms of compatibility and equilibrium constraints at the interface. Modal substructuring, also known as component-mode synthesis (CMS), refers to the subdomain of dynamic substructuring, where the substructures' dynamics are represented in the modal domain [1]. The accuracy of modal substructuring is known to be sensitive to the selection of modes used to describe the dynamics of the individual substructures. Due to the reduction of physical degrees of freedom (DoFs) to a smaller set of generalized coordinates, the established physical constraints generally cannot be satisfied exactly, as it often occurs that the total number of the available substructures' modes is lower than the number of constraints. In addition, for multi-point connections, some of the physical constraints may be redundant, resulting in interface stiffening and spurious eigenfrequencies [2]. Both issues are addressed by weakening the interface compatibility and equilibrium constraints. Weakening is performed by transforming the constraint equations using a suitable reduction basis, which may consist of vibration modes [3] or dominant singular vectors [4]. Regardless of the weakening procedure, the accuracy of modal substructuring remains subject to the modes considered in the modal reduction and weakening procedures. In this work, a framework for assessing whether the selected modes are appropriate for estimating coupled dynamics is proposed. The framework is based on neglecting one constraint at a time and performing modal substructuring considering the remaining constraints. By analyzing the stability of the calculated eigenfrequencies, the appropriateness of the selected modes for modal substructuring can be assessed. The proposed approach is demonstrated on the Round Robin benchmark structure with the objective of estimating the dynamics of the assembled Frame & Thick Wing by decoupling the Thin Wing from the assembled Frame & Thin Wing and coupling the Thick Wing. Two distinct representations of interface connection are considered, namely the equivalent multi-point connection, where the constraints are defined for the physical DoFs, and the connection via virtual points, where the constraints refer to the virtual point DoFs.

Jure Korbar · Miha Pogačar · Gregor Čepon
Faculty of Mechanical Engineering, University of Ljubljana, Aškerčeva cesta 6, 1000 Ljubljana
e-mail: jure.korbar@fs.uni-lj.si; miha.pogacar@fs.uni-lj.si; gregor.cepon@fs.uni-lj.si



Walter D'Ambrogio, et al. (eds.), *Dynamic Substructuring & Transfer Path Analysis, Vol. 4*, Conference Proceedings of the Society for Experimental Mechanics Series, https://doi.org/10.13052/9788743801498-9

## Primal Formulation of Modal Substructuring

Dynamic substructuring in the modal domain allows to combine individual component models to estimate assembly dynamics or remove parts of the structure based on the individual substructures' modal parameters. The governing linear equation of motion for a substructure $s$ in the physical domain can be written as:

$$\mathbf{M}^{(s)}\ddot{\boldsymbol{u}}^{(s)} + \mathbf{C}^{(s)}\dot{\boldsymbol{u}}^{(s)} + \mathbf{K}^{(s)}\boldsymbol{u}^{(s)} = \boldsymbol{f}^{(s)} + \boldsymbol{g}^{(s)}, \tag{1}$$

where $\mathbf{M}^{(s)}$, $\mathbf{C}^{(s)}$, and $\mathbf{K}^{(s)}$ are the mass, damping, and stiffness matrices, respectively, $\boldsymbol{u}^{(s)}$ is the displacement vector, $\boldsymbol{f}^{(s)}$ is the external force vector, and $\boldsymbol{g}^{(s)}$ is the connecting interface force vector. The velocity and acceleration vectors are denoted as the first and second time derivatives $\dot{\boldsymbol{u}}^{(s)}$ and $\ddot{\boldsymbol{u}}^{(s)}$ of displacements $\boldsymbol{u}^{(s)}$. The equations of motion for $N$ individual substructures can be combined into a single uncoupled equation of motion:

$$\mathbf{M}\ddot{\boldsymbol{u}} + \mathbf{C}\dot{\boldsymbol{u}} + \mathbf{K}\boldsymbol{u} = \boldsymbol{f} + \boldsymbol{g}, \tag{2}$$

where $\mathbf{M} = \mathrm{diag}\left(\mathbf{M}^{(1)}, \ldots, \mathbf{M}^{(N)}\right)$, $\mathbf{C} = \mathrm{diag}\left(\mathbf{C}^{(1)}, \ldots, \mathbf{C}^{(N)}\right)$, $\mathbf{K} = \mathrm{diag}\left(\mathbf{K}^{(1)}, \ldots, \mathbf{K}^{(N)}\right)$, $\boldsymbol{u} = \left\{\boldsymbol{u}^{(1)\top}, \ldots, \boldsymbol{u}^{(N)\top}\right\}^\top$, $\boldsymbol{f} = \left\{\boldsymbol{f}^{(1)\top}, \ldots, \boldsymbol{f}^{(N)\top}\right\}^\top$, and $\boldsymbol{g} = \left\{\boldsymbol{g}^{(1)\top}, \ldots, \boldsymbol{g}^{(N)\top}\right\}^\top$. Coupling the individual dynamic models requires satisfying the constraints regarding the compatibility of interface displacements and equilibrium of interface forces:

$$\mathbf{B}\boldsymbol{u} = \mathbf{0}, \tag{3}$$

$$\mathbf{L}^\top\boldsymbol{g} = \mathbf{0}. \tag{4}$$

The physical DoFs are reduced to a smaller set of generalized DoFs $\boldsymbol{\eta}$ using a block-diagonal reduction matrix $\mathbf{R}_{\mathrm{m}}$:

$$\boldsymbol{u} = \mathbf{R}_{\mathrm{m}}\boldsymbol{\eta}, \tag{5}$$

where $\mathbf{R}_{\mathrm{m}} = \mathrm{diag}\left(\mathbf{R}_{\mathrm{m}}^{(1)}, \ldots, \mathbf{R}_{\mathrm{m}}^{(N)}\right)$, $\boldsymbol{\eta} = \left\{\boldsymbol{\eta}^{(1)\top}, \ldots, \boldsymbol{\eta}^{(N)\top}\right\}^\top$, and $\mathbf{R}_{\mathrm{m}}^{(s)}$ denotes the reduction matrix. In this study, a set of mass-normalized mode shapes $\boldsymbol{\Phi}^{(s)}$ is considered as a reduction matrix. Pre-multiplying Eq. (2) with $\mathbf{R}_{\mathrm{m}}^\top$ and considering the reduction in Eq. (5) yields the following equation of motion and corresponding compatibility and equilibrium constraints:

$$\mathbf{M}_{\mathrm{m}}\ddot{\boldsymbol{\eta}} + \mathbf{C}_{\mathrm{m}}\dot{\boldsymbol{\eta}} + \mathbf{K}_{\mathrm{m}}\boldsymbol{\eta} = \boldsymbol{f}_{\mathrm{m}} + \boldsymbol{g}_{\mathrm{m}}, \tag{6}$$

$$\mathbf{B}_{\mathrm{m}}\boldsymbol{\eta} = \mathbf{0}, \tag{7}$$

$$\mathbf{L}_{\mathrm{m}}^\top\boldsymbol{g}_{\mathrm{m}} = \mathbf{0}, \tag{8}$$

where $\mathbf{M}_{\mathrm{m}} = \mathbf{R}_{\mathrm{m}}^\top\mathbf{M}\mathbf{R}_{\mathrm{m}}$, $\mathbf{C}_{\mathrm{m}} = \mathbf{R}_{\mathrm{m}}^\top\mathbf{C}\mathbf{R}_{\mathrm{m}}$, $\mathbf{K}_{\mathrm{m}} = \mathbf{R}_{\mathrm{m}}^\top\mathbf{K}\mathbf{R}_{\mathrm{m}}$, $\mathbf{B}_{\mathrm{m}} = \mathbf{B}\mathbf{R}_{\mathrm{m}}$, $\mathbf{L}_{\mathrm{m}} = \mathrm{null}\left(\mathbf{B}_{\mathrm{m}}\right)$, $\boldsymbol{f}_{\mathrm{m}} = \mathbf{R}_{\mathrm{m}}^\top\boldsymbol{f}$, and $\boldsymbol{g}_{\mathrm{m}} = \mathbf{R}_{\mathrm{m}}^\top\boldsymbol{g}$. Eq. (7) enforces exact/strong compatibility of physical DoFs in the reduced DoF space. If the number of constraints exceeds the combined number of the individual substructures' modes, the set compatibility equations is overdetermined and generally cannot be satisfied exactly. In addition, the number of modes which can be calculated for the coupled structure decreases with the increasing number of constraints and decreasing number of modes of the individual substructures. Therefore, it can be reasonable to reduce the number of constraints by weakening the compatibility conditions, i. e. approximately satisfying Eq. (7) by pre-multiplication with a weakening matrix $\mathbf{W}$:

$$\widetilde{\mathbf{B}}_{\mathrm{m}}\boldsymbol{\eta} = \mathbf{0}, \tag{9}$$

$$\widetilde{\mathbf{L}}_{\mathrm{m}}\boldsymbol{g}_{\mathrm{m}} = \mathbf{0}, \tag{10}$$

where $\widetilde{\mathbf{B}}_{\mathrm{m}} = \mathbf{W}\mathbf{B}_{\mathrm{m}}$ and $\widetilde{\mathbf{L}}_{\mathrm{m}} = \mathrm{null}\left(\mathbf{W}\mathbf{B}_{\mathrm{m}}\right)^\top$. Substructuring in the primal formulation is performed by expressing the dual set of generalized coordinates $\boldsymbol{\eta}$ using a unique set of generalized coordinates $\boldsymbol{\xi}$:

$$\boldsymbol{\eta} = \widetilde{\mathbf{L}}_{\mathrm{m}}\boldsymbol{\xi}, \tag{11}$$

therefore, the coupled equation of motion is written as:

$$\widetilde{\mathbf{M}}_{\mathrm{m}}\ddot{\boldsymbol{\xi}} + \widetilde{\mathbf{C}}_{\mathrm{m}}\dot{\boldsymbol{\xi}} + \widetilde{\mathbf{K}}_{\mathrm{m}}\boldsymbol{\xi} = \widetilde{\boldsymbol{f}}_{\mathrm{m}}, \tag{12}$$

where $\widetilde{\mathbf{M}}_{\mathrm{m}} = \widetilde{\mathbf{L}}_{\mathrm{m}}^\top\mathbf{M}_{\mathrm{m}}\widetilde{\mathbf{L}}_{\mathrm{m}}$, $\widetilde{\mathbf{C}}_{\mathrm{m}} = \widetilde{\mathbf{L}}_{\mathrm{m}}^\top\mathbf{C}_{\mathrm{m}}\widetilde{\mathbf{L}}_{\mathrm{m}}$, $\widetilde{\mathbf{K}}_{\mathrm{m}} = \widetilde{\mathbf{L}}_{\mathrm{m}}^\top\mathbf{K}_{\mathrm{m}}\widetilde{\mathbf{L}}_{\mathrm{m}}$, and $\widetilde{\boldsymbol{f}}_{\mathrm{m}} = \widetilde{\mathbf{L}}_{\mathrm{m}}^\top\boldsymbol{f}_{\mathrm{m}}$.

## Modal Domain Virtual Point Transformation

Performing experimental dynamic substructuring requires collocated response measurements on the adjacent sides of the interface, which proves to be challenging. Furthermore, coupling rotational DoFs is subject to one's ability to measure the rotational DoFs. By coupling multiple closely spaced translational DoFs to implicitly account for rotational DoFs, one may encounter over-stiffening of interface. The virtual point transformation (VPT) addresses all of the above issues by performing a geometric transformation of the measured DoFs into a single virtual point (VP) [5]. The VPT assumes that the local motion of the interface can be described using a set of interface deformation modes (IDMs). This allows to couple experimental models with collocated DoFs and directly couple rotational DoFs.[1]

The VPT is typically applied when performing frequency-based substructuring, however, there are no assumptions regarding modal substructuring that would prohibit applying the VPT in the modal domain. Furthermore, modal-domain VPT (M-VPT) can introduce the advantages of the classical VPT to modal substructuring. Denoting by $\boldsymbol{u}_b$ the subset of displacements pertaining to the measured interface DoFs, which are to be transformed to the VP DoFs $\boldsymbol{q}_b$, the relation between the measured and VP DoFs can be written as:

$$\boldsymbol{u}_b = \mathbf{R}_b \boldsymbol{q}_b, \tag{13}$$

where $\mathbf{R}_b$ is the IDM matrix transforming the VP DoFs to the measured DoFs. Eq. (13) is typically overdetermined (the number of measured DoFs exceeds the number of IDMs) to mitigate measurement errors. The objective of VPT is to transform measured DoFs to VP DoFs, therefore, the VP DoFs $\boldsymbol{q}$ are calculated as a least-squares solution of Eq. (13):

$$\boldsymbol{q}_b = \mathbf{T}_b \boldsymbol{u}_b, \tag{14}$$

where $\mathbf{T}_b = \mathbf{R}_b^+$, with $\star^+$ denoting the Moore-Penrose inverse of $\star$. Since the interface DoFs are being replaced with VP DoFs, the reduction of physical DoFs to generalized DoFs is now modified at interface:

$$\boldsymbol{q}_b = \mathbf{T}_b \boldsymbol{u}_b = \mathbf{T}_b \mathbf{R}_{\mathrm{m},b} \boldsymbol{\eta}, \tag{15}$$

where $\mathbf{R}_{\mathrm{m},b}$ denotes the section of the reduction matrix $\mathbf{R}_{\mathrm{m}}$ pertaining to the interface DoFs.

## Mode Selection Framework

Successfully performing modal substructuring is greatly influenced by the selection of the individual substructures' modes involved in the reduction to generalized coordinates and the weakening formulation. Without prior knowledge of the coupled structure's modal parameters, it is difficult to know whether a certain combination of selected modes is appropriate. This work proposes a framework for assessing whether a particular subset of substructures' modes is appropriate for modal substructuring.

A prerequisite for applying the described framework is that the number of physical interface DoFs is larger than the minimum number of interface DoFs required to sufficiently couple the substructures. This can be achieved by selecting enough measurement locations at the interface to achieve some level of redundancy. A six step framework for assessing the validity of the selected modes for modal substructuring is described in the following:

1. Select a subset of modes for each substructure to perform DoF reduction.
2. Establish compatibility equations for interface displacements.
3. Remove compatibility equations pertaining to a single DoF at the interface.[2]
4. Perform modal substructuring by considering the remaining compatibility equations.
5. Obtain a set of eigenfrequencies for all possible combinations of compatibility equations with one missing DoF by repeating steps 2–4 for each interface DoF.
6. Assess the stability of the set of all eigenfrequencies calculated in step 5.

If the calculated eigenfrequencies do not significantly deviate within the frequency range of interest, the selected modes can be considered as a valid subset of modes for modal substructuring.

[1]The VPT typically assumes the motion of the interface can be described using rigid IDMs, therefore, the measured interface DoFs are transformed into a set of three translational and three rotational DoFs. However, the set of IDMs can be extended to include flexible interface motion [6].

[2]When performing substructuring using a transmission simulator, there are two coupling equations per interface DoF; one for decoupling the transmission simulator and one for coupling the substructure missing from the assembly.

## Round Robin Structure Case Study

A case study of the proposed framework was performed on the Round Robin structure, addressing the Thin & Thick Wing challenge, where the objective is to estimate the dynamics of the Frame & Thick Wing assembly by decoupling the Thin Wing from the Frame & Thin Wing assembly and coupling the Thick Wing. For brevity, the Frame, Thin Wing and Thick Wing substructures are denoted as A, TS, and B, respectively. Frame & Thin Wing and Frame & Thick Wing assemblies are respectively denoted as ATS and AB. The substructuring procedure can be conceptually outlined in the equation form: ATS − TS + B = AB.

The substructures involved in this case study are shown in Fig. 1, along with the sensor placement. Both Thin and Thick Wing are mounted to the Frame at four connection points. Three triaxial accelerometers are placed in the proximity of each connection point to facilitate the application of M-VPT. A VP is located at each connection point for a total of four VPs, depicted as green spheres in Fig. 1, with six rigid IDMs per VP. The four triaxial accelerometers in the corners of TS and B were also included in the substructuring procedure, therefore, the extended interface formulation was considered.

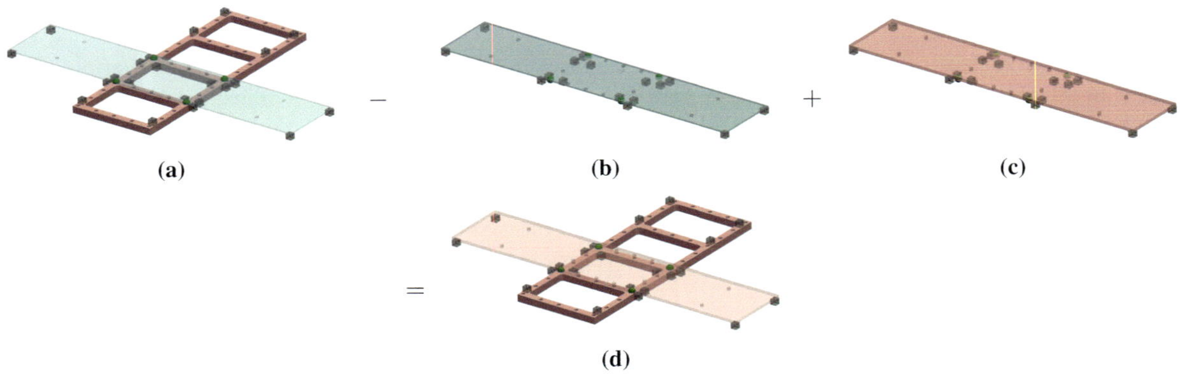

**Fig. 1** Substructures used in the Thin & Thick Wing challenge: (a) ATS, (b) TS, (c) B, (d) AB.

To clarify the substructuring procedure, the matrices $\mathbf{W}$, $\mathbf{B}_{\mathrm{m}}$, and generalized coordinates $\boldsymbol{\eta}$ are written in a block form as:

$$\mathbf{W} = \begin{bmatrix} \boldsymbol{\Phi}_b^{(\mathrm{TS})+} & \mathbf{0} \\ \mathbf{0} & \boldsymbol{\Phi}_b^{(\mathrm{TS})+} \end{bmatrix}, \quad \mathbf{B}_{\mathrm{m}} = \begin{bmatrix} \boldsymbol{\Phi}_b^{(\mathrm{ATS})} & -\boldsymbol{\Phi}_b^{(\mathrm{TS})} & \mathbf{0} \\ \mathbf{0} & \boldsymbol{\Phi}_b^{(\mathrm{TS})} & -\boldsymbol{\Phi}_b^{(\mathrm{B})} \end{bmatrix}, \quad \boldsymbol{\eta} = \begin{Bmatrix} \boldsymbol{\eta}^{(\mathrm{ATS})} \\ \boldsymbol{\eta}^{(\mathrm{TS})} \\ \boldsymbol{\eta}^{(\mathrm{B})} \end{Bmatrix}. \tag{16}$$

The proposed mode selection framework is performed by removing a single row in $\boldsymbol{\Phi}_b^{(\mathrm{TS})}$ and the corresponding rows in $\boldsymbol{\Phi}_b^{(\mathrm{ATS})}$ and $\boldsymbol{\Phi}_b^{(\mathrm{B})}$. The number of modes taken into consideration for modal substructuring for substructures ATS, TS, and B, are denoted as $M^{(\mathrm{ATS})}$, $M^{(\mathrm{TS})}$, and $M^{(\mathrm{B})}$, respectively . The results of the mode selection framework are shown in Fig. 2, where two distinct subsets of modes $S_1$ and $S_2$ are tested, which are defined as follows:

$$S_1 = \begin{cases} M^{(\mathrm{ATS})} = 25 \\ M^{(\mathrm{TS})} = 22 \\ M^{(\mathrm{B})} = 22 \end{cases} \quad \text{and} \quad S_2 = \begin{cases} M^{(\mathrm{ATS})} = 30 \\ M^{(\mathrm{TS})} = 18 \\ M^{(\mathrm{B})} = 18 \end{cases}. \tag{17}$$

Modal substructuring is performed on each subset of modes with two distinct interface representations, namely the equivalent multi-point connection (EMPC), where the constraints are defined directly on the physical set of coordinates, as well as the M-VPT representation, where the constraints refer to the VP DoFs. It can be seen that the more stable eigenfrequency results are generally in line with the reference eigenfrequencies, which are not known in practice. The stability of the eigenfrequencies can therefore be used as a guide whether the subset of modes used within modal substructuring is appropriate to estimate the assembly dynamics in the frequency range of interest.

The results also allow to compare the EMPC and the M-VPT approaches for describing the interface DoFs. One significant advantage of M-VPT over the EMPC is the improved accuracy of rigid-body dynamics. All substructures in this study have free boundary conditions, therefore, each substructure has six rigid-body vibration modes. The M-VPT approach accurately predicts six rigid-body modes for the coupled structure, while the EMPC approach typically predicts only four eigenfrequencies at 0 Hz. The M-VPT also generally outperforms the EMPC in terms of more accurate estimation of the coupled eigenfrequencies.

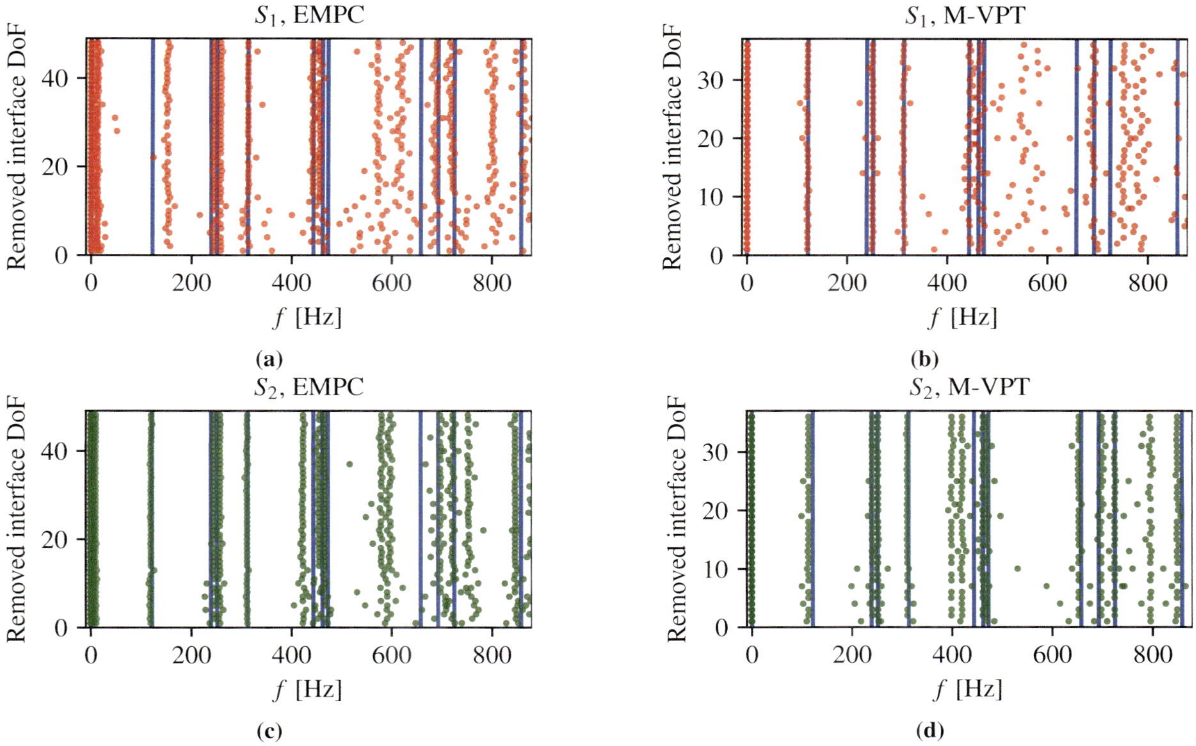


**Fig. 2** Application of the mode selection framework: (a) mode subset $S_1$ with EMPC , (b) mode subset $S_1$ with VPT, (c) mode subset $S_2$ with EMPC, (d) mode subset $S_2$ with VPT.

## Conclusion

This work proposes a framework for assessing whether the selected subset of substructures' modes is appropriate for performing modal substructuring. By removing a single interface DoF, the stability of the coupled eigenfrequencies can be assessed, with higher stability indicating that the subset of modes considered for substructuring is appropriate.

## References

1. De Klerk, D., Rixen, D.J., and Voormeeren, S.N. "General framework for dynamic substructuring: history, review and classification of techniques". *AIAA journal*, 46(5):1169–1181 (2008).
2. Rixen, D.J. "How measurement inaccuracies induce spurious peaks in frequency based substructuring". In *Proceedings of the Twenty Sixth International Modal Analysis Conference, Orlando, FL. Society for Experimental Mechanics, Bethel, CT* (2008).
3. Allen, M.S., Mayes, R.L., and Bergman, E.J. "Experimental modal substructuring to couple and uncouple substructures with flexible fixtures and multi-point connections". *Journal of Sound and Vibration*, 329(23):4891–4906 (2010).
4. Pogačar, M., Čepon, G., and Boltežar, M. "Weakening of the multi-point constraints in modal substructuring using singular value decomposition". *Mechanical Systems and Signal Processing*, 163:108109 (2022).
5. van der Seijs, M., van den Bosch, D., Rixen, D.J., and de Klerk, D. "An improved methodology for the virtual point transformation of measured frequency response functions in dynamic substructuring". *COMPDYN* (2013).
6. Pasma, E.A., van der Seijs, M., Klaassen, S.W.B., and van der Kooij, M.W. "Frequency based substructuring with the virtual point transformation, flexible interface modes and a transmission simulator". In *Dynamics of Coupled Structures, Volume 4*, pages 205–213. Springer (2018).

# Chapter 10
# Investigation of the Use of Commercial Robotic Arms for Real-Time Hybrid Substructuring

**Arian Kist, David Stadler, Rok Belšak, Vasja Plesec, Timi Karner, Gregor Harih and Daniel Rixen**

**Abstract** Real-Time Hybrid Substructuring (RTHS) allows components of dynamic systems to be tested under realistic boundary conditions from an early stage of development by coupling the prototype of the component in real-time with a co-simulation of the remaining structure of the dynamic system. Real-time synchronization of the interfaces using actuators and sensors is a prerequisite for stable experiments and high-fidelity results. Since these hardware components suffer from imperfect transfer behavior and delays, additional controllers are required in most RTHS setups. This paper presents an RTHS framework specifically designed to use off-the-shelf robotic arms as actuators. A combination of Iterative Leaning Control and Normalized Passivity Control acts as a pure outer-loop controller in the robot's task-space, making it independent of the robot used and avoiding integration into the robot's low-level controllers. We demonstrate the ability of the control framework to provide high-fidelity results in virtual, i.e. fully simulated, RTHS experiments. We also present a first experimental realization of the framework using a KUKA® KR16 robot arm as an actuator. Although a stable RTHS experiment could be performed with this hardware, the fidelity improvement through iterative learning could not be experimentally validated yet.

**Keywords** Real-time hybrid substructuring · Robotic arm · Iterative learning control · Passivity control · Delay compensation

## Introduction

Dynamic Substructuring is a well-established method in the field of structural dynamics [1]. In this approach, the dynamics of an overall system is examined by independently analyzing the dynamic behavior of several substructures that compose the overall system. This can be done both experimentally and numerically. The dynamics of the overall system is then emulated by coupling the individual subcomponents. To do this, the *compatibility* of the displacements and the *equilibrium* of the forces at each of the interfaces between the substructures must be satisfied. This concept is visualized in fig. 1a. The overall system is decomposed into two substructures (A and B). To preserve the overall dynamics when investigating the dynamics of the substructures individually, the compatibility condition requires the interface displacements to be equal, i.e. $\boldsymbol{z}^A = \boldsymbol{z}^B$, and the equilibrium condition requires the resulting interface force to be zero, i.e. $\boldsymbol{F}^A + \boldsymbol{F}^B = \boldsymbol{0}$. Based on these conditions, the dynamics of the whole system can be reconstructed from the individual results of the substructures.

Real-Time Hybrid Substructuring (RTHS) is an efficient and economical testing approach to investigate the dynamic behavior of systems under realistic conditions by coupling a numerical co-simulation to physical components in real-time [2, 3, 4]. The approach therefore makes use of the concept of Dynamic Substructuring and the system under test is

Arian Kist · David Stadler · Daniel Rixen
Chair of Applied Mechanics, TUM School of Engineering and Design, Munich Institute of Robotics and Machine Intelligence (MIRMI), Technical University of Munich, Boltzmannstr. 15, 85748 Garching, Germany
e-mail: arian.kist@tum.de; david.stadler@tum.de; rixen@tum.de

Rok Belšak · Timi Karner
Laboratory for Robotics, Faculty of Mechanical Engineering, University of Maribor, Smetanova Ulica 17, 2000 Maribor, Slovenia
e-mail: rok.belsak@um.si; timi.karner@um.si

Vasja Plesec · Gregor Harih
Laboratory for Integrated Product Development and CAD, Faculty of Mechanical Engineering, University of Maribor, Smetanova Ulica 17, 2000 Maribor, Slovenia
e-mail: vasja.plesec@um.si; gregor.harih@um.si



Walter D'Ambrogio, et al. (eds.), *Dynamic Substructuring & Transfer Path Analysis, Vol. 4*, Conference Proceedings of the Society for Experimental Mechanics Series, https://doi.org/10.13052/9788743801498-10

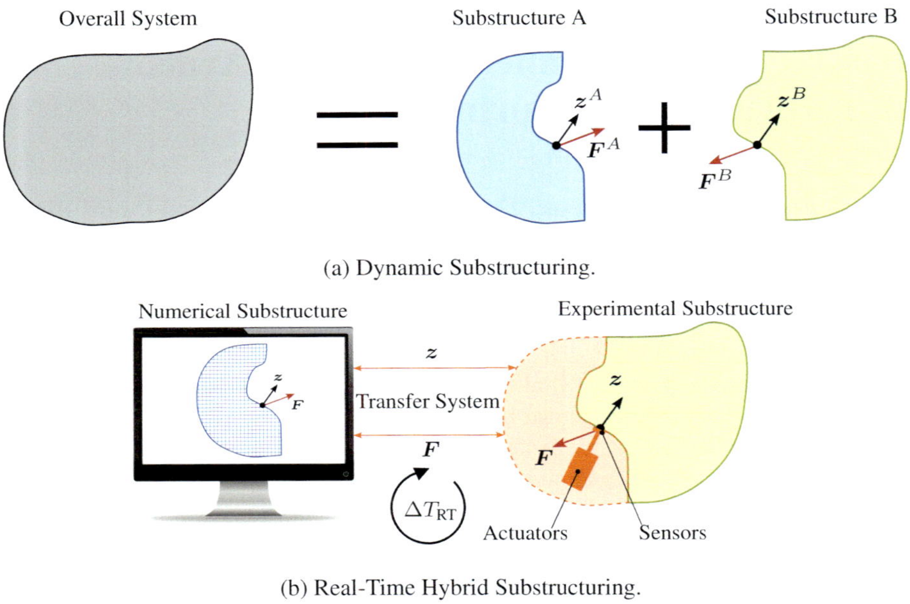


**Fig. 1** Illustration of Real-Time Hybrid Substructuring (b) in the context of Dynamic Substructuring (a). In order to couple the dynamics of substructures A and B to satisfy the dynamics of the overall system, the compatibility of the interface displacements ($\boldsymbol{z}^A = \boldsymbol{z}^B$) and the equilibrium of the interface forces ($\boldsymbol{F}^A + \boldsymbol{F}^B = \boldsymbol{0}$) must hold. In Real-Time Hybrid Substructuring, one substructure is simulated numerically and is therefore referred to as the Numerical Substructure. This is usually the well-defined part of the overall system, i.e. it can be modeled with high fidelity by numerical methods. For the part of the system where simulation is not feasible, a physical component is realized and investigated experimentally. This part is called the Experimental Substructure. To emulate the dynamics of the overall system, both substructures are coupled in real-time (denoted by $\Delta T_{\mathrm{RT}}$ in the figure). To synchronize the interface motion $\boldsymbol{z}$ and the interface force $\boldsymbol{F}$ and thus to satisfy the compatibility and equilibrium conditions, a Transfer System is used. The Transfer System aims to mimic the interface to the numerical substructure using actuators and sensors.

divided into different substructures. An illustration of the RTHS approach is given in fig. 1b. The specific challenge of RTHS is that it involves both numerically simulated as well as experimentally investigated substructures, i.e. physical components, and that the coupling of these substructures is performed in real-time. A Transfer System aims at satisfying the compatibility and equilibrium condition at the interfaces between the numerical and experimental substructures under real-time conditions. This requires the use of sensors and actuators to measure and apply interface motions and forces. Perfect emulation of the reference system is only possible if these hardware components do not introduce synchronization errors, i.e., violate the compatibility and equilibrium condition. In addition, imperfect interface synchronization can even make the test unstable and damage the test setup. The stability issues of RTHS experiments due to interface synchronization errors and delays are discussed, e.g., in [5, 6, 7, 8]. Therefore, special attention must be paid to the selection of the hardware components involved and control strategies must be developed to mitigate synchronization errors and ensure stability. Over the last decades intensive research has been put in designing appropriate control approaches, among many others, ranging from simple polynomial extrapolation [5] over model based controllers [9] to learning [10, 11] and passivity-based [12, 13] approaches. An overview can be found, e.g., in [14].

In this contribution, we share our experience in using a commercial robotic arm, a KUKA® KR16 [15] with a KR C4 robot controller [16], as an actuator to synchronize the interface between a Numerical and an Experimental Substructure in an RTHS test setup. In our previous work, see e.g. [17, 18, 19], we used as an actuator a Stewart platform built in house. This Stewart platform allowed direct access to the joint motors, minimizing potential communication delays. It also allowed the design of advanced joint controllers to reduce tracking errors and compensate for actuator dynamics. When considering industrial RTHS applications, the use of application-specific actuators may not always be an economically efficient approach. Since industrial robot arms are well established in many industries, their use as actuators for RTHS may be the more feasible option in many cases. In addition, as shown in [19], the capabilities of our Stewart platform are limited in terms of workspace and performing highly dynamic motions, which could be overcome by using an industrial robot arm. Therefore, we compare the performance of the KUKA® KR16 [15] in a simple RTHS experiment with the performance of our Stewart platform.

Since the user typically does not have access to the low-level control of individual joints on industrial robots, we adapted our learning-based joint-space control scheme [18] to directly adjust the desired task-space trajectory to mitigate synchronization errors. We demonstrate the performance and robustness of this adapted control scheme through simulation, i.e., in a virtual RTHS experiment, by treating the controlled actuator as a black box and inducing different amounts of artificial delay.

## RTHS Test Setup

For the investigations of this contribution, we analyze the system shown in fig. 2 using RTHS. It consists of a one-dimensional two-mass oscillator including a contact scenario. The upper mass $m_{\text{NUM}}$ is connected to a moving point by a spring with stiffness $k_{\text{NUM}}$ and a damper with damping ratio $d_{\text{NUM}}$. The lower mass $m_{\text{EXP}}$ is attached to the upper mass by a spring with stiffness $k_{\text{EXP}}$. To excite the system, a cosine trajectory $z_{\text{ex}}(t)$ with frequency $f_{\text{ex}}$ is prescribed for the moving point so that the lower mass intermittently contacts the ground. For the investigation using RTHS, the moving point, the upper spring-damper and the upper mass are chosen as the Numerical Substructure, while the lower spring, the lower mass and the ground contact are chosen as the Experimental Substructure. The interface displacement is denoted as $z(t)$. A detailed description of this system has been given in our previous work, e.g.[20, 10, 13, 18], and further description is omitted here. The simplicity of the system allows us to compute a reference solution to assess the fidelity of the RTHS experiments, which is generally not possible for RTHS experiments.

The fig. 3 illustrates the signal flow of the entire RTHS loop executed in real-time. It includes the control scheme to ensure a robust and high-fidelity experiment, which we adapted from our previous work [18] to serve as a pure outer-loop control scheme. The numerical time-integration of the Numerical Substructure provides as output for each hybrid simulation step the interface displacement $z$ (and its corresponding velocity $\dot{z}$). This serves as the desired task-space motion command for the robot to synchronize the interface displacements and satisfy the compatibility condition. The Experimental Substructure is mounted on the interface to the robot's end-effector. In this work, we investigate the use of a commercial robot as an actuator, so we assume that the low-level control of this robot is not accessible and the controlled actuator is treated as a black box. Possible processing and communication delays are accounted for using delay blocks as shown in fig. 3. The robot executes the motion $z'$, $\dot{z}'$ and moves the Experimental Substructure. This motion is not the same as the commanded motion due to the preprocessing delay block (which accounts for delays in the robot for receiving and processing the motion command) and imperfect trajectory tracking of the low-level actuator controller. For our outer-loop control, a measurement of $z'$ is required. Since acquiring this measurement may introduce additional delay compared to the actual motion of the robot, we introduce a postprocessing delay block that accounts for delays due to acquiring, processing, and transmitting the actual task-space position of the robot. The measured task-space motion of the robot is denoted as $z'_{\text{meas}}$, and it is generally assumed that $z \neq z' \neq z'_{\text{meas}}$.

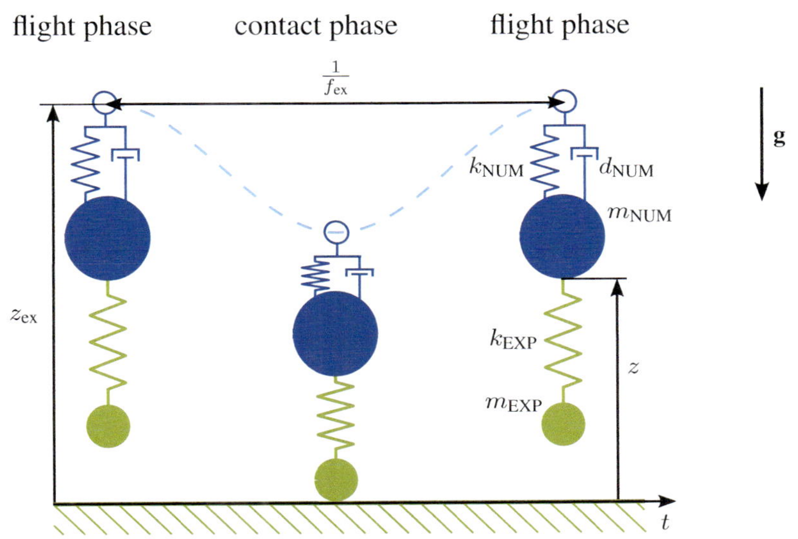


**Fig. 2** Overall system whose dynamics are investigated with RTHS in this work. The Numerical Substructure is colored in blue, while the Experimental Substructure is colored in green. Figure adapted from [21].

In order to drive the total measurable tracking error of the robot $e_{\text{track}} = z - z'_{\text{meas}}$ to zero, we implement an Iterative Learning Controller (ILC). In [10, 18] we used the ILC as a feedforward controller within the joint controller of the Stewart platform. In this work, we use the ILC scheme to directly adjust the task-space motion command $z$ to best compensate $e_{\text{track}}$. ILC is used when a motion command is executed several times in a row. It stores and processes the tracking error from the current trial $j$ to improve the tracking error in the next trial $j+1$ by learning to generate a control signal $\Delta_{\text{ILC}}$ to compensate for the plant dynamics. The updating law for this control signal over the trials is:

$$\Delta_{\text{ILC},j+1} = Q \cdot (L \cdot e_{\text{track},j} + \Delta_{\text{ILC},j}) \tag{1}$$

Note that $e_{\text{track},j}(t)$ and $\Delta_{\text{ILC},j}(t)$ are the full time signals during the whole motion task of trial $j$, i.e. $t \in \left[0, \frac{1}{f_{\text{ex}}}\right]$. The learning function $L$ is used to process the error signal of trial $j$. To ensure the robustness and convergence of the ILC, the entire signal is processed with a robustness filter $Q$. In this work, the learning function is chosen as a P-controller, i.e.: $L = L_{\text{p}}$ and the robustness filter is chosen as a zero-phase lowpass filter with a cutoff frequency $f_{\text{Q,cut}}$. Our implementation of the ILC is a purely data-based approach that does not require modeling of the system. However, the parameters $L_{\text{p}}$ and $f_{\text{Q,cut}}$ must be tuned to achieve stable convergence of the learning. A further discussion of our ILC implementation and its convergence analysis is omitted here and the reader is referred to our previous publications [10, 18]. The adjusted task-space motion command for the robot is thus $z_{\text{ad}} = z + \Delta_{\text{ILC}}$.

To enforce the equilibrium condition of the interface forces, the force $F$ resulting from the robot's motion $z'$ is measured at the interface of the Experimental Substructure using a force-torque sensor (FTS). The measured force $F'$, which is generally not equal to the actual interface force due to measurement errors in the FTS, is fed back to the Numerical Substructure. Note that we assume that this signal is not affected by the postprocessing delays in the robot, since we assume to use an additional FTS and do not rely on internal force/torque measurements in the robot.

To ensure stable RTHS experiments, a Normalized Passivity Controller (NPC) is implemented[1]. The NPC monitors the power flow between the substructures and the Transfer System. If the power at the interface to the Experimental Substructure $P_{\text{EXP}} = \dot{z}'_{\text{meas}} \cdot F'$ [2] is greater than the power at the interface to the Numerical Substructure $P_{\text{NUM}} = \dot{z} \cdot F'_{\text{ad}}$, i.e. $P_{\text{error}} = P_{\text{EXP}} - P_{\text{NUM}} \geq 0$, the experiment is considered active and therefore unstable. In this case, artificial damping force $F_{\text{d}}$ is added to the force feedback to restore the passivity of the experiment. It is calculated as follows:

$$F_{\text{d}} = \begin{cases} G_{\text{d}} \cdot \frac{\tilde{P}_{\text{error}}}{|\tilde{P}_{\text{tot}}|} \cdot \dot{z} & \text{if } \tilde{P}_{\text{error}} > 0 \\ 0 & \text{if } \tilde{P}_{\text{error}} \leq 0 \end{cases} \tag{2}$$

Where $P_{\text{tot}} = P_{\text{EXP}} + P_{\text{NUM}}$, i.e. the total power, is used to normalize the error power $P_{\text{error}}$ for the calculation of the damping gain. $G_{\text{d}}$ is a tuning parameter. Note also that $\tilde{\bullet}$ indicates filtering with a first order lowpass filter with time constant $T_{\text{LP}}$ to smooth the signals. Since the NPC already acts on the outer control loop and not within the robot's low-level controllers, we have not modified our implementation of [13, 18] and it is referred to that publication for further information on our NPC scheme. The adapted measured interface force $F'_{\text{ad}} = F' + F_{\text{d}}$ is then finally used in the next time step of the numerical simulation.

Compared to our previous implementation of the scheme, the adapted control scheme presented here has the advantage of acting completely as an outer-loop controller. No interaction nor integration with internal computations in the hardware components, such as low-level actuator controllers, is required. The only requirements for the used robot/actuator are that its desired task-space motion can be commanded by an external device and that a measurement of the actual task-space motion is available. Apart from these requirements, with this control scheme the hardware components for the robot and the FTS are interchangeable. However, if different hardware components are used, the control parameters of the ILC and NPC may need to be adjusted, but a model of either of these components is not required. In addition, it should be noted that the ILC now aims to drive the total delay induced by the robot/actuator used to zero, thus including processing and communication delays and not just improving low-level actuator tracking performance.

[1]In our control approach, ILC aims to improve the test fidelity and NPC ensures stability [18]. Even though a high fidelity RTHS experiment is less likely to become unstable, since ILC reduces the tracking error iteratively, in the first iterations the error is usually still large and additional approaches are needed to ensure stability. In our case, this is addressed by the NPC. On the other hand, NPC alone cannot improve the fidelity of the test, as it only prevents instability. Hence the combination of NPC and ILC.

[2]The actual power at the interface to the Experimental Substructure would require using the actual interface force $F$. However, in a real experimental setup, only the measured force $F'$ is available, and we use this force to calculate the power at this interface.

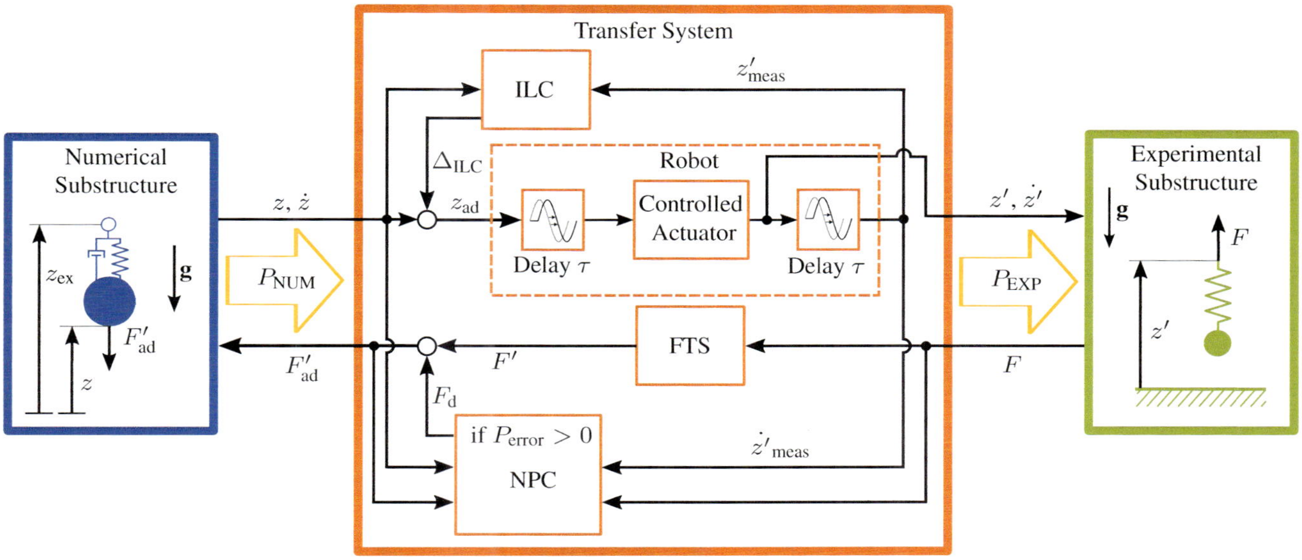


**Fig. 3** Signal flow of the whole RTHS loop used in this work. It includes the control scheme using a combination of ILC and NPC to ensure robust and high-fidelity RTHS experiments, which we adapted from our previous work outlined in [18] to act as a pure outer-loop controller. Figure adapted from [21].

## Simulative Investigation

To validate the performance of the adapted control scheme presented in the previous section, we performed simulations, namely purely virtual RTHS experiments. Thereby, the Transfer System and the Experimental Substructure are also simulated numerically. The virtual RTHS experiment is implemented in MATLAB®/Simulink® (version R2024a, MathWorks®). We use three different dynamic models for the actuator. To model the dynamics of industrial robotic arms, we use the *Robotic System Toolbox* in MATLAB®. In addition to the available functions for calculating the robot kinematics, we use the *Joint Space Motion Model* block in *Independent Joint Motion* mode with its default parameters to simulate the closed-loop motion of the robot. The algorithms of this toolbox are based on the definition of a *rigidBodyTree* object in MATLAB®. To simulate the KUKA® KR16 robotic arm, we custom-built the *rigidBodyTree* object from CAD data. We also import a *rigidBodyTree* object for the ABB® IRB 120 robot arm, which is already importable from the *Robotic System Toolbox*. For our in-house Stewart platform, a dynamic and kinematic model is available from previous work. The dynamics of its legs have been identified up to 100 Hz and a standard cascaded individual joint controller with velocity feedforward is used for joint control. See [17, 21] for more information on the Stewart platform, its system identification, and the implemented joint controller. To also model the effects of delays, two *Transport Delay* blocks are implemented within the signal flow as shown in fig. 3. These blocks delay a signal by the time delay $\tau$. In our simulation we assume that the delay for the pre- and postprocessing is the same. We also assume ideal transfer behavior of the FTS, i.e. $F' = F$. The control scheme is implemented as described in the previous section.

In the virtual RTHS experiment, the Experimental Substructure is also simulated. The simplicity of the system design also allows for easy modeling of the Experimental Substructure as it is a simple one-dimensional oscillator. For a more stable numerical simulation, we added a small damper with damping ratio $d_{EXP}$ in parallel to the spring $k_{EXP}$. To model ground contact, we activate a ground reaction force on $m_{EXP}$ as soon as it touches the ground. This reaction force is based on a nonlinear spring-damper model as described in [22] (note that we have increased the vertical ground stiffness of this model to $10^6\,\mathrm{kg\,m^{-2}}$ to minimize ground penetration). In table 1 all relevant parameters for the simulation can be found, including the chosen Simulink® solver and the sample time $\Delta T_{RT}$.

Figure 4 shows the result of a virtual RTHS experiment with a delay value[3] of $\tau = 10\,\mathrm{ms}$ and the KUKA® KR16 as actuator without ILC (Trial 1) and after ILC convergence (Trial 20) compared to the reference solution. The reference solution is the actual behavior of the system in fig. 2 and what the RTHS experiment should ideally predict. In fig. 4a the interface displacement $z'_{meas}$ of the virtual RTHS experiment is shown compared to the reference solution $z_{ref}$. The result

[3]Note that we implement two delays, one for receiving and preprocessing and one for postprocessing and sending, with the same value $\tau$, cf. fig. 3. Thus, the total artificial delay in addition to the robot's trajectory tracking errors is always $2\tau$.

**Table 1** Parameters used in the simulation. For an exact formulation of the initial conditions of the system shown in fig. 2 and the trajectory $z_{ex}(t)$ see [18, 21]. We have also added a small damping to the Experimental Substructure to allow a more stable numerical simulation of it.

| variable | value | variable | value |
|---|---|---|---|
| $f_{ex}$ | $0.25\,\mathrm{Hz}$ | $g$ | $9.81\,\mathrm{m\,s^{-2}}$ |
| $m_{NUM}$ | $9.6187\,\mathrm{kg}$ | $m_{EXP}$ | $0.3813\,\mathrm{kg}$ |
| $k_{NUM}$ | $10\,000\,\mathrm{kg\,s^{-2}}$ | $d_{NUM}$ | $500\,\mathrm{kg\,s^{-1}}$ |
| $k_{EXP}$ | $8650\,\mathrm{kg\,s^{-2}}$ | $d_{EXP}$ | $10\,\mathrm{kg\,s^{-1}}$ |
| $L_p$ | $0.5$ | $f_{Q,cut}$ | $4\,\mathrm{Hz}$ |
| $G_d$ | $800\,\mathrm{kg\,s^{-1}}$ | $T_{LP}$ | $0.01\,\mathrm{s}$ |
| Simulink® Solver | ode3 | $\Delta T_{RT}$ | $0.001\,\mathrm{s}$ |

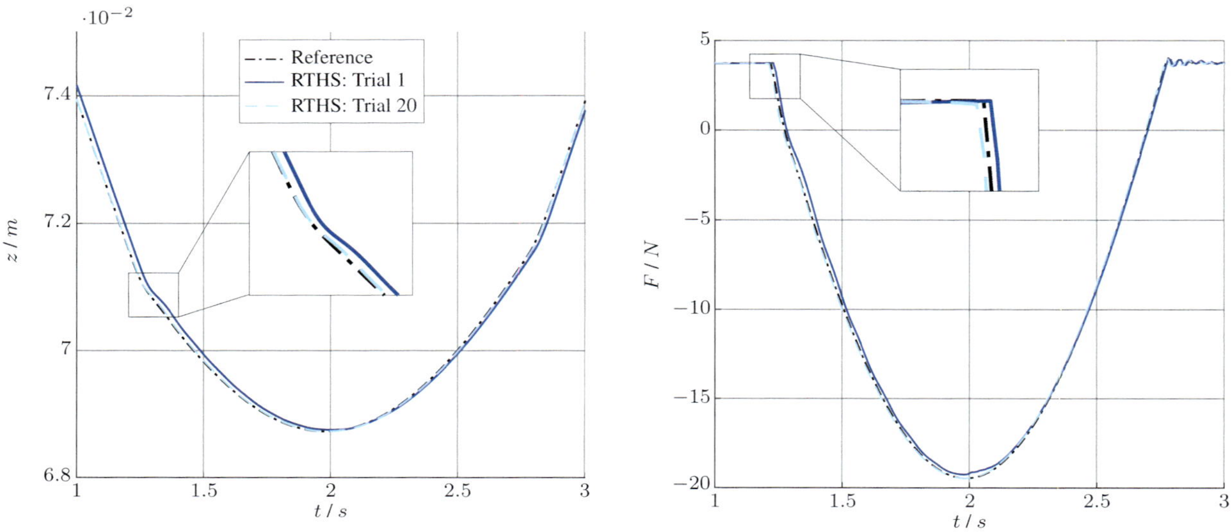


(a) The interface displacement $z'_{meas}$ of the virtual RTHS experiment before and after convergence of the ILC compared to the reference solution $z_{ref}$.

(b) The interface force $F'_{ad}$ of the virtual RTHS experiment before and after convergence of the ILC compared to the reference solution $F_{ref}$.

**Fig. 4** Result of a virtual RTHS experiment with a delay value of $\tau = 10\,\mathrm{ms}$ and the KUKA® KR16 as actuator without ILC (Trail 1) and after ILC convergence (Trial 20) compared to the reference solution.

clearly shows that our adapted control scheme achieves the desired behavior and proves its applicability, as the RTHS solution after convergence of the ILC is almost identical to the reference solution. Thus, the ILC is able to almost completely mitigate the total error induced by the transfer system on the interface displacement. In fig. 4b the interface force $F'_{ad}$ of the virtual RTHS experiment is shown compared to the reference solution $F_{ref}$. The results show that the RTHS experiment predicts a slightly earlier contact after convergence of the ILC compared to the reference solution. The reason is that the ILC aims at shifting the motion command for the robot to compensate the total measured error on the interface displacement $e_{track} = z - z'_{meas}$, which includes an additional delay for measuring, postprocessing, and transmitting the actual motion of the robot $z'$. The Experimental Substructure and the force measurement by the FTS, which is assumed not to be part of the robot, are not affected by this additional delay. Since the ILC takes this delay into account, it shifts the motion command too much from the perspective of the Experimental Substructure, resulting in the early contact compared to the reference solution. This aspect must be taken into account when interpreting the results when the RTHS experiment and the control scheme are implemented as presented. In addition, we mention that even if the NPC was in use, it barely intervened, since the RTHS loop was hardly active[4] during the whole (virtual) experiment (The additional damping force was below $0.2\,\mathrm{N}$ during the whole experiment).

[4] In terms of passivity analysis, 'active' refers to the case where the Transfer System induces additional energy into the RTHS loop, which was hardly the case here.

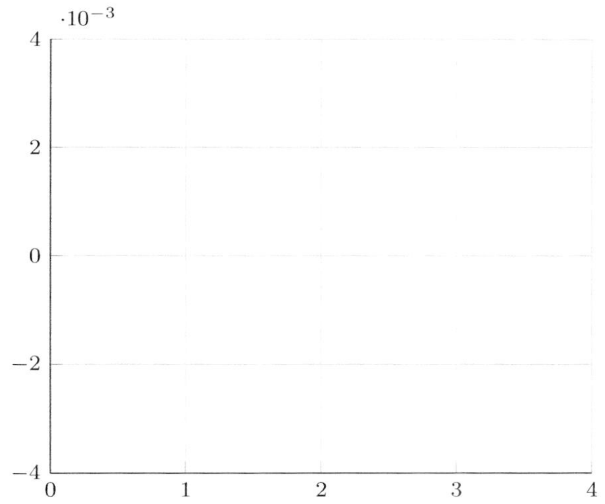
·10−3
4
2
0
−2
−4
0
1
2
3
4

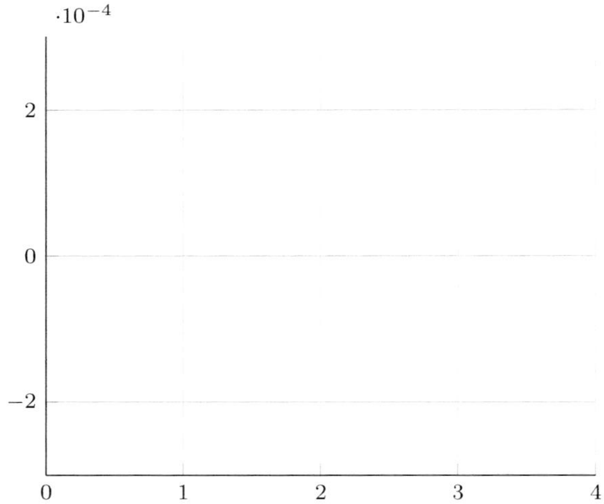
·10^−4
2
0
−2
0
1
2
3
4

P-controller in the robot's task-space is used to further drive the tracking error to zero (not visualized in fig. 7). So the motion command for the robot is: $z_{\text{ad}} = z + \Delta_{\text{ILC}} + K_{\text{p,Kuka}}\,(z - z'_{\text{meas}})$. As robot we use a KUKA® KR16 with a KR C4 robot controller. To move the robot based on an external motion command, the software KUKA.RobotSensorInterface® (RSI) [24] is used. RSI runs on the KUKA® KR C4 controller and handles communication with external devices.

In this work, we use UDP/IP communication via Ethernet. To send the motion command $z_{\text{ad}}$ from the MicroLabBox to the KR C4 controller, we implemented a UPD server in Simulink® based on [25] and utilizing the RTI Ethernet Blockset [26] provided by dSpace®. The UDP server is compiled with the rest of the software and executed on the MicroLabBox. The RSI, which acts as the UDP client in our structure, initiates the communication by sending the current measured position of the robot's end effector $z'_{\text{meas}}$. As soon as the UPD server on the MicroLabBox receives a data packet, it processes it, calculates the current motion command $z_{\text{ad}}$ and responds by sending this command back to the UPD client on the KR C4. The RSI processes the received data packet, forwards it to the internal KUKA® controller as the desired end-effector position and waits until the next communication cycle starts before sending the next data packet. Before using it in the RTHS setup, we validated the UDP communication in a simple benchmark test similar to [27]. In this test, a 3D sine trajectory in the robot's task-space is commanded to the robot from the MicroLabBox via the UDP communication and recorded for 90 seconds. During this time, no packet loss was detected and we estimated the total delay to be 32 ms. This delay includes delays for receiving and preprocessing the motion command, trajectory tracking errors, and delays for measuring, postprocessing, and transmitting the actual motion of the robot[5].

The actual motion of the KR16 robot $z'$ is transferred to the Experimental Substructure, which consists of a compression spring and a 3D-printed half-sphere attached to the robot's end effector. A table is used as a ground to which the half-sphere intermittently makes contact. A SCHUNK® FTS 330/30 [28] is placed between the compression spring and the robot end effector to measure the interface force. The amplified force signal is fed to one of the analog inputs of the MicroLabBox where, after analog-to-digital conversion (ADC) and adaptation by the NPC, it is used as input for the numerical time-integration in the Numerical Substructure.

As shown, in fig. 7, different cycle rates are used in the experimental setup. The major reason is that the smallest possible cycle time of the RSI is 4 ms. Thus, the time between two data packets sent by the UPD client is 4 ms and consequently a motion command can only be sent with this cycle time. Therefore, we run the ILC and the UDP server on the MicroLabBox with a cycle rate of 250 Hz, while the rest of the software on the MicroLabBox, namely the sampling of the force measurement, the NPC, and the numerical time-integration of the Numerical Substructure, is run with a cycle rate of 1 kHz. So every fourth time step of the time-integration is used to generate the motion command for the robots. This procedure is done to be able to use a small time step for the numerical time-integration in order to obtain a stable and high fidelity numerical simulation. Note that the NPC requires the current measured velocity of the robot's end effector $\dot{z}'_{\text{meas}}$, which is only available at the rate of the UDP communication. To overcome this, $\dot{z}'_{\text{meas}}$ is held constant for three time steps until the next measurement is available. It is also assumed that the robot's internal motion control operates at a much faster cycle rate than the RSI, which is indicated by $f_{\text{KUKA}}$ in fig. 7.

In the following, we compare the results of a first RTHS experiment using the setup from fig. 7 with the performance of the setup from our previous work using the Stewart platform as actuator, which is described in detail in [18, 21]. To be able to compare the results, we also used the SCHUNK ® FTS 330/30 for the measurements with the Stewart platform presented here. However, for the Stewart platform measurements, we used the ILC integrated in the joint controllers as described in [18] and not the version adapted for the commercial robots. It should be noted that in the setup for the Stewart platform also the low-level control is implemented on the MicroLabBox and a desired motor current is commanded to the current controlled joint motors via the analog outputs of the MicroLabBox. Therefore, in this setup no serial communication with additional hardware was required keeping all delays except trajectory tracking errors to a minimum. Table 2 summarizes the important parameters for the experimental investigations.

An RTHS experiment using the KUKA® KR16 robot as actuator was successfully performed with the presented setup. In fig. 8 the results are shown without the ILC (Trial 1) and after 25 ILC trials. They are compared with the reference solution and with the results of an experiment using the Stewart platform as actuator. Although the interface quantities could be roughly replicated using the KUKA setup, the deviation from the reference solution is significantly larger compared to the Stewart platform setup. This is especially true for the period $t \in [1.2\,\text{s}, 2\,\text{s}]$, where not only a delayed trajectory but also additional unwanted oscillations are visible for the measured interface displacement $z'_{\text{meas}}$ in fig. 8a. Such unwanted oscillations are typically observed when the RTHS experiment tends to become unstable due to an interface synchronization error. Since the NPC was in operation, the instability and further growth of unwanted oscillations was prevented. Note that the RTHS experiment could not be performed in the presented KUKA setup without the use of NPC due to stability problems, i.e., rapidly growing unwanted oscillations during the contact phase. Apart from that, it can be observed in fig. 8b

[5]We assume that delays due to processing on the MicroLabBox are well below 1 ms and thus do not affect the results

**Table 2** Parameters used for the experimental investigations. For an exact formulation of the initial conditions of the system shown fig. 2 and the trajectory $z_{\text{ex}}(t)$ refer to [18, 21]. Note that there are slight adaptions compared to table 1.

| variable | value | variable | value |
|---|---|---|---|
| $f_{\text{ex}}$ | $0.25\,\text{Hz}$ | $g$ | $9.81\,\text{m}\,\text{s}^{-2}$ |
| $m_{\text{NUM}}$ | $9.6187\,\text{kg}$ | $m_{\text{EXP}}$ | $0.0995\,\text{kg}$ |
| $k_{\text{NUM}}$ | $10\,000\,\text{kg}\,\text{s}^{-2}$ | $d_{\text{NUM}}$ | $100\,\text{kg}\,\text{s}^{-1}$ |
| $k_{\text{EXP}}$ | $8650\,\text{kg}\,\text{s}^{-2}$ | $K_{\text{p,Kuka}}$ | $0.25$ |
| $L_{\text{p,Kuka}}$ | $0.04$ | $Q_{\text{Kuka}}$ | $0.85$ |
| $L_{\text{p,Stewart}}$ | $25$ | $f_{\text{Q,cut,Stewart}}$ | $4\,\text{Hz}$ |
| $G_{\text{d}}$ | $800\,\text{kg}\,\text{s}^{-1}$ | $T_{\text{LP}}$ | $0.01\,\text{s}$ |
| Simulink® Solver | ode8 | $\Delta T_{\text{RT}}$ | $0.001\,\text{s}$ |

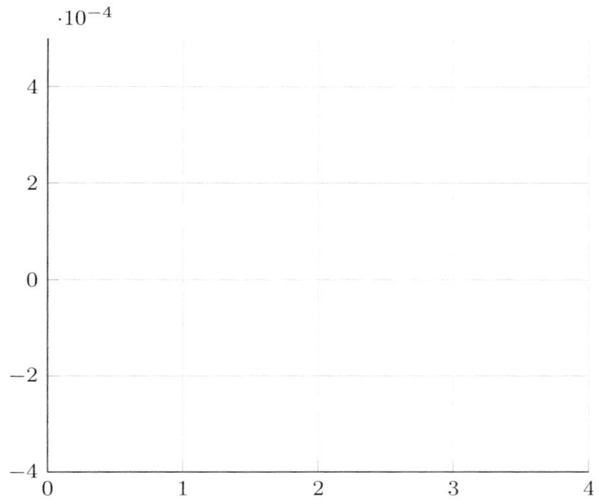
·10^−4
4
2
0
−2
−4
0
1
2
3
4

## References


1. de Klerk, D., Rixen, D. J., and Voormeeren, S. N. "General Framework for Dynamic Substructuring: History, Review and Classification of Techniques". *AIAA Journal* 46.5 (2008), pp. 1169-1181. DOI: 10.2514/1.33274.
2. Nakata, N., Dyke, S. J., Zhang, J., Mosqueda, G., Shao, X., Mahmoud, H., Head, M. H., Bletzinger, M. E., Marshall, G. A., Ou, G., and Song, C. *Hybrid Simulation Primer and Dictionary.* Tech. rep. George E. Brown, Jr. Network for Earthquake Engineering Simulation, 2014.
3. Blakeborough, A., Williams, M. S., Darby, A. P., and Williams, D. M. "The development of real-time substructure testing". *Philosophical Transactions of the Royal Society of London. Series A: Mathematical, Physical and Engineering Sciences* 359.1786 (Sept. 2001). Ed. by Chapman, S. J., pp. 1869-1891. DOI: 10.1098/rsta.2001.0877.
4. Williams, M. S. "Real-time hybrid testing in structural dynamics". *5th Australasian Congress on Applied Mechanics, ACAM 2007.* 2007.
5. Horiuchi, T., Inoue, M., Konno, T., and Namita, Y. "Real-time hybrid experimental system with actuator delay compensation and its application to a piping system with energy absorber". *Earthquake Engineering & Structural Dynamics* 28.10 (1999), pp. 1121-1141. DOI: 10.1002/(SICI)1096-9845(199910)28:10⟨1121::AID-EQE858⟩3.0.CO;2-O.
6. Darby, A. P., Williams, M. S., and Blakeborough, A. "Stability and Delay Compensation for Real-Time Substructure Testing". *Journal of Engineering Mechanics* 128.12 (Dec. 2002), pp. 1276-1284. DOI: 10.1061/(ASCE)07339399(2002)128:12(1276).
7. Botelho, R. M. and Christenson, R. E. "Robust Stability and Performance Analysis for Multi-actuator Real-Time Hybrid Substructuring". *Dynamics of Coupled Structures, Volume 4.* Ed. by Allen, M., Mayes, R. L., and Rixen, D. J. Cham: Springer International Publishing, 2015, pp. 1-7. DOI: 10.1007/978-3-319-15209-7_1.
8. Bartl, A., Insam, C., and Rixen, D. J. "Stability Issues in Hardware-in-the-Loop Tests of Flexible Components". *PAMM* 18.1 (2018). DOI: 10.1002/pamm.201800361.
9. Carrion, J. E. and Spencer, B. F. "Model-based Strategies for Real-time Hybrid Testing". *Newmark Structural Engineering Laboratory Report Series 006* (Dec. 2007). URL: http://hdl.handle.net/2142/3629.
10. Insam, C., Kist, A., and Rixen, D. J. "High Fidelity Real-Time Hybrid Substructure Testing Using Iterative Learning Control". *ISR 2020; 52th International Symposium on Robotics.* Dec. 2020.
11. Hochrainer, M. J. and Puhwein, A. M. "Investigation of Nonlinear Dynamic Phenomena Applying Real-Time Hybrid Simulation". *Nonlinear Structures and Systems, Volume 1.* Ed. by Kerschen, G., Brake, M. R. W., and Renson, L. Cham: Springer International Publishing, 2020, pp. 125-131. DOI: 10.1007/978-3-030-12391-8_16.
12. Hashan Peiris, L. D., Plummer, A. R., and Du Bois, J. L. "Passivity Control in Real-Time Hybrid Testing". *2018 UKACC 12th International Conference on Control (CONTROL).* Sept. 2018, pp. 317-322. dOI: 10.1109/CONTROL.2018.8516814.
13. Insam, C., Peiris, L. D. H., and Rixen, D. J. "Normalized passivity control for hardware-in-the-loop with contact". *International Journal of Dynamics and Control* 9.4 (Mar. 2021), pp. 1471-1477. DOI: 10.1007/s40435-021-00790-8.
14. Palacio-Betancur, A. and Gutierrez Soto, M. "Recent Advances in Computational Methodologies for Real-Time Hybrid Simulation of Engineering Structures". *Archives of Computational Methods in Engineering* 30.3 (Apr. 2023), pp. 16371662. DOI: 10.1007/s11831-022-09848 -y.
15. KUKA Roboter GmbH. *Technical Data KR 16.* Augsburg, 2005.
16. KUKA Roboter GmbH. *KR C4 Betriebsanleitung.* Augsburg, 2018.
17. Insam, C., Göldeli, M., Klotz, T., and Rixen, D. J. "Comparison of Feedforward Control Schemes for Real-Time Hybrid Substructuring (RTHS)". *Dynamic Substructures, Volume 4.* Ed. by Linderholt, A., Allen, M., and D'Ambrogio, W. Conference Proceedings of the Society for Experimental Mechanics Series. Cham: Springer International Publishing, 2021, pp. 1-14. DOI: 10.1007/978-3-030-47630-4_1.
18. Insam, C., Kist, A., Schwalm, H., and Rixen, D. J. "Robust and high fidelity real-time hybrid substructuring". *Mechanical Systems and Signal Processing* 157 (Aug. 2021), p. 107720. DOI: 10.1016/j.ymssp.2021.107720.
19. Insam, C., Ballat, L.-M., Lorenz, F., and Rixen, D. J. "Hardware-in-the-Loop Test of a Prosthetic Foot". *Applied Sciences* 11.20 (Oct. 2021), p. 9492 . DOI: 10.3390/app11209492.
20. Insam, C., Bartl, A., and Rixen, D. J. "A Step Towards Testing of Foot Prostheses Using Real-Time Substructuring (RTS)". *Special Topics in Structural Dynamics & Experimental Techniques, Volume 5.* Ed. by Dervilis, N. Conference Proceedings of the Society for Experimental Mechanics Series. Cham: Springer International Publishing, 2020, pp. 1-9. DOI: 10.1007/978-3-030-12243-0_1.
21. Insam, C. "Fundamental Methods for Real-Time Hybrid Substructuring with Contact: Enabling Testing of Prosthetic Feet". PhD thesis. Technical University of Munich, 2022.
22. Geyer, H. and Herr, H. "A Muscle-Reflex Model That Encodes Principles of Legged Mechanics Produces Human Walking Dynamics and Muscle Activities". *IEEE Transactions on Neural Systems and Rehabilitation Engineering* 18.3 (June 2010), pp. 263-273. DOI: 10.1109/tnsre.2010.2047592.
23. dSpace GmbH. *MicroLabBox Product Brochure.* 2020.
24. KUKA Roboter GmbH. *KUKA.RobotSensorInterface 3.2 Documentation - Version: KST RSI 3.2 V1.* Augsburg, 2013.
25. Leland, J. *kukaslxctrl. GitHub repository.* 2017. URL: https://github.com/mitmedialab/kukaslxctrl(visitedon03/12/2024).
26. dSPACE GmbH. *RTI Ethernet Blockset. Reference for RTI Ethernet Blockset 1.2.1.* 2016.
27. Lisowski, W. and Miekina, L. "Exemplary application of a virtual sensor with a manipulating robot". *The 18th Conference "Mechatronic Design Workshops".* May 2018.
28. Fritz Schunk GmbH. *Montage- und Bedienungsanleitung für 6 Achsen Kraft-/Momentensensor Type FTS.* 1992.

# Chapter 11
# Model Initialization in Real Time Hybrid Testing for Experimental Detection of Isolated Branches

**A. Mario Puhwein and Markus J. Hochrainer**

**Abstract** Due to complex interactions nonlinear structures exhibit a variety of difficult and unexpected phenomena. Isolated branches are of technical relevance and they typically occur at moderate to high vibration amplitudes. In a numerical model, their systematic detection assumes steady state conditions, and is e.g. based on the systematic detection of bifurcation points until the isolated branch appears. However, since a nonlinear system cannot be investigated experimentally at instable operating points, this approach is not feasible in experimental testing. In this work, a physical Duffing type absorber is constructed and coupled to a simulated Duffing type host structure, in a real-time hybrid testing configuration. The experimental study is limited to steady state conditions, however, satisfactory results are obtained in this configuration. Only small and slow excitation adjustments are possible to keep the coupled system in steady state. Therefore, it takes a long time to guide the system to an isolated branch along a predefined trajectory. In this work, an alternative to this cumbersome process is proposed. If the desired vibration amplitudes of the coupled systems are known, the physical absorber is driven to the desired configuration, before it is coupled to the virtual model. Similarly, the simulated structure can be brought directly into the corresponding state by prescribing correct initial conditions, before permanently linking the models. The developed switchover technique allows to explore isolated branches experimentally without time-consuming path following strategies. It is demonstrated that RTHT experiments can be used to investigate stable parts of isolated branches, which can be reached fast and reliably.

**Keywords** Real-time hybrid testing · Nonlinear oscillation · Isolated branches · Dynamic vibration absorber

## Introduction

Nonlinear mechanical systems exhibit a wide variety of effects which are often analyzed in numerical studies because they are difficult to reproduce in real physical experiments. These include bifurcations, multiple solutions at different energy levels, hysteresis, quasi-stationary oscillation ranges, sub- and superharmonic as well as chaotic solutions. Due to material or structural optimizations many technical systems have non-linear behavior, and if they posess low damping, non-linear resonant vibrations can occur. This results in high vibration amplitudes, which are desirable in the case of energy harvesting, but should be avoided in the majority of applications e.g. by vibration damping measures like the traditional vibration absorber. Since linear absorbers are inefficient in non-linear systems, the dynamic absorbers must also behave non-linear [1].

Since the construction and investigation of non-linear systems with predefined properties is complex, expensive and sometimes dangerous, experimental studies use either model tests or significantly simpler alternatives such as real-time hybrid testing (RTHT). The latter allows accurate testing of physical components, and has become popular in research and industry due to advances in simulation techniques and the availability of real-time hardware. The basic idea is to construct only a part of the overall system (physical model), and implement the remaining parts in a computer simulation (virtual model). Coupling the models at carefully selected interfaces is crucial, because all kinematic and dynamic boundary conditions must be fulfilled to reproduce the correct system dynamics. For mechanical applications the interface often requires a

A. Mario Puhwein
Technical University of ViennaKarlsplatz 13, 1040 Vienna, Austria
e-mail: antonmario.puhwein@gmail.com

Markus J. Hochrainer
University of Applied SciencesJohannes Gutenberg-Strasse 3, 2700 Wiener Neustadt, Austria
e-mail: markus.hochrainer@fhwn.ac.at



Walter D'Ambrogio, et al. (eds.), *Dynamic Substructuring & Transfer Path Analysis, Vol. 4*, Conference Proceedings of the Society for Experimental Mechanics Series, https://doi.org/10.13052/9788743801498-11

"transfer system", which transmits the simulated forces or movements to the physical model. It typically uses actuators with robust and high-precision control, which also limit the possible frequency and amplitude range of the RTHT experiment. The transfer system and its control directly influence the quality and even the stability of the entire setup, see. e.g. [2–5]. Since dynamic properties are often unknown, different control schemes have been proposed, e.g. phase shift [6],[7] and inverse compensation [8],[9], adaptive control with actuator identification [10]-[13], or AI based approaches [14]. Another way to reduce the adverse effects of the transfer system is the restriction to repeatable processes. To reduce inaccuracies due to coupling errors, the entire experiment can be repeated while adapting the transfer system control until the required accuracy is achieved, see e.g. [15]. If the experiment is focused on quasi-stationary periodic processes, iterative learning control (ILC) or repetitive control (RC) [16]-[19] can be applied successfully. ILC used in this project works very accurately and robustly, does not require prior knowledge of the dynamics of physical model and transfer system, dead times are perfectly compensated and it has proven to operate reliably for quasi-stationary processes.

In RTHT the focus is generally on the physical model, but the simulation model plays an equally important role as it must represent all relevant components that are not tested experimentally. In recent years the development of virtual models is well supported by high end simulation tool which have developed significantly. Today several reliable tool chains are commercially available that automatically convert the simulation model into the required representation for the real-time application.

To study nonlinear dynamic absorption, the physical oscillator used in this work is of Duffing type, see Fig 1a). The nonlinear restoring forces result from the interaction of a strong base magnet with a small magnet attached to the absorber mass, see Fig 1b). For the host structure, it is assumed that it can also be represented as a Duffing type SDOF oscillator, see again Fig 1a). According to the RTHT methodology, host structure and absorber are separated, the absorber is tested physically, whereas the host structure (virtual model, simulated model) is simulated in real time.

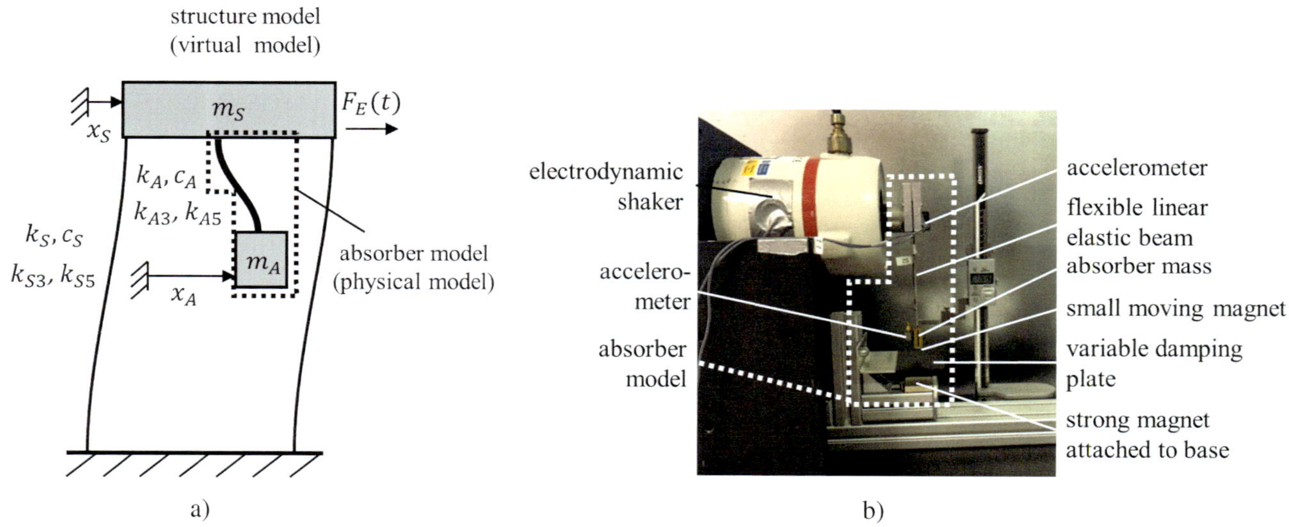


**Fig. 1** a) Model of the coupled system b) laboratory setup of test bench

The simulated host structure can be regarded as "digital twin" of the real structure which can be configured easily in a wide parameter range. Any dynamics or nonlinearity can be implemented, as long as the motion can be calculated in real time. Therefore, the selection of nonlinear host structure parameters which reflect a proper physical system is crucial. For the dynamic absorber, in the contrary, it is rather difficult to develop a setup with adjustable dynamic parameters. For oscillators of Duffing type this can be achieved either by geometric or material nonlinearities or the application of magnetic interaction. The latter approach is quite convenient, because permanent magnets are commercially available in any size and shape. It has been demonstrated in [20], that a duffing type absorber can be obtained by adding repelling or attracting magnets to the mass of a linear oscillator. Since magnetic interaction forces depend on the magnetic field strength, it is convenient to attach a small light magnet onto the moving mass while providing a strong magnetic field by a much larger permanent magnet which is rigidly attached to the base. Furthermore, almost linear damping can be added to the oscillator by inserting an aluminum plate close to the moving magnet. It creates eddy current damping effects which are proportional to the relative velocity of the magnet with respect to the damping plate. By changing the distance between moving magnet and plate the almost linear viscous damping can be adjusted.

As already discussed, the coupling of virtual and physical model is a challenge in RTHT. In the given configuration a transfer system is required to transform the motion of the simulated model into a real displacement of the physical absorber's base. To keep the coupled structure in dynamic equilibrium, the interaction forces must be determined and applied to the virtual model as additional forcing. If both effects are incorporated correctly, the RTHT setup behaves exactly like the corresponding physical model. Considering the setup in some more detail two cascaded feedback loops become apparent. One is formed by the feedback control necessary for the transfer system. It assures that the shaker exciting the dynamic absorber exactly behaves like the simulated structure. This inner control loop must be robust, fast and with negligible phase lag, because otherwise any additional delay might lead to instabilities for lightly damped resonant mechanical systems. The second, much slower control loop results from the RTHT force feedback. It contains the host structure reference acceleration $\ddot{x}_{Sref}$, the electrodynamic shaker excitation $\ddot{x}_S$, and the interaction feedback force $F_A$ which is calculated from the measured absorber acceleration, see Fig 2. Because a properly configured RTHT setup is able to reproduce any stable dynamics of the corresponding physical system, it can, e.g. be used to follow stable paths to investigate isolated branches if the periodically excited system response remains stable and quasi stationary. Therefore, it can take quite a time to experimentally drive the coupled system to a desired configuration. Isolated branches, e.g. can only be reached via rather complicated paths on the frequency-excitation response surface. However, if a desired steady state response is known from simulations, it can also be reached directly in a RTHT experiment by first treating the physical and virtual model as separated systems which are independently guided to the predefined steady state. Only if both systems are in equilibrium, they are coupled, thereby closing the RTHT loop. Using this methodology, the physical absorber is treated as SDOF system, and stable points on the nonlinear frequency response curve can be reached by varying either excitation force or frequency. For the simulated model, any desired model state can directly be specified at the instant of coupling, by prescribing system states as initial conditions.

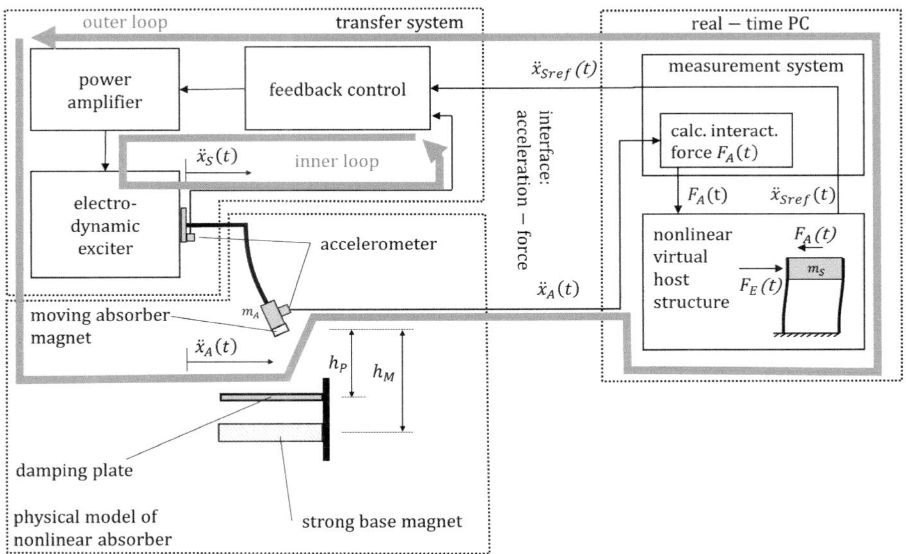


**Fig. 2** Schematics of the RTHT setup with feedback loops and relevant interface quantities.

Nonlinear systems in e.g. resonant regimes can be very sensitive to parameter variations. To obtain a desired vibration absorption behavior it is essential that numerical and physical models are known and described accurately, and that the absorber parameters are selected carefully. Thus, several steps are suggested to obtain a reliable RTHT configuration, see Fig 3. Starting point is the definition of the host structure properties like mass, damping and linear as well as nonlinear stiffnesses. Based on these assumptions, the absorber can be designed according to published design rules, see e.g. [1]. After its construction the absorber is tested experimentally to identify its actual parameters. Then, the RTHT experiment is performed to measure the coupled system response. If it deviates from the desired equal peak behavior, the physical absorber is modified and tuned until the coupled response behaves as desired. Then, isolated branches and a possible stable path to them can be established from numerical simulations. If the Isolas identified turn out to be outside the physical limits of the test bench, the absorber model must be redesigned e.g. for a different mass ratio or according to the expected vibration amplitudes. In case of a redesign, however, the entire process must be repeated again. Only if the numerical simulations provide a possible stable path to Isolas, the RTHT experiment can be carried out successfully. Even if the proposed procedure appears rather complicated, it has proven effective in laboratory operation because the nonlinear absorbtion investigated has a high parameter sensitivity. Studies in the laboratory confirm, that the iterations indicated in Fig 3 typically must be carried several times before the performance of the test-setup turns out to be optimal.

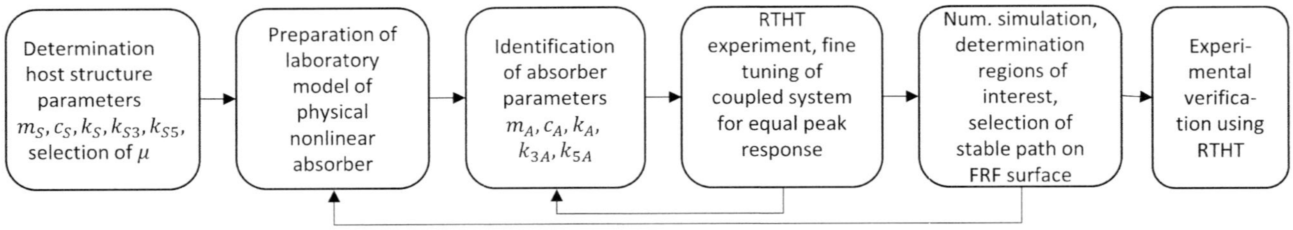


**Fig. 3** Flowchart for successful implementation of RTHT; modified from [22].

## Analytical Model and Coupled Simulation

When analyzing the motion of complex structures, it is common to decompose the response of complex structures into modal components, thereby identifying the dominant degrees of freedom in a frequency range of interest. In many situations it is even possible to approximate the system dynamics using only a single degree of freedom. If this simplification is acceptable for host structure and absorber, their coupled dynamics can be described by Duffing type oscillators with cubic and quintic stiffness terms [21]

$$\begin{aligned} m_S\ddot{x}_S + c_S\dot{x}_S + k_S x_S + k_{S3}{x_S}^3 + k_{S5}{x_S}^5 - c_A\left(\dot{x}_A - \dot{x}_S\right) - k_A\left(x_A - x_S\right) - k_{A3}(x_A - x_S)^3 - k_{A5}(x_A - x_S)^5 \\ = F_E\, cos(\omega t) \\ m_A\ddot{x}_A + c_A\left(\dot{x}_A - \dot{x}_S\right) + k_A\left(x_A - x_S\right) + k_{A3}(x_A - x_S)^3 + k_{A5}(x_A - x_S)^5 = 0 \end{aligned} \tag{1}$$

$m_S, c_S, k_S, k_{S3}, k_{S5}, F_E$ describe the mass, damping coefficient, linear, cubic and quintic stiffness as well as the excitation force of the host structure and $m_A, c_A, k_A, k_{A3}, k_{A5}$ denote the mass, damping coefficient, linear, cubic and quintic stiffness of the absorber. Introducing the the equivalent static deformation $x_{S,stat} = F_E/k_S$, structure and absorber displacements can be expressed dimensionless as $q_S = x_S/x_{S,stat}$ and $q_A = (x_S - x_A)/x_{S,stat}$, the external forcing amplitude is commonly described by the equivalent acceleration amplitude $a_F = F_E/m_S$. The natural frequencies $\omega_{nS} = \sqrt{k_S/m_S}$, $\omega_{nA} = \sqrt{k_A/m_A}$, and the damping ratios $\zeta_S = c_S/(2m_S\omega_{nS})$, $\zeta_A = c_A/(2m_A\omega_{nA})$ of host structure and absorber can be used to describe the underlying linear two degree of freedom dynamics. The normalized excitation frequency is given by $\gamma = \omega/\omega_{nS}$ and the mass ratio is denoted $\mu = m_A/m_S$. Although the equations of motion, eq. (1), can be solved numerically by time integration, analytical solutions can be given by the harmonic balance method (HBM) for steady state conditions. Comparisons with exact numerical results have revealed, that in the current work single term approximations for both, structure and absorber, are sufficiently accurate. The HBM approach yields a set of nonlinear coupled equations, which must be solved numerically to obtain the desired response amplitudes.

In [21] a systematic method is described to discover and identify Isolas by evaluating a turning point condition to generate and follow the turning point curve (TP-curve), see Fig 4. Numerically, it is also possible to determine the closed curve of an isolated branch using a path following algorithm for constant excitation force (FRF-Isola). However, these results cannot be reproduced experimentally because instable branches of the coupled system cannot be followed. Consequently, it is also not possible to follow the turning point curve experimentally. An Isola can only be reached by keeping the system states on the stable regions of the generally complex shaped response surfaces. This often requires a simultaneous adjustment of amplitude $a_F$ and frequency $\gamma$ of the structural excitation. Typically, a desired isolated state can be reached experimentally by starting from an initial stable configuration with a higher excitation force, followed by force reductions and frequency corrections until the desired state is reached. In the experiment, however, the actual time to drive the system to an isolated state can take up to serval minutes, because the dynamic system must remain in quasi periodic conditions during the entire process. This time can be reduced significantly by the proposed variable system configuration.

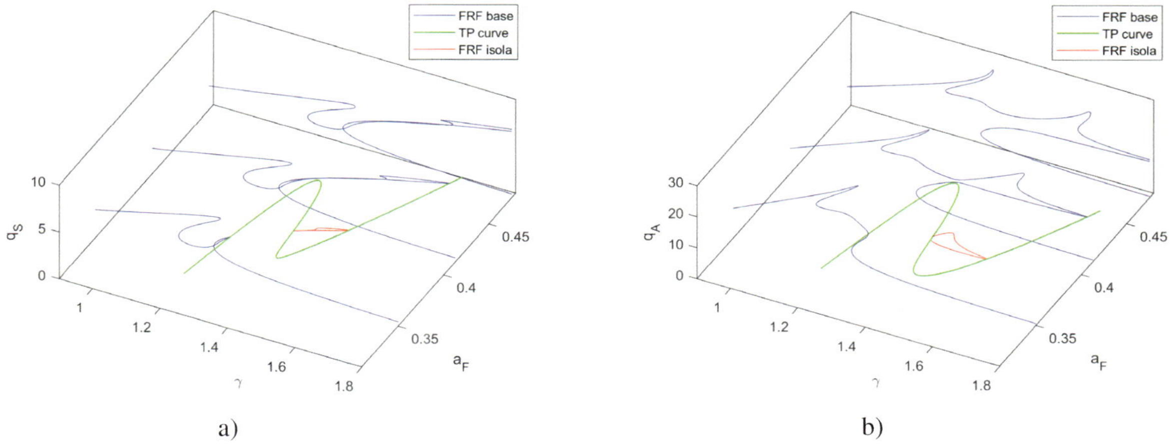


**Fig. 4** Isolated branches (red) detected by numerical computations of the turning point curve (TP-curve) a) host structure b) dynamic absorber [22].

## Variable System Configuration

The path following just described must be strictly followed in a traditional laboratory experiment of the coupled 2DOF nonlinear structure. It is not possible to leave or circumvent the stable path to get to an isolated branch. However, in the RTHT setup only the absorber is physically tested, the host structure is simulated numerically and consequently its initial conditions can be prescribed. If it is possible to guide the isolated physical absorber to a desired vibration state while completely neglecting the virtual model, an accurate RTHT experiment can still be carried out if the corresponding initial conditions are prescribed to the virtual model before coupling both systems. The only prerequisite for a valid experiment is knowledge of the true states of the virtual model. They must be either determined beforehand or calculated while setting up the vibration of the physical absorber. When implementing this concept, it becomes clear that the moment of coupling the systems is critical, since a very smooth transition is essential. This requires both subsystems to be in almost perfect equilibrium before the switchover, otherwise transient beating phenomena will be induced. Once the physical model has reached the desired vibration, the effective virtual host structure forcing $F_A+F_E$ can be derived from eq. (1)

$$F_A+F_E=m_S\ddot{x}_S+c_S\dot{x}_S+k_Sx_S+k_{S3}{x_S}^3+k_{S5}{x_S}^5, \tag{2}$$

and the structure-absorber interaction force $F_A$ can be defined by

$$F_A=m_A\,\ddot{x}_A=c_A\left(\dot{x}_A-\dot{x}_S\right)+k_A\left(x_A-x_S\right)+k_{A3}(x_A-x_S)^3+k_{A5}(x_A-x_S)^5. \tag{3}$$

Since only the accelerations $\ddot{x}_S$ and $\ddot{x}_A$ are recorded, the velocities and displacements must be determined from analytical integration. Since steady state conditions are considered, the measured signal is least squares approximated by a Fourier series up to the fifth harmonics. This is a smart way to filter periodic signals, see [23], [24], with the advantage of eliminating noise and drift while maintaining excellent accuracy due to the exact integration. Thus the r.h.s. of eq. (2) is known and since $F_A$ can be determined using eq. (3) the unknown structural force excitation $F_E$ can be estimated

$$F_E=m_S\ddot{x}_S+c_S\dot{x}_S+k_Sx_S+k_{S3}{x_S}^3+k_{S5}{x_S}^5-m_A\,\ddot{x}_A. \tag{4}$$

This signal is again least squares approximated

$$F_E=A_E\cos\omega t\;+B_E sin\omega t, \tag{5}$$

to minimize measurement noise and errors as well as inaccuracies caused by transient processes. With the variable system configuration the absorber can be directly forced into its final condition. During this process, eqs. (2)–(5) are permanently updated and the switchover to the coupled RTHT configuration can take place once the absorber has reached the intended steady state operating point. The simulation of the virtual model only begins just after the switchover at time $t_s$, with the initial conditions $x_S\left(t_s\right)$ and $\dot{x}_S\left(t_s\right)$ derived from the filtered shaker acceleration $\ddot{x}_S$. If all tasks are carried out correctly, the switching is smooth, and the system states do not change significantly. However, if the coupled system is not in a stable equilibrium, it will converge to a different state. All major steps of the developed switching process are illustrated in Fig 5.

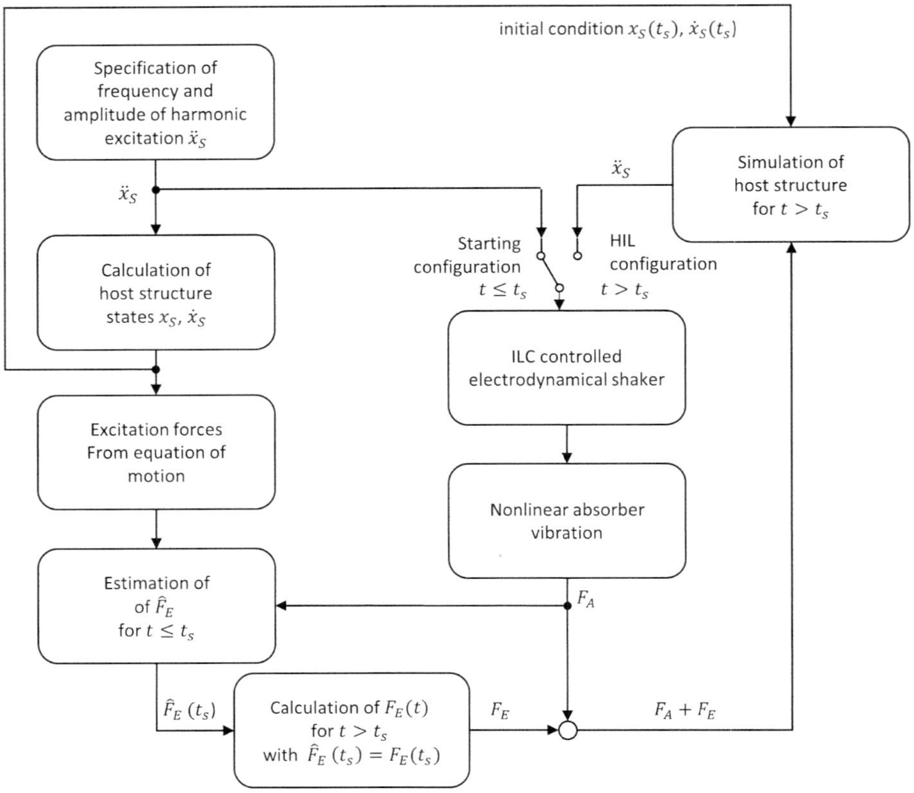


**Fig. 5** Flowchart of variable system configuration. For $t \leq t_s$ the absorber is driven to the predefined steady state, before switching to the RTHS configuration for $t > t_s$.

## Experimental Results

The setup and identification of the RTHT experiment is described in [22]. For the frequency range of interest $6\pi \leq \omega \leq 10\pi\ rad/s$, the essential parameters can be summarized by $m_S = 3.78kg$, $c_S = 4.68Ns/m$, $k_S = 1.68 \cdot 10^3 N/m$, $k_{S3} = 8.20 \cdot 10^7 N/m^3$, $k_{S5} = 0N/m^5$, $\mu = 1/45$, $m_A = 0.084kg$, $k_A = 32.4313N/m$, $c_A = 0.256Ns/m$, $k_{A3} = 1.54 \cdot 10^5\ N/m^3$ and $k_{A5} = -1.20 \cdot 10^8 N/m^5$. With this configuration, the nonlinear dynamic system has isolated branches within the amplitude range $0.15 < a_F < 0.215m/s^2$ and the frequency range $4.7 < f < 6.6Hz$ which is appropriate for the RTHT experiment. For this setup, Fig 6 illustrates the experimental determination of the Isola surface for both, structure and absorber by combining the results of 18 path following experiments.

Starting from the stable point A the Isolas can be reached by reducing the excitation force amplitude to move to point B. Then, the excitation frequency is decreased (experiment 1) or increased (experiment 2) until the stable surface is left. To obtain the remaining response curves 3 to 18 the experiment is repeated with the separation point B shifted slightly for each run. The isolated surface is clearly indicated by the amplitude jumps which occur at the border of the surface. Once the Isola has been left by jumping to a lower amplitude stable solution, it cannot be reached again by varying the excitation frequency only. For the experiments shown in Fig 6 a single run typically takes about 15min because the system has to be driven through the quasi stable region quite carefully. The approach proposed in this work is much faster, because the Isola can be reached directly. In Fig 7 the developed technique appears as a direct shortcut from an initial condition $I$ the desired point $i$ on the isolated surface. Subsequently, it is possible to navigate on the Isola surface by varying the excitation amplitude and frequency. To demonstrate the flexibility and robustness of the method, the experiment is repeated for slightly different initial conditions, see point $II$ in Fig 7. Again the same Isola is reached at a different point $ii$ and in this experiment a different path is prescribed and followed.

The "shortcut" paths directly connecting point $I$ and $i$ as well as $II$ and $ii$, demonstrate, that the proposed switchover works as intended. However, to assess its quality, the time response of the accelerations should be considered. In Fig.8 two different switching processes are illustrated for the same operating point. Up to the switching time $t_S = 19s$ the absorber is excited with a predefined acceleration $\ddot{x}_S$.

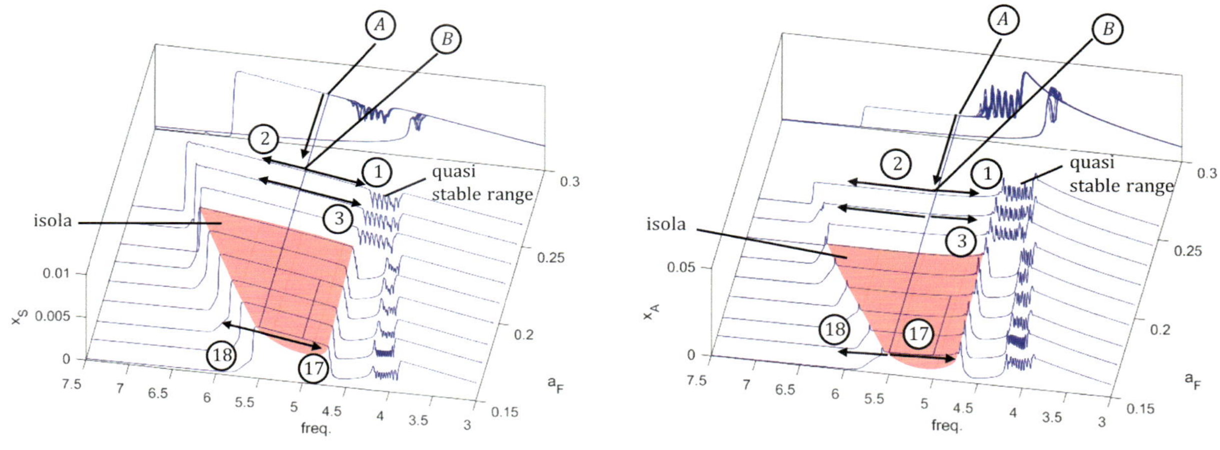


a) b)

**Fig. 6** RTHT experiment to confirm isolated branches experimentally a) structural response, b) dynamic absorber response; modified from [22].

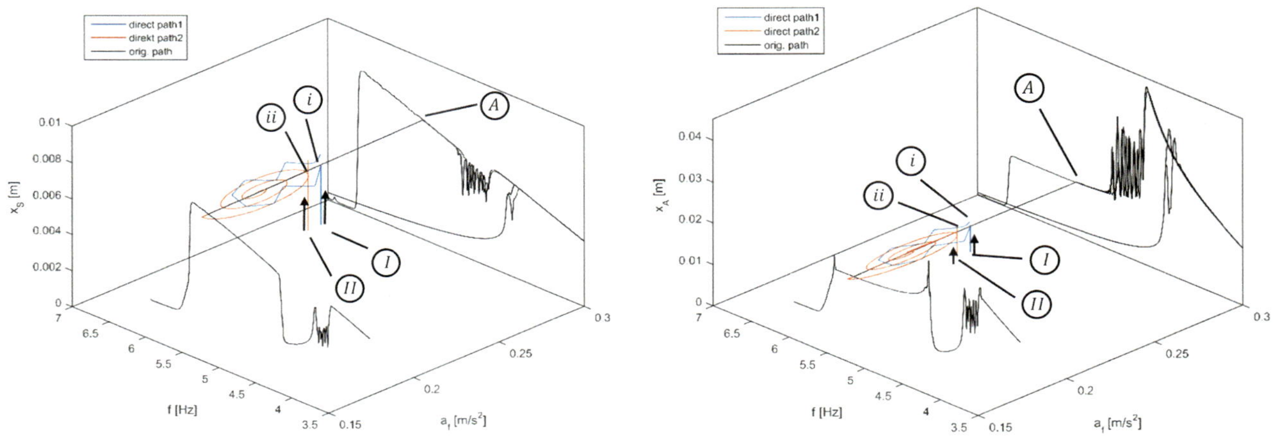


a) b)

**Fig. 7** Comparison of traditional and proposed paths to reach the isolated branch. Displacements of a) the simulated host structure b) the physical absorber.

After the dynamic absorber has reached steady state conditions, the switchover can take place. In Fig 8 a)-b) a suboptimal vibration pattern results from an inaccurate estimation of the external structural excitation $F_E$ using eq. (4). Apparently it takes about $10s$ until steady state conditions are established again. Obviously, rather large fluctuations occur in the measured accelerations, which are tolerable because the desired state on the Isola can still be maintained. In general, however, such errors result in a jump of the system to a stable state with much lower vibration energy and amplitudes. In Fig 8 c)-d) the same switching situation is illustrated for favorable conditions. Again, a transient beating process is recognizable after switching, but with significantly reduced settling times and amplitude variations. It takes about $5s$ to reach stable steady state conditions on the isolated branch. Although improvements are still possible in the experimental results presented, they are already satisfactory, because the estimation of the excitation force from eq. (4) is difficult in the case of resonant vibrations. It turns out that transient effects from the switchover are depending on the chosen system state, and they reduce for smaller vibration amplitudes and energy levels. The experiments have revealed, that the structural acceleration $\ddot{x}_S$ of the host structure must be known with the best possible accuracy to guide the coupled system to a desired state. $\ddot{x}_S$ can be either determined from the numerical model or from measurements, if a condition of interest has already been reached experimentally before.

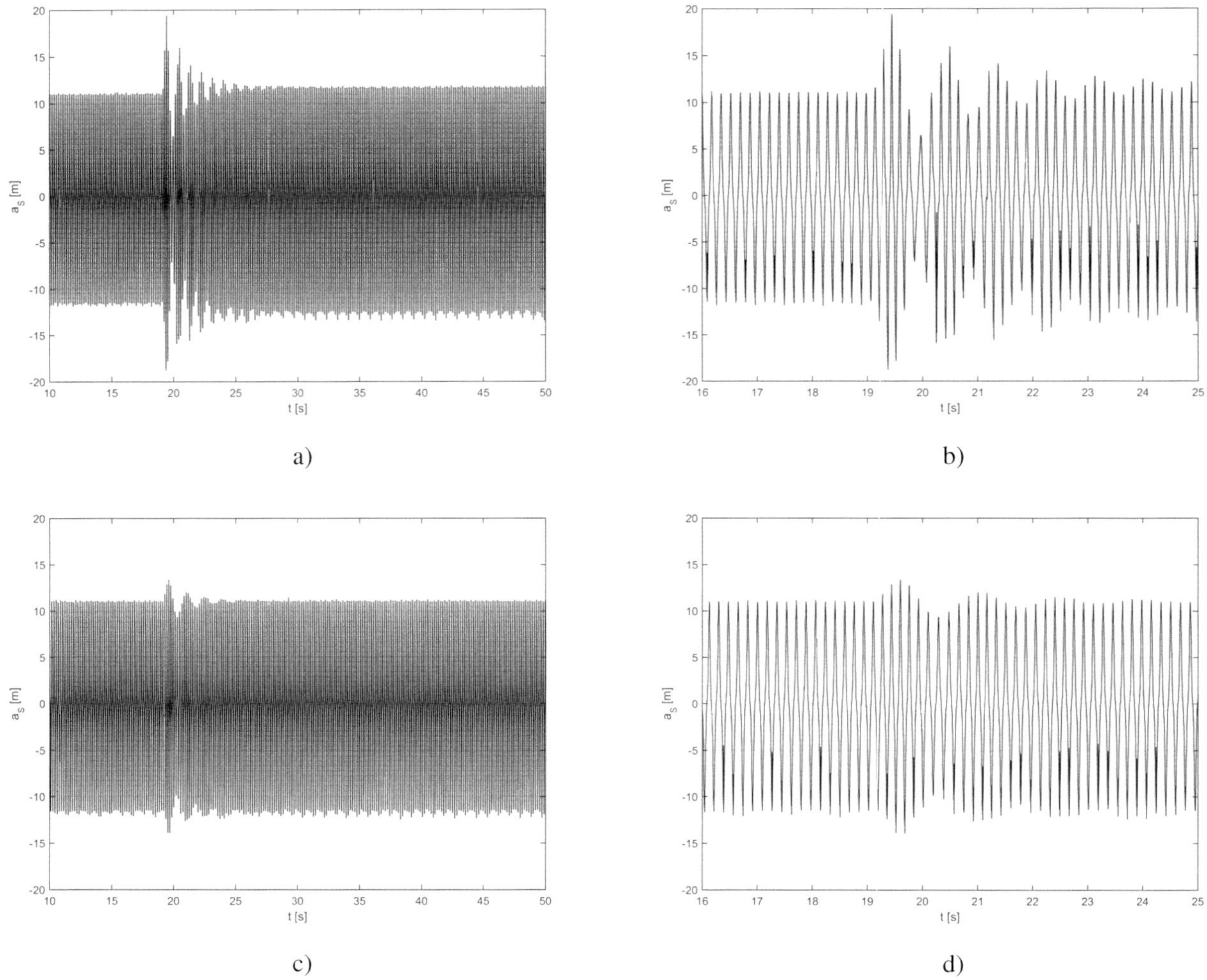


**Fig. 8** Transient effect (RTHT experiment) of the switchover at time $t_s = 19s$ on the acceleration $a_S = \ddot{x}_S$ a) inaccurate determination of the external excitation for $10s < t < 50s$ and b) for $16s < t < 25s$ c) favorable switching conditions with minor transient effects for $10s < t < 50s$ and d) for $16s < t < 25s$.

## Conclusion

Isolated branches of nonlinear systems can be of interest for several studies. Generally, they are investigated numerically using path following strategies which slowly guide the dynamic system in to a desired condition, often following turning point curves or other sections of instable paths. However, this approach is not possible in an experimental setup, because close to boundaries of stability, any physical system immediately jumps to a stable low-energy configuration. Once an isolated branch has been detected, it is always possible to reproduce steady state vibrations in a numerical simulation using proper initial conditions. A similar approach, which avoids path following methods by directly guiding the dynamic system into its final state, is proposed for RTHT. In the hybrid experiment a physical nonlinear dynamic absorber is attached to a virtual nonlinear host structure to mitigate its vibration. Such a coupled system creates fairly complex response characteristics including bifurcations, internal resonances and isolated branches. The latter can be investigated experimentally using a two-step procedure if the desired steady state conditions are known with sufficient accuracy. First, the absorber is driven to its desired point of operation, before it is coupled to the virtual system in a matching vibration state. If the configurations of the individual systems at the switching moment are close to a stable branch, the experiment will converge to the desired state within several oscillations. When subsequently guiding the oscillators along a predefined trajectory on the isolated branch, the system must always remain in steady state, and consequently only small and slow adjustments in the excitation are permissible. In this work it is undoubtedly demonstrated that RTHT experiments can be used to investigate stable parts of isolated branches, which can be reached fast and reliably.

**Acknowledgments** This work was carried out as part of the COMET Centre InTribology (FFG no. 906860), a project of the "Excellence Centre for Tribology" (AC2T research GmbH). InTribology is funded within the COMET – Competence Centres for Excellent Technologies Programme by the federal ministries BMK and BMAW as well as the federal states of Niederösterreich and Vorarlberg based on financial support from the project partners involved. COMET is managed by The Austrian Research Promotion Agency (FFG).

## References

1. Habib, G., Detroux, T., Viguié, R., Kerschen, G., „Nonlinear generalization of Den Hartog?s equal-peak method", Mech. Syst. Signal Process., Bd. 52–53, S. 17–28, Feb. 2015, doi:10.1016/j.ymssp.2014.08.009.
2. Hochrainer, M. J., "Real-Time Hybrid Testing: Challenges and Experiences from a Teaching Point of View". In: Mains M., Dilworth B. (eds) Topics in Modal Analysis & Testing, vol. 9, Conference Proceedings of the Society for Experimental Mechanics Series. Springer, 2019.
3. Hochrainer, M. J., Puhwein, A. M., "Investigation of Nonlinear Dynamic Phenomena Applying Real-Time Hybrid Simulation." In: G. Kerschen (ed.), Nonlinear Structures and Systems, Volume 1, Conference Proceedings of the Society for Experimental Mechanics Series, 2019, https://doi.org/10.1007/978-3-030-12391-8_16.
4. Saouma, V., Sivaselvan, M., Hybrid Simulation: Theory, Implementation and Applications. Taylor & Francis Ltd., 2008.
5. Bursi, O.S., Wagg, D., "Modern testing techniques for structural systems." In: Dynamics and Control, CISM International Centre for Mechanical Sciences, vol. 502. Springer, 2008.
6. Zhao, J., French, C., Shield, C., Posbergh, T., "Considerations for the development of real-time dynamic testing using servo-hydraulic actuation", Earthquake Engineering & Structural Dynamics vol.32/11, pp.1773–1794, 2003.
7. Gawthrop, P.J., Wallace, M.I., Neild, S.A., Wagg, D.J., "Robust real-time substructuring techniques for under-damped systems", Structural Control & Health Monitoring, vol. 14/4, pp.591–608, 2007.
8. Bonnet, P.A. , Lim, C.N., Williams, M.S., Blakeborough, A., Neild, S.A. Stoten, D.P., Taylor, C.A.,"Real-time hybrid experiments with Newmark integration MCSmd outer-loop control and multi-tasking strategies", Earthquake Engineering & Structural Dynamics vol. 36/1, pp.119–141, 2007.
9. Chen, C., Ricles, J.M., "Analysis of actuator delay compensation methods for real-time testing", Engineering Structures vol. 31/11, pp. 2643–2655, 2009.
10. Chae, Y., Kazemibidokhti, K. , Ricles, J.M. , "Adaptive time series compensator for delay compensation of servo-hydraulic actuator systems for real-time hybrid simulation", Earthquake Engineering & Structural Dynamics, vol. 42/11, pp.1697–1715, 2013.
11. Chen, P.-C. , Chang, C.-M. , Spencer, B.F., Tsai, K.-C. , "Adaptive model-based tracking control for real-time hybrid simulation", Bull. Earthquake Engineering vol. 13/6, pp. 1633–1653, 2015.
12. Najafi, A., Spencer, B.F., "Adaptive model reference control method for real-time hybrid simulation", Mechanical Systens & Signal Processing, vol.132 pp.183–193, 2019.
13. Tsokanas N., Pastorino, R. , Stojadinović, B., "Adaptive model predictive control for actuation dynamics compensation in real-time hybrid simulation", Mechanism and Machine Theory, vol.172, https://doi.org/10.1016/j.mechmachtheory.2022.104817
14. Insam, C., Göldeli, M., Klotz, T., Rixen, D. J., „Comparison of Feedforward Control Schemes for Real-Time Hybrid Substructuring (RTHS)", in *Dynamic Substructures, Volume 4*, pp. 1–14, 2021. doi:10.1007/978-3-030-47630-4_1.
15. Witteveen, W., Koller, L., Penninger, D., "Non-simultaneous real-time hybrid simulation of a numerical and experimental mechanical system with moderate nonlinearities via iterative coupling based on Frequency Response Functions", Mechanical Systems and Signal Processing, vol. 163, 2022. https://doi.org/10.1016/j.ymssp.2021.108055
16. Insam, C., Kist, A., Rixen, D. J., "High Fidelity Real-Time Hybrid Substructure Testing Using Iterative Learning Control", ISR 2020; 52th International Symposium on Robotics, 2020. ISBN - 978-3-8007-5428-1
17. Wang, Y., Gao, F., Doyle, F. J., "Survey on iterative learning control, repetitive control, and run-to-run control", Journal of Process Control vol. 19, pp. 1589–1600, 2009.
18. Xu, J.-X., Tan, Y., "Linear and Nonlinear Iterative Learning Control", Springer, 2003.
19. Bristow, D.A., Baron, K.L., Alleyne, A.G., Iterative Learning Control, in Levine, W.S. (ed.), The control handbook, CRC press, 2011.
20. Hochrainer, M.J., Puhwein, A.M., Design and Characterization of a Multi-purpose Duffing Oscillator with Flexible Parameter Selection, In: Epp, D.S. (eds) Special Topics in Structural Dynamics & Experimental Techniques, Volume 5. Conference Proceedings of the Society for Experimental Mechanics Series. Springer, 2021, https://doi.org/10.1007/978-3-030-47709-7_7
21. Puhwein, A.M., Hochrainer, M.J., Investigation of Multiple Branches in Nonlinear Oscillators Using Real-Time Hybrid Testing. In: Allen, M., D'Ambrogio, W., Roettgen, D. (eds) Dynamic Substructures, Volume 4. SEM 2023. Conference Proceedings of the Society for Experimental Mechanics Series. Springer, Cham. https://doi.org/10.1007/978-3-031-36694-9_4
22. Puhwein, A.M., Hochrainer, M.J., Investigation of Isolated Branches in Nonlinear Oscillators Using Real-Time Hybrid Testing, W. D'Ambrogio et al. (eds.), Dynamic Substructures, Vol. 4, SEM 2025 Conference Proceedings of the Society for Experimental Mechanics, Series, https://doi.org/10.1007/978-3-031-68897-3_6
23. Nelles, O., "Nonlinear System Identification", Springer, 2010.
24. Eykhoff, P., "System Identification", Wiley, 1974.

# Chapter 12
# Extrapolating Dynamic Transfer Functions from Multi-Input Multi-Output Vibration Testing and Simulation

**E. J. Perez, E. E. Regula, C. J. Wynn, J. D. Blessinger, J. L. Davis, E. P. Dawson, and S. J. Zimmerman**

**Abstract** Finding transfer functions between an applied force and a measured output is well defined for dynamic systems. These transfer functions, which assume a force input, can have many names depending on the output they measure, such as receptance for displacement, mobility for velocity, or accelerance for acceleration. Similarly, the transfer path is a type of transfer function, which is simply the ratio of two outputs from a system in the frequency domain. For single-input linear systems, finding the transfer path between two outputs is a well-understood calculation. These techniques become more complex when dealing with multi-input multi-output tests. While these multi-input vibration tests are more complicated, they provide a more complete understanding of the system being analyzed. These complex systems have well-defined methods for obtaining transfer functions if a forcing function is known. However, when the input force is not known, the methods to find the transfer path between two locations is less understood. This work explores methods for finding unknown transfer paths using accelerations. The tests and validations are done using the Box Assembly, a common structure used in dynamics testing. Experimentally measured and finite-element approximated transfer functions are used to create a transmissibility matrix. This matrix allows for estimation of unknown accelerations based on measured accelerations at different locations on the box assembly.

**Keywords** Transfer Paths · Transmissibility · FRF · MIMO · FEA

## Introduction

Frequency response functions (FRFs) are well understood in dynamic systems and are useful for system identification and structural health monitoring. Finding the FRF for a single-input single-output (SISO) case is straightforward, as only scalar values are used. However, these calculations become more complex with multi-input multi-output (MIMO) testing because the scalar values are replaced with matrices. Not only are the FRFs of MIMO harder to obtain, but so are the transfer paths. Similar to FRFs, which are ratios of output to input, transfer paths are ratios of two system outputs. This relationship

E. J. Perez
Department of Engineering, University of California Riverside, Riverside, CA 92524
e-mail: epere194@ucr.edu

E. E. Regula
Department of Engineering, Messiah University, Mechanicsburg, PA 17055
e-mail: er1324@messiah.edu

C. J. Wynn
Department of Engineering, University of Utah, Salt Lake City, UT 84112
e-mail: cjwynn7@outlook.com

J. D. Blessinger
Los Alamos National Laboratory, Los Alamos, NM 87545
e-mail: jbless@lanl.gov

S. J. Zimmerman
Department of Engineering, University of Southern Indiana, Evansville, IN 47712
e-mail: edawson@lanl.gov



Walter D'Ambrogio, et al. (eds.), *Dynamic Substructuring & Transfer Path Analysis, Vol. 4*, Conference Proceedings of the Society for Experimental Mechanics Series, https://doi.org/10.13052/9788743801498-12

defines the movement of one location relative to another for any set forcing scenario. These transfer paths can be placed into a transmissibility matrix that defines the contribution of any output to any other distinct output. These transmissibility matrices are different for all independent forcing scenarios. Using these methods, it may be possible to also find theoretical FRFs from locations that have no input during testing but do have measured accelerations. This has the potential to save test engineers time and money on large tests, allowing them to measure acceleration at multiple points while not having to force at those points, and predicting FRFs from those locations.

The box assembly (BA), which can be seen in Figure 1, was utilized for this research. This subassembly of the box assembly with removable component (BARC) structure is commonly used as a test object in multi-axis dynamic testing. Abaqus CAE was used to model the BA for finite-element analysis (FEA). This software allows for the extraction of mass and stiffness matrices which can then be imported into MATLAB to be used to calculate the FRFs and transfer paths used in modeling.

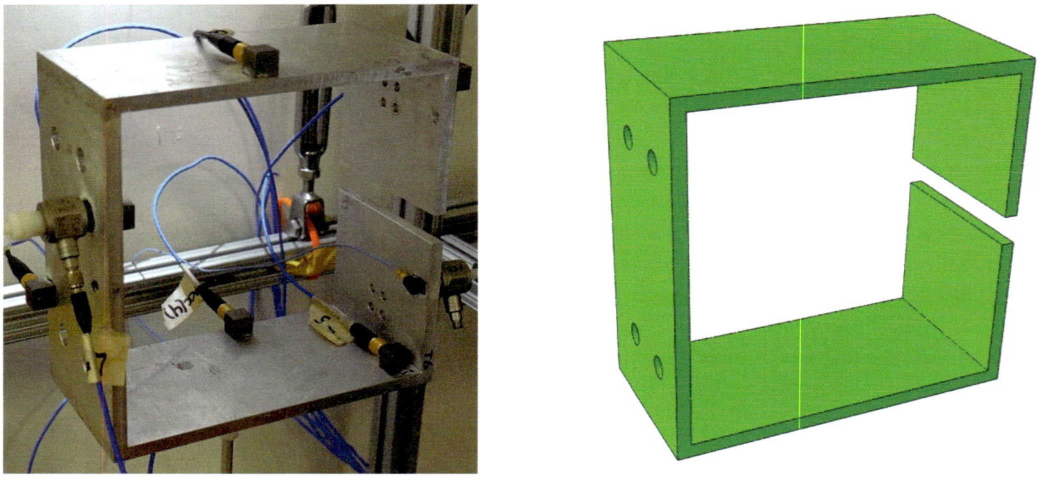

(a) Experimental test setup (b) Abaqus model

**Fig. 1** Box assembly configuration used in both experimental and numerical approach

The concept of a transfer path has been previously explored as a method of defining the relationships between different points in a MIMO system. The basic expression, as seen in Equation 1, follows previous calculations [1, 2, 3]. Where $T$ is the transfer path while $H_u$ and $H_k$ are the unknown (u) and known (k) frequency response functions of interest.

$$T = \frac{H_u}{H_k} \tag{1}$$

The transfer path in Equation 1 is a scalar one to one relationship, as it is dependent on two single FRFs. Another method of expressing the transfer path between two outputs of a system is to use the pure responses. Using Equations 2 and 3, the response of the system can be obtained, where $f_k$ is the known forcing magnitude of the system and $x_u$ and $x_k$ are responses of the system in the frequency domain. The transfer path from $x_u$ to $x_k$ can be obtained as shown in Equations 2, 3, and 4.

$$x_u = H_u \cdot f_k \tag{2}$$

$$x_k = H_k \cdot f_k \tag{3}$$

$$T = \frac{x_u}{x_k} \tag{4}$$

The transmissibility matrix is then derived from evaluating the transfer path of multiple outputs to multiple outputs at once. If $[H_k]$ is a $n \times n$ matrix and $[H_u]$ is a $m \times n$ matrix, then Equation 5 shows the transmissibility matrix obtained from $[H_u]$ and $[H_k]$.

$$[T] = [H_u][H_k]^{-1} \tag{5}$$

It is important to note that when computing the transmissibility matrix shown in Equation 5, it does not hold the same information as the one to one transfer path obtained with Equation 1. That is to say, the individual elements in the transmissibility matrix are not one to one transfer paths. Rather, an element in the transmissibility matrix is the contribution weighting of a degree of freedom (DOF) from the the set of knowns to a DOF from the set of unknowns. Using similar methods described above, the transmissibility matrix can be computed using a set of unknown and known responses.

$$[x_u] = [\{x_u\}_1\{x_u\}_2 \ldots \{x_u\}_n] = [H_u][\{f_k\}_1\{f_k\}_2 \ldots \{f_k\}_n] = [H_u][F] \tag{6}$$

$$[x_k] = [\{x_k\}_1\{x_k\}_2 \ldots \{x_k\}_n] = [H_k][\{f_k\}_1\{f_k\}_2 \ldots \{f_k\}_n] = [H_k][F] \tag{7}$$

$$[T] = [\{x_u\}_1\{x_u\}_2 \ldots \{x_u\}_n][\{x_k\}_1\{x_k\}_2 \ldots \{x_k\}_n]^{-1} = [x_u][x_k]^{-1} \tag{8}$$

Here, $[x_u]$ and $[x_k]$ are obtained from the known and unknown FRF matrix, $[H]$, as well as the force matrix, $[F]$. Here the force matrix is comprised of multiple forcing vectors as seen in Equations 6 and 7, where each column vector indicates a forcing case. These forcing cases are necessary to expand the size of the $[x_k]$ matrix. Equation 8 is satisfied as long as the number of columns in $[x_u]$ is equal to the number of columns in $[x_k]$, all while $[x_k]$ is maintained a square matrix. Which means the number of forcing cases must be equal to the number of known outputs. This indicates that the size of the forcing matrix is dependent on the number of known outputs and number of forcing locations. To be explicit, the forcing matrix in Equation 7 and 6 is a $m \times n$ matrix where, $m$ = number of forcing locations onto the system and $n$ = number of known outputs. It is also important that the forcing matrix is full ranked to successfully obtain the full contributions within the transmissibility matrix.

However, these processes have some limitations, as the necessary step of inverting the known response of the system requires a square matrix. Maintaining a square matrix while increasing the number of known responses requires an equal increase in the load vector's length. Thus, when increasing the size of the known square matrix, it is also necessary to increase the number of columns of the unknown response matrix to match the size of the square matrix. If the known response matrix to be non-square, a pseudo-inverse operator must be employed.

For example, computing the transmissibility matrix with $n$ known responses with only two load vectors can be illustrated as $[T] = [x_u]_{2\times 2} \cdot [x_k]^{\dagger}_{n\times 2}$. Where $[x_k]^{\dagger}$ is the pseudo-inverse of the non-square matrix $[x_k]$. Even though $[x_k]$ has $n$ rows and two columns, the pseudo-inverse (†) can be used to compute the transmissibility matrix. To verify equivalence, computation of an FRF, $[H_u]_{11}$, was made using the pseudo-inverse operator on a non-square matrix and the traditional inverse for a square matrix. In general, using Equation 5 and the fact that no information is lost calculating $[H_u]_{11}$ from either a square or non-square matrix, then Equation 9 holds.

$$H_u = \sum_{i=1}^{2} T_{1i} \cdot H_{i1_k} = \sum_{j=1}^{n} T_{1j} \cdot H_{j1_k} \tag{9}$$

Figure 2 shows the overlay of an FRF ($[H_u]_{11}$) gathered using two different methods, but for the same system. The first method is by using $[H]_k$ of size $2 \times 2$, while the second uses $[H]_k$ of size $2 \times 100$. This second method required the use of a pseudo-inverse if no other changes were made to the $[H]_u$ matrix and forcing vectors. In Figure 2, it shows that both methods are equal.

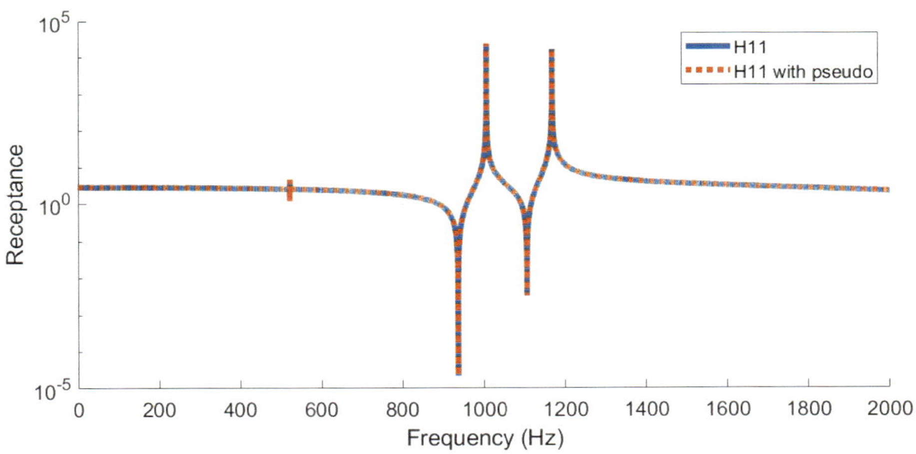


**Fig. 2** Receptance calculated with square matrices vs with pseudo inverse.

These methods allow the transmissibility matrix to be calculated with solely experimental data, with solely FEA, or with a combination of the two. For example, if there is a limited number of sensors for experimentation, the acceleration data of one location can be experimentally gathered and FEA can be used to estimate the acceleration of another location. These two sets of data can be used to find a transfer path between the two locations.

The transmissibility matrix may also be acquired in other ways, including construction using the results from Spectral Power Density (SPD) and Singular Value Decomposition (SVD) [4]. Several different methodologies by which the accuracy of the transmissibility matrix may be evaluated using power spectrum density or H-estimators have also been investigated [5, 6].

Many methods of gathering MIMO FRFs and transmissibility matrices require known forcing inputs. This research seeks to establish reliable and accurate methodologies which would allow for FRFs to be estimated between measured accelerations without the need for knowledge about forcing. An added benefit of this approach would be the development of a method by which the forcing can actually be discovered and characterized by measurement of the accelerations at specific points. The benefits of expanding this technique into the realm of MIMO FRFs were also evaluated by their accuracy and efficacy against techniques with similar goals using SISO FRFs.

These techniques were evaluated numerically through simulation as well as experimentally through physical measurement and verification using a BA. The numerical techniques were also used to attempt to predict the unknown FRF matrix seen in the earlier calculations in order to help derive a transmissibility matrix.

## Analysis

The BA was analyzed in two different ways, numerically and experimentally. Numerical evaluation of the system was conducted first in order to gather an informed prediction of sensor and shaker placement, as well as an overall idea of the dynamics of the structure. The first step in the numerical evaluation of the BA was the creation of an FEA model and the validation of its accuracy. To achieve this goal, a model based on measurements of the test structure was created, this model was created in Abaqus using C3D8R elements. Once the model was made, a frequency analysis was performed to gather all the natural frequencies under 2000 Hz. This process was repeated several times with slightly different variations of the model. For example, in one model, the bolt holes in the bottom of the structure were modeled, in the other they were left out. Also, different global meshing sizes were used until there was a convergence in the predicted values, as shown in Figure 3. This Figure also shows the time to compute the natural frequencies for the structure, demonstrating the dramatic increase in computation time relative to mesh seed size. The BA is made from Aluminum 6061-T6 and was modeled as such in Abaqus. The model used a density of 2700 kg/m$^3$, a Young's modulus of 68.9 GPa, and a Poisson's ratio of 0.33.

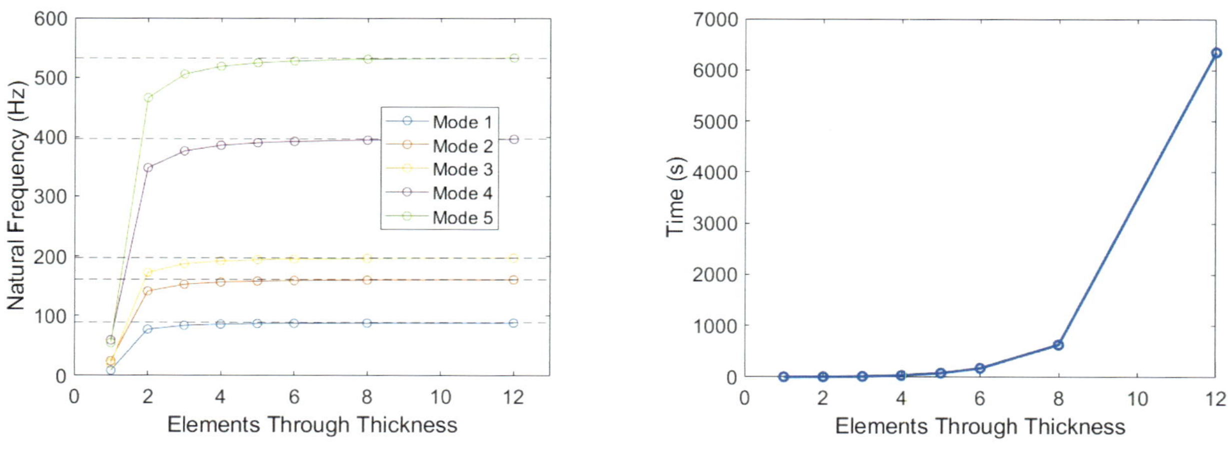


(a) Convergence of natural frequencies for varying mesh seed sizes (b) Time taken for computation to complete based on mesh seed sizes

**Fig. 3** Mesh convergence study performed on the numerical BA.

The values of some of the larger mesh sizes were then compared with the smallest size to find a more computationally efficient model that retained most of its accuracy. To do this, percent differences were calculated between the finest mesh size, and the next three smallest mesh sizes. Five elements through the thickness was chosen to model the BA as there was only a

~1.5% error to the highest fidelity model. Additionally, FEA tests indicated that the bolt holes in the bottom of the structure were significant in the model results, but the bolt holes in the top of the BA were insignificant. Finally, point masses equal to 0.42 g were placed at each of the nodes corresponding to the triaxial accelerometer locations. This significantly helped the model match experimental results as well.

Both the mass and stiffness matrices were then exported from Abaqus into MATLAB to have a greater control of the underlying information created by the model. This process required a reduction to make the matrices more easily handled in later calculations. The method used for this process was Guyan Reduction, which reduced the number of nodes by a factor of ~225 while retaining much of the information about the dynamics. This reduction was verified by solving for the natural frequencies of the BA in MATLAB using the exported mass and stiffness matrices. These reduced results were then compared to the high-fidelity model, and were found to match closely with each other, having a mean average percent error of 0.34%.

A key requirement for testing and model validation was the proper selection of sensor and stinger locations such that the greatest number of modes would be excited. The sensor location was chosen using engineering judgement to obtain signal of a large enough magnitude that any noise might be overcome as well as avoiding symmetry which would generally result in redundant information. The location of the stingers and accelerometers can be seen in Figure 4.

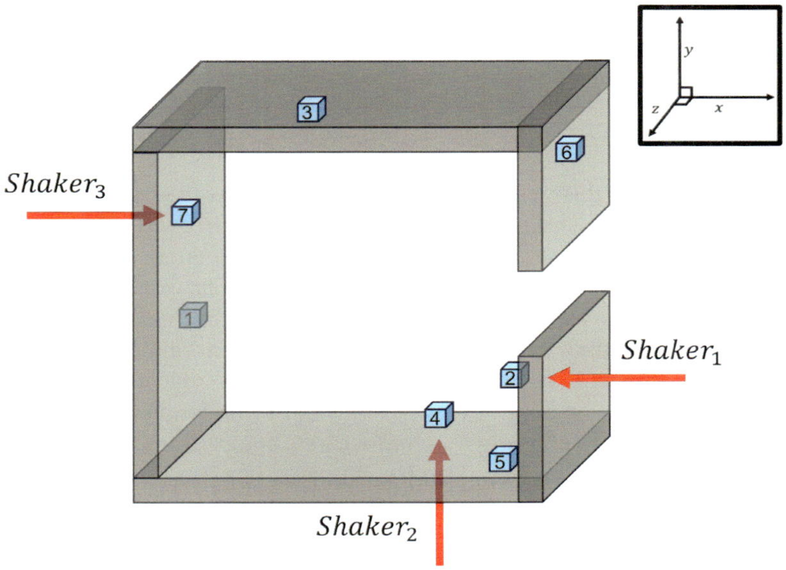


**Fig. 4** Sensor and shaker placement for BA.

For the experimental setup, due to limited sensor availability as well as a limits on the number of channels from the Data Acquisition System (DAQ), a collection of six triaxial accelerometers, two load cells, and one uniaxial accelerometer were used to obtain all measurements. In order to excite the BA, three shakers were mounted to the structure seen in Figure 4 with stingers to minimize torsional loads. Since the DAQ had only two input channels, one shaker had to remain off for any specific testing scenarios. Even though only two shakers were used, keeping the third shaker attached was crucial to maintain consistent boundary conditions while switching between different testing scenarios seen in Figure 5.

Two of the triaxial accelerometers had a noisy z-direction, so those signals were neglected. The total number of individual signals that can be processed was nineteen. From the accelerometers, seventeen acceleration signals were obtained: seven signals in the x-direction, six signals in the y-direction, and four signals in the z-direction. The two active shakers each had load cells that were attached to the BA, which yielded two forcing signals.

For each test the two active shakers were given a burst random input signal to excite the BA. The burst random signal is preferable to randomly excite the BA with a span of 1-2000 Hz. Using a chirp input signal would have yielded similar results, but the burst random input signal was more time efficient, and no windowing was needed as long as our system's response decayed before the next burst.

There were six tests conducted in order to obtain the experimental data for both MIMO and SIMO systems; these setups can be seen in Figure 5. In the first MIMO test setup (MIMO 1), shaker one and shaker two excited the BA while shaker three remained off. The output of this stage yielded thirty-four FRFs, since there are two forcing locations and seventeen accelerometer channels. The second MIMO test setup (MIMO 2) had shaker two and shaker three exciting the BA while

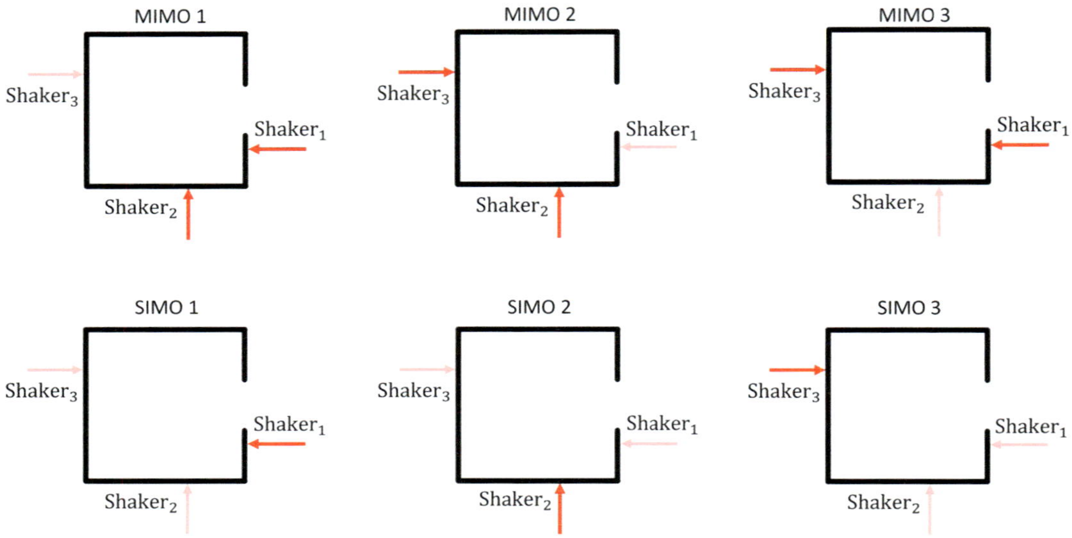


**Fig. 5** The three MIMO and three SIMO experiments conducted.

shaker one remained off. The output of this stage yielded thirty-four FRFs with respect to forcing at shakers two and three. For the final MIMO test setup (MIMO 3), shakers one and three excited the BA while shaker two was off. The output of this stage yielded thirty-four FRFs with respect to forcing from shakers one and three. There were also three SIMO test setups following a similar pattern. In the first SIMO test setup (SIMO 1), shaker one excited the BA while shakers two and three were off. In the second test setup (SIMO 2), shaker two excited the BA while shakers one and three remained off. In the last SIMO test setup (SIMO 3), shaker three excited the BA while shakers one and two remained off. For each of the SIMO tests, 17 FRFs were created, with respect to forcing by the indicated shaker. Conclusions about differences in transfer paths between MIMO and SIMO systems can be drawn experimentally only with frequency ranges where the FRFs match. This indicates a similar system between setups, allowing for more accurate comparisons.

From these experiments the comparison between SIMO and MIMO FRFs have been made in Figure 6. These comparisons are necessary as they will help guide the conclusions of comparing the differences in transfer paths between MIMO and SIMO. In theory the FRFs from the MIMO and SIMO systems will be identical as long as the system and boundary conditions remain the same throughout each experiment. This means that all modes of the structure will excited and captured between different configurations. However, as displayed in Figure 6, the MIMO and SIMO FRFs do not match up exactly in some frequency ranges. It is important that no conclusions about the transfer paths are made based on the frequency ranges where the FRFs do not match. Since the SIMO experiments have only a single shaker receiving the burst random signal to excite the BA, some of the mode shapes can be misrepresented as the box assembly is unable to be fully excited from a single shaker. Conversely, the MIMO experiments have two active shakers exciting the box assembly. This allows for the system to be excited more equally across more modes, allowing better full system identification.

Clear deviations in Figure 6 for Shaker 1 and Shaker 3 Drive Points can be seen for MIMO3. For the MIMO3 setup, shakers 1 and 3 are driving, as seen in Figure 5. Where there is no excitation in the y-direction, and the excitation in the x-direction oppose each other, the response of the system changes slightly. This can also be seen for Shaker 2 Drive Point where the FRFs match each other much more closely up to about 600 Hz and no MIMO3 test occurs.

When evaluating the differences in transfer paths between SIMO and MIMO system experimentally, conclusions can only be made for regions where MIMO and SIMO FRFs are approximately equal. On the other hand, analytical FRFs are always equal regardless of the number of inputs.

As a result, this makes analyzing the difference in transfer paths easy as any conclusion made are valid for any frequency range. Figure 7 shows an 8-degree-of-freedom spring-mass linear system used in previous studies [1]. Evaluating the transfer paths for a SIMO system where the forcing is on mass one, the transfer paths can be obtained using Equation 10. Where $\{x_k\} = \{x_1\}$ and $\{x_u\} = \{x_5\}$. For the rest of the examples discussing different MIMO and SIMO systems, all of the transfer paths will follow Equation 10, while the only changes being made between the systems are forcing locations.

$$\{T\} = \{x_u\}\{x_k\}^{-1} = \{x_5\}\{x_1\}^{-1} = \{T_{51}\} \tag{10}$$

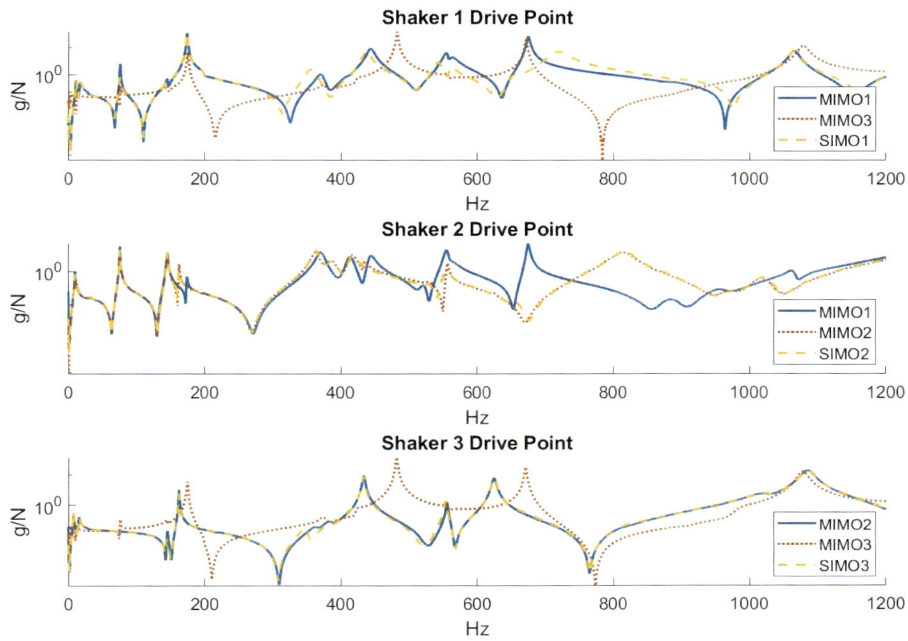


**Fig. 6** Comparison of MIMO FRFs to SIMO FRFs with the same accelerometer to the same shaker.

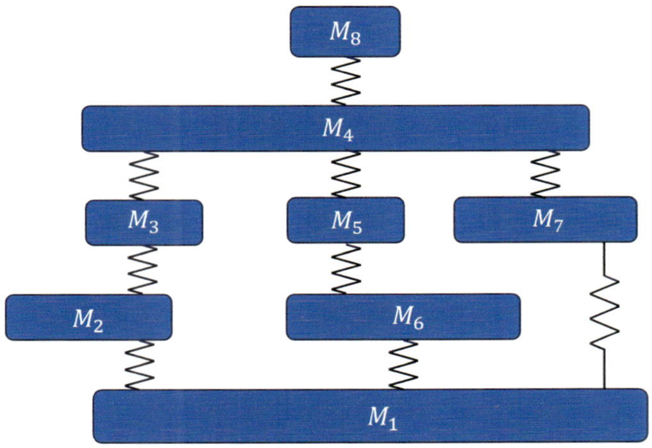


**Fig. 7** Spring mass system used to evaluated transfer paths between MIMO and SIMO cases.

From the transfer path obtained from the SIMO case of forcing only on mass one, it is important to note that if the forcing on mass one is scaled up or down by some constant, the transfer path $\{T_{51}\}$ will not change. That is to say, the contribution of mass one to mass five will remain constant regardless of forcing magnitude. Which also means for a SIMO system where $\{x_k\}$ and $\{x_u\}$ are held constant, the transfer path will always be the same when forcing at a given location with any magnitude. For instance if $\{x_k\}$ and $\{x_u\}$ remain the same, then there could only be eight SIMO transfer paths obtained from individual forcing at all eight masses. For the analysis where $\{x_k\}$ and $\{x_u\}$ are kept constant, the eight transfer paths that can be obtained from the system in Figure 7 will be associated with each individual SIMO case.

When evaluating transfer paths for MIMO systems, it is important to note that the same rules for SIMO systems do not apply. Using the same $\{x_k\}$ and $\{x_u\}$, the transmissibility matrix can be obtained in a similar fashion to SIMO system. The difference is that there are now multiple forcing locations with different magnitudes. Here is where the MIMO transfer paths start to differ from the SIMO systems. If the transfer path $\{T_{51}\}$ is computed from forcing at multiple locations, then the transfer path will change depending on the magnitudes of the individual forcing. Letting the forcing magnitude be equal for each of the forcing locations, a balanced MIMO transfer path is obtained. As long as the forcing magnitudes are the same for any MIMO case and thus the magnitude ratio, $\{x_u\}/\{x_k\}$, is kept constant, then the balanced MIMO transfer paths will always be obtained. This can be a baseline for MIMO transfer paths as the balanced transfer paths reveal the responses from equal weighting from forcing at different locations. On the other hand, if forcing magnitudes start to differ from one location to another, then the transfer path obtained will start to diverge from the balanced MIMO transfer path. The resultant transfer path will then start to converge towards the SIMO transfer path corresponding to the force with the greatest magnitude in a MIMO system.

Let $\{x_k\}$ and $\{x_u\}$ remain the same from Equation 10 and have forcing on masses one, two, and three from Figure 7. The balanced transfer path $\{T_{51}\}$ can be obtained assuming $F_1 = F_2 = F_3$. Now three more transfer paths $\{T_{51}\}$ can be computed from the SIMO systems that make up the MIMO system. That is, $\{T_{51}\}$ forcing only on mass one, $\{T_{51}\}$ forcing only on mass two and $\{T_{51}\}$ forcing only on mass three. In Figure 8 the balanced transfer path $\{T_{51}\}$ obtained from the MIMO system is displayed, as well as the three SIMO transfer paths $\{T_{51}\}$ obtained from the three SIMO cases stated above. As stated previously, the balanced MIMO transfer path can be obtained only when multi-forcing magnitudes are equal. As forcing magnitudes deviate from each other, the MIMO transfer path will diverge from the balanced transfer path. In the limit where the relative forcing magnitude at one location exceeds the magnitude of forcing at other locations, the MIMO transfer path will start to converge towards the SIMO transfer path of the large forcing magnitude. This can also be shown in Figure 8, which shows the balanced transfer path $\{T_{51}\}$ obtained from the MIMO system stated above and the three SIMO convergences possibilities. As the forcing magnitude at one location exceeds the forcing magnitudes of the other locations, the figure shows the convergence of the MIMO balanced transfer path to a SIMO transfer path. This can also be reasoned out mathematically by recognizing that if one force was made large enough in Equation 11, the others would have increasingly little effect.

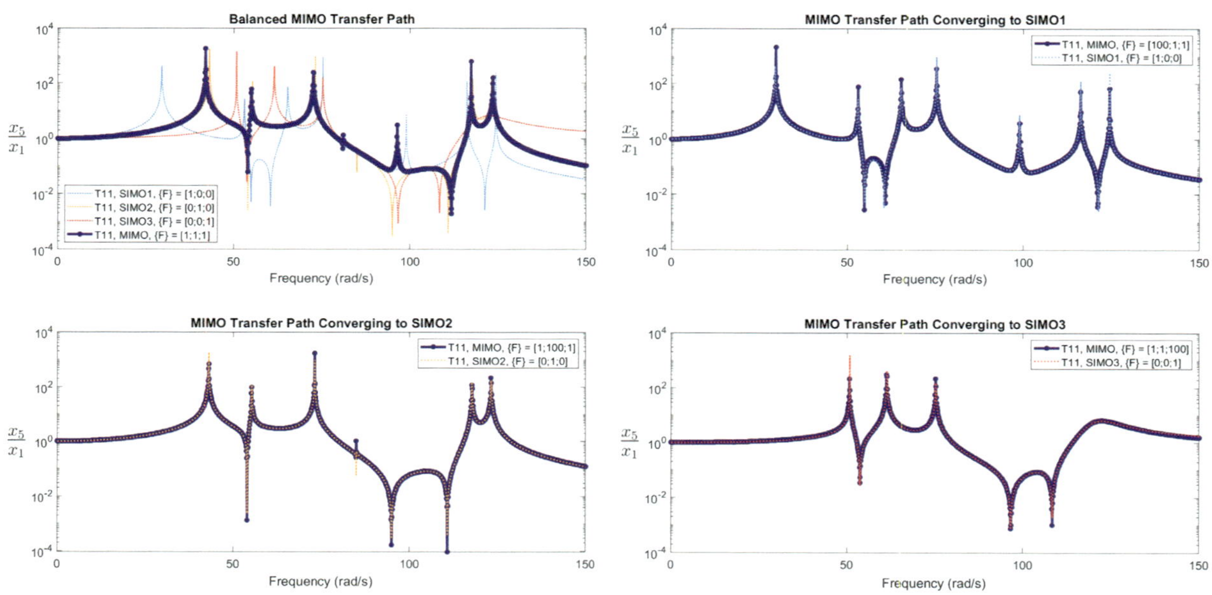


**Fig. 8** MIMO transfer path converging to SIMO counterparts.

In other words, MIMO transfer paths are not complete linear combinations of the SIMO transfer paths as the relative multi forcing magnitude contributes significantly in computing the MIMO transfer path.

$$T_{12} = \frac{H_{11} \cdot F_1 + H_{12} \cdot F_2 + \cdots + H_{1n} \cdot F_n}{H_{21} \cdot F_1 + H_{22} \cdot F_2 + \cdots + H_{2n} \cdot F_n} \neq A \cdot \frac{H_{11} \cdot F_1}{H_{21} \cdot F_1} + B \cdot \frac{H_{12} \cdot F_2}{H_{22} \cdot F_2} + \cdots + X \cdot \frac{H_{1x} \cdot F_x}{H_{2x} \cdot F_x} \tag{11}$$

The correct calculation of transfer paths were key to the secondary goal, to predict the FRF between two unforced locations on the physical BA. Two methods for this goal were evaluated. The first approach that was attempted was the multiplication of a transfer path and an FRF to reach the desired transfer function. This operation can be seen in Equation 12, which relates frequency domain information and where subscript notation $t$ is target, $u$ is unknown and $k$ is known.

$$\frac{X_t}{F_u} = \frac{X_t}{X_k} \cdot \frac{X_u}{F_k} \tag{12}$$

The FRF in this equation relating the force at the unknown point to the acceleration of the known point can be gathered without actually forcing at the unknown point by utilizing the concept of reciprocity of a linear structure and finding the FRF by forcing at the known point. Methods for gathering the transfer path have already been explored at length in this paper. Attempting this prediction method on the analytical spring mass system shown earlier results in Figure 9.

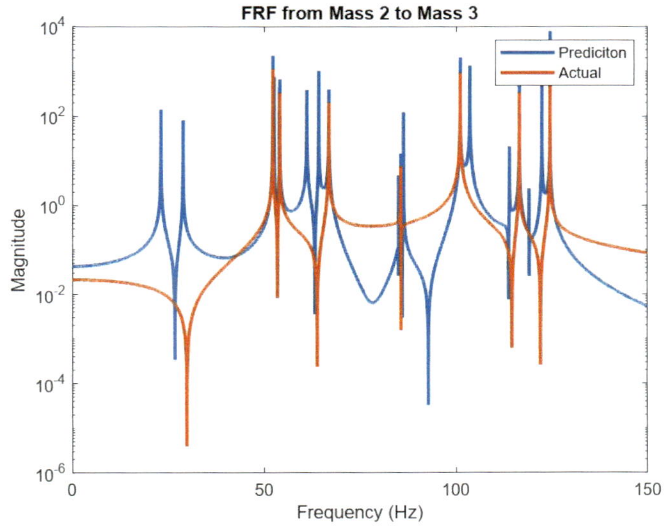


**Fig. 9** FRF Prediction using Transfer Paths and FRFs

In an effort to improve the results of the prediction as well as to make it more applicable to the Box Assembly, numerical simulation data was used to predict the transfer paths which were then combined with experimental FRFs. The results of those calculations can be seen in Figure 10.

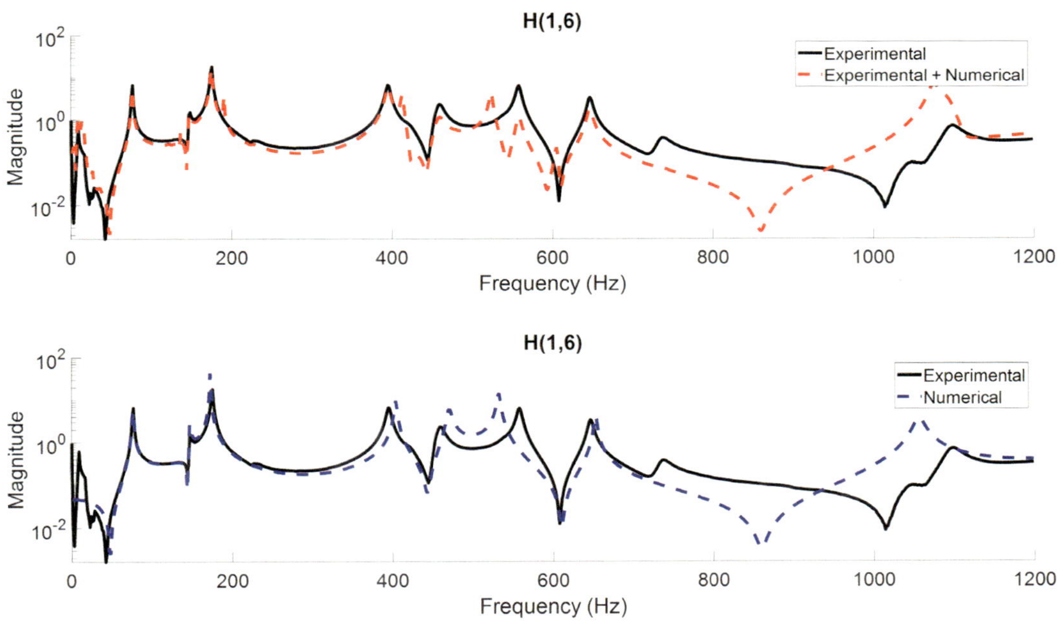


**Fig. 10** Comparison of predictions of a mix of numerical and experimental data.

The second method used to predict the unknown FRF involved taking the matrix of FRFs gathered during a SIMO or MIMO test and retrieving modal parameters using the least squares rational function algorithm. The parameters that were retrieved were the mode shape matrix, the damping ratios for each mode, and the natural frequencies. Using these values, any FRF between two measured accelerometer locations could be predicted using Equations 13 and 14 as a template for the construction of FRFs where $[\Phi]$ is the mode shape matrix for an n-dof system with m-modes, and $[M]$ represents the inverse pole matrix. The results of our analysis built from Equation 13 are shown in figure 11 as our final conclusion for the predictability of these transfer functions.

$$[H] = [\Phi][M][\Phi]^T = \begin{bmatrix} \Phi_{11} & \Phi_{12} & \cdots & \Phi_{1m} \\ \Phi_{21} & \Phi_{22} & \cdots & \Phi_{2m} \\ \vdots & \vdots & \ddots & \vdots \\ \Phi_{n1} & \Phi_{n2} & \cdots & \Phi_{nm} \end{bmatrix} \begin{bmatrix} M_1 & 0 & \cdots & 0 \\ 0 & M_2 & \cdots & 0 \\ \vdots & \vdots & \ddots & \vdots \\ 0 & 0 & \cdots & M_x \end{bmatrix} \begin{bmatrix} \Phi_{11} & \Phi_{21} & \cdots & \Phi_{n1} \\ \Phi_{12} & \Phi_{22} & \cdots & \Phi_{n2} \\ \vdots & \vdots & \ddots & \vdots \\ \Phi_{1m} & \Phi_{2m} & \cdots & \Phi_{nm} \end{bmatrix} \tag{13}$$

$$M_x = \frac{-\omega^2}{\omega_{nx}^2 - \omega^2 + 2i\zeta_x\omega_{nx}\omega} \tag{14}$$

## Conclusion

One of the key purposes of this paper was to investigate the differences between the transfer paths and transmissibility matrices of MIMO and SIMO systems. Towards this goal there were two main findings. The first was that as one of the forces applied in a MIMO system became an order of magnitude larger than any of the other forces, the transfer path between any two points converged to the path that would be seen in a SIMO case where just the larger force was present. The second key finding was that the MIMO transfer paths could not be constructed as a linear combination of SIMO cases.

In the effort to predict FRFs at locations that were never forced, the first method used measured transfer paths and FRFs in a calculation like Equation 12. This method did predict some peaks but was often vastly inaccurate. The error in this method was found to be in the assumption that the transfer path would be the same between the measured and predicted cases. This method necessitates a transfer path based on the frequency information from the unknown forcing case, however obtaining this data would be contrary to the original goal of not forcing at the location. Therefore, this method was unsuitable for experimental only prediction without the integration of some numerical simulation data.

The failure of the first method due to the inability to gather the needed transfer paths inspired an attempt to integrate numerical simulation predictions for transfer path into the calculation. The results of this method were seen in Figure 10. This method produced predictions that were improved from the prior method. However it was still erroneous, likely due to FE model inaccuracies used for numerical transfer path prediction or boundary condition differences between the model and the experiment. Another shortcoming of this method is that its requirement for highly accurate modeling makes solving for FRFs using simulation more reasonable than the method proposed.

The third attempt at FRF prediction followed roughly the method outlined in Equation 13. This method produced many of the best results as can be seen in Figure 11.

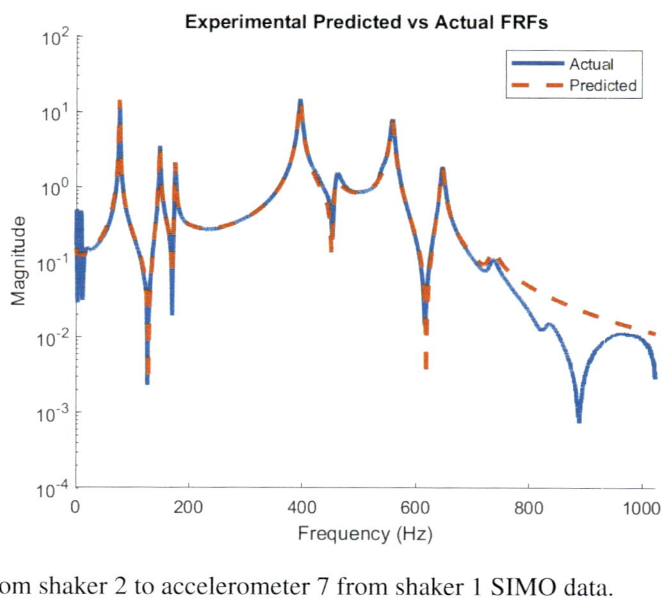


**Fig. 11** Prediction of FRF from shaker 2 to accelerometer 7 from shaker 1 SIMO data.

This method provides good predictions for frequencies below 700 Hz, however there are some qualities that can improve its accuracy. The method of fitting the input FRFs can affect how well the overall system parameters can be gathered. It may also have some correlation to the quality of fitting the two accelerometer channels directly related to the desired FRF.

An analysis of the capabilities and limitations of this method for predicting FRFs as well as an analysis on what might improve these results is a topic deserving of further study. The benefits of a further refined and automated system for predictions would be well worth study as they would reduce time and cost for MIMO testing. This paper presents a novel approach for estimating the frequency response function at locations where only acceleration is measured. This method demonstrates reliability and accuracy up to approximately 700 Hz with our system. The methodology of estimating FRFs under these constraints offers a valuable tool to be used to reduce labor intensive experimentation.

**Acknowledgments** This research was funded by Los Alamos National Laboratory (LANL) through the Engineering Institute's Los Alamos Dynamics Summer School. The Engineering Institute is a research and education collaboration between LANL and the University of California San Diego's Jacobs School of Engineering. This collaboration seeks to promote multidisciplinary engineering research that develops and integrates advanced predictive modeling, novel sensing systems, and new developments in information technology to address LANL mission relevant problems.

## References

1. N.M.M. Maia. J.M.M. Silva, A.R. "The transmissibility concept in multi-degree-of-freedom systems". *Mechanical Systems and Signal Processing*, 15:129–137 (2001)
2. António P.V. Urgueira, N.M.M., Raquel A.B. Almeida. "On the use of the transmissibility concept for the evaluation of frequency response functions". *Mechanical Systems and Signal Processing*, 25(3):940–951 (2011)
3. M. Fontul, J.S.N.M., A.M.R. Ribeiro. "Transmissibility matrix in harmonic and random processes". *Shock and Vibration*, 11:563–571 (2004)
4. Iván Gómez Araújo, J.E.L. "Operational modal analysis using svd of power spectral density transmissibility matrices". *Mechanical Systems and Signal Processing*, 46(1):129–145 (2014)
5. Leclère, Q., Roozen, N.B., and Sandier, C. "On the use of the hs estimator for the experimental assessment of transmissibility matrices". *Mechanical systems and signal processing*, 43(1-2):237–245 (2014)
6. X.Z. Li, X.D.Z.P., X.B. Yue. "Response transmissibility versus power spectrum density transmissibility: Dynamic property analysis and comparison". *Journal of Sound and Vibration*, 454:32–50 (2019)

# Chapter 13
# Revisiting the Dual Admittance-Based Quasi-Static Formulation for the Identification of Linear Joint Dynamics with Dynamic Substructuring

**Francesco Trainotti, M. Brons, S. Klaassen, D. J. Rixen**

**Abstract** Accurately capturing joint characteristics in mechanical systems is essential for structural analysis, design optimization and condition monitoring. Substructuring decoupling offers a straightforward procedure to isolate the linear(-ized) joint dynamics in terms of frequency-based dynamic stiffness or admittance functions. Assuming a quasi-static representation of the joint, identified therein as no mass and no cross-coupling effects, a variety of formulations with (seemingly) different roots can be found in literature that differ in several aspects, such as the type, location and amount of impedances/transfer functions employed, as well as the (sub)-components required. This article presents a concise review of popular quasi-static identification formulations. Additionally, it provides an analysis of the discrepancies and equivalence between these formulations, by revisiting the dual admittance-based framework.

**Keywords** Joint identification · Experimental substructuring · Inverse substructuring · Quasi-static joint

## Introduction

Mechanical joints are fundamental components in assembled mechanical systems, significantly impacting their dynamic performance, structural integrity, and noise and vibration characteristics. Accurate identification of joint properties is essential for effective structural analysis, optimized design, and reliable condition monitoring. With advancements in measurement technology and the adoption of Experimental Substructuring techniques [1, 2], there has been a growing interest in efficient procedures to isolate and identify joint dynamics.

A key consideration in joint modeling is the choice between dynamic and quasi-static representations. Dynamic joint modeling treats the joint as a full dynamic component. In contrast, quasi-static joint modeling simplifies the joint to a compliant linkage between interfaces, neglecting inertial effects. This assumption has gained popularity due to its simplicity and practicality, making it suitable for a wide range of engineering applications. One important advantage of quasi-static modeling is its applicability to measurements of the joint in its assembled configuration, often referred to as 'in-situ' measurements.

Another important aspect is the formulation approach, which can be categorized into primal and dual methods. Primal formulations impose compatibility constraints by assuming that the interface degrees of freedom are identical and perfectly aligned. On the other hand, dual formulations allow discrepancies between the interface degrees of freedom and aim to minimize these gaps, providing greater flexibility, especially in systems where perfect alignment is difficult to achieve.

Focusing on quasi-static identification approaches, several studies have contributed to the development and refinement of these methods. Primal quasi-static joint identification techniques have been explored in works such as [3, 4, 5, 6, 7, 8]. Dual quasi-static methods have been extensively investigated in works like [9, 10, 11, 12, 13, 14].

Despite the variety of formulations available, discrepancies often arise due to differences in the types, locations, and quantities of impedances or transfer functions used, associated with interface or internal measurements of the required (sub)-components. Such variations make it challenging to select the most appropriate method for a given application.

Francesco Trainotti · M. Brons · D. J. Rixen
Technical University of Munich, Chair of Applied Mechanics, Department of Mechanical Engineering, TUM School of Engineering and Design, Boltzmannstr. 15, 85748 Garching, Germany
e-mail: francesco.trainotti@tum.de; maribr@dtu.dk; rixen@tum.de

S. Klaassen
Vibes Technology, Molengraaffsingel 14, 2629 JD Delft, The Netherlands
e-mail: sklaassen@vibestechnology.com

 
Walter D'Ambrogio, et al. (eds.), *Dynamic Substructuring & Transfer Path Analysis, Vol. 4*, Conference Proceedings of the Society for Experimental Mechanics Series, https://doi.org/10.13052/9788743801498-13

This paper presents a short review of popular quasi-static joint identification formulations, with a focus on both primal and dual approaches. By analyzing the discrepancies and commonalities between these methods, along with a novel revisitation of the dual admittance-based framework, the aim is to provide a unified perspective on these techniques.

The paper is structured as follows: section 2 provides the theoretical background on quasi-static identification, with a focus on two popular formulations within the primal and dual domains. Then, section 3 introduces insights into the equivalence of these methods, including a novel derivation of the joint isolation process in admittance form in the dual sense. Finally, section 4 offers concluding remarks.

## Quasi-static Substructuring Identification

The goal of the substructuring characterization is to isolate the dynamic effects of a joint component $J$ from the connecting components $A$ and $B$ defining the linear(-ized) time invariant assembly $AJB$. This is illustrated in fig. 1. A partitioning of the measurements between internal and interface is realized. For simplicity, we assume that the 'experimental' interface is defined such that there is a compatible set of degrees of freedom between the joint and the two components that ensures full controllability and observability of the interface and joint dynamics. Furthermore, we assume that the joint is not easily accessible in practice, so that the extraction of its dynamical properties should be done via indirect measurements of the assembled system.

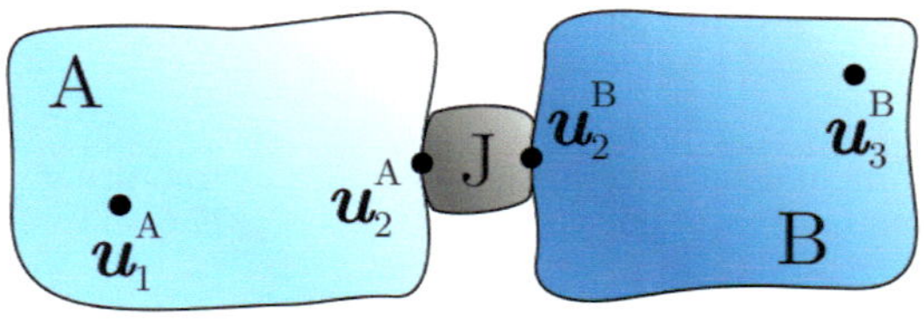


**Fig. 1** Illustration of the assembled system connecting components $A$ and $B$ to a joint $J$. $\mathbf{u}$ refers to displacement, where the subscripts $\star_1$, $\star_3$ indicate internal degrees of freedom and $\star_2$ represents the interface.

The substructuring isolation can be classified on the basis of the following general features:

- Dynamic joint or quasi-static joint
- Primal approach or dual approach
- Using interface measurements or internal measurements

The focus of this article is on a quasi-static identification performed via a primal or a dual formulation with solely interface measurements. A quasi-static joint is herein defined as a mass-less compliant mechanism with no cross-coupling between the connected interfaces' degrees of freedom.

A quasi-static joint parametrized with a simple spring-damper mechanism is represented in fig. 2.

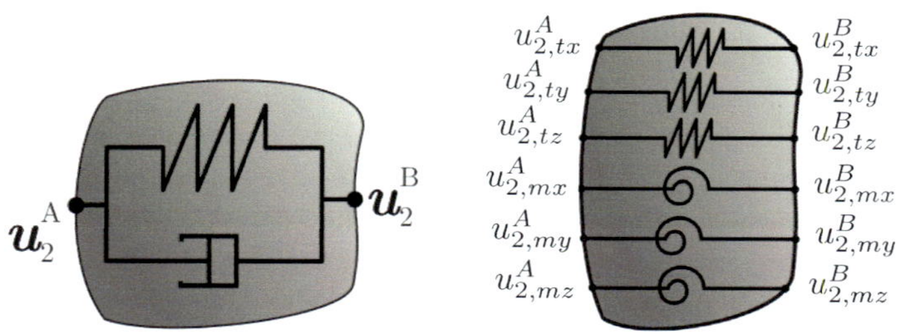


**Fig. 2** Illustration of the quasi-static joint. Left: Representation of the joint as a massless component. Right: Example illustrating the DoF to DoF topology of the joint, i.e. the no cross-coupling assumption.

Let us define the admittances of the assembled system and sub-components at the defined collocated interface degrees of freedom as $\mathbf{Y}_{22}^{AJB}$ and $\mathbf{Y}_{22}^{A|B} = \begin{bmatrix} \mathbf{Y}_{22}^{A} & \mathbf{0} \\ \mathbf{0} & \mathbf{Y}_{22}^{B} \end{bmatrix}$ respectively. Note that both matrices have the same dimension, given the quasi-static assumption. The corresponding impedances are obtained by simple inversion assuming the same number of output and input degrees of freedom as $\mathbf{Z}_{22}^{AJB} = \left(\mathbf{Y}_{22}^{AJB}\right)^{-1}$ and $\mathbf{Z}_{22}^{A|B} = \left(\mathbf{Y}_{22}^{A|B}\right)^{-1}$.

A global set of ordered interface degrees of freedom $\mathbf{u}_2$, external forces $\mathbf{f}_2$ and interface forces $\mathbf{g}_2$ are also introduced, stacking together the corresponding quantities from components $A$ and $B$.

## *Quasi-static primal formulation with impedances: Inverse Substructuring*

A primal formulation of substructuring, incorporating the equations of motion and interface conditions, and assuming exact compatibility, is expressed in impedance form as follows [1]:

$$\mathbf{Z}_{22}^{A|B}\mathbf{u}_2 = \mathbf{f}_2 + \mathbf{g}_2 \tag{1a}$$

$$\mathbf{u}_2 = \mathbf{L}\mathbf{q}_2 \tag{1b}$$

In this formulation, the compatibility condition is automatically satisfied by reducing the global interface coordinate, $\mathbf{u}_2$, to a minimal matching set of solving coordinates ,$\mathbf{q}_2$, through the localization Boolean operator $\mathbf{L}$.

Integrating eq. (1b) into eq. (1a) and ensuring equilibrium by left-projecting the equations of motion using $\mathbf{L}$ leads to the assembled configuration.

A primal formulation naturally favors the use of impedances, as it exposes the topological connections between components. By incorporating a dynamic joint component $\begin{bmatrix} \mathbf{Z}^J_{2A2A} & \mathbf{Z}^J_{2A2B} \\ \mathbf{Z}^J_{2B2A} & \mathbf{Z}^J_{2B2B} \end{bmatrix}$ into the global uncoupled impedance of eq. (1a), the dynamic stiffness of the assembled system can be expressed as:

$$\mathbf{Z}_{22}^{AJB} = \begin{bmatrix} \mathbf{Z}^A_{22} + \mathbf{Z}^J_{2A2A} & \mathbf{Z}^J_{2A2B} \\ \mathbf{Z}^J_{2B2A} & \mathbf{Z}^B_{22} + \mathbf{Z}^J_{2B2B} \end{bmatrix} \tag{2}$$

The joint dynamics are directly accessible from the off-diagonal terms of eq. (2). In an experimental setting, the isolation of the joint properties is thus implicitly achieved by inverting the assembled admittance model $\mathbf{Y}_{22}^{AJB}$. This allows for direct access to the joint dynamics within the system. This identification procedure is referred to here as Inverse Substructuring.

Notably, no assumptions regarding the joint properties have been made so far. However, by explicitly imposing the characteristics discussed in the introduction of section 2, namely the absence of mass and cross-coupling, the following quasi-static joint model can be reconstructed:

$$\left(\mathbf{Z}^J_{2A2A} \approx -\mathbf{Z}^J_{2A2B} \approx -\mathbf{Z}^J_{2B2A} \approx \mathbf{Z}^J_{2B2B}\right) = \mathbf{Z}^J_c, \qquad \mathbf{Z}^J = \begin{bmatrix} \mathbf{Z}^J_c & -\mathbf{Z}^J_c \\ -\mathbf{Z}^J_c & \mathbf{Z}^J_c \end{bmatrix} \tag{3}$$

The approximation symbol $\approx$ denotes the application of symmetry in the context of experimental measurements, while the property itself arises from the topological construction at the theoretical level.

Note that the literature distinguishes between isolating the joint dynamics via eq. (2) and fully determining a joint model as in eq. (3). Due to the topological construction, the dynamics of the joint, as well as those of the sub-components, are isolated and identified solely based on the dynamics of the assembled system.

## *Quasi-static dual formulation with admittances: LM-FBS with weakened interface*

A dual formulation of the substructuring operation is expressed in admittance form as follows [1]:

$$\mathbf{u}_2 = \mathbf{Y}_{22}^{A|B}(\mathbf{f}_2 - \mathbf{B}^T\boldsymbol{\lambda}) \tag{4a}$$

$$\mathbf{B}\mathbf{u}_2 = \mathbf{0} \tag{4b}$$

In this formulation, the compatibility condition is strictly enforced (no interface flexibility) using a Boolean operator $\mathbf{B}$, while the equilibrium condition is inherently satisfied by introducing Lagrange multipliers $\boldsymbol{\lambda}$. These multipliers represent the intensity of the interface forces $\mathbf{g}_2$ and serve as solving variables.

A massless joint is introduced as a relaxation of the connection between the components. Mathematically, this is expressed as:

$$\mathbf{B}\boldsymbol{u}_2 = \Delta\boldsymbol{u}^J_2 = \mathbf{Y}^J_c\boldsymbol{\lambda} \tag{5}$$

A gap $\Delta\boldsymbol{u}$ is now relaxing the compatibility condition via the quasi-static action of the joint dynamics $\mathbf{Y}^J_c$. Note that quasi-staticity at this point is solely referred to the absence of a mass term disrupting the equilibrium between the two sides of the interface.

By solving eq. (4a) using the relaxed compatibility in eq. (5), the expression for the assembled system admittance is retrieved:

$$\mathbf{Y}_{22}^{AJB} = \left[\mathbf{I} - \mathbf{Y}_{22}^{A|B}\mathbf{B}^T\left(\mathbf{B}\mathbf{Y}_{22}^{A|B}\mathbf{B}^T + \mathbf{Y}_c^J\right)^{-1}\mathbf{B}\right]\mathbf{Y}_{22}^{A|B} \tag{6}$$

which is referred as Lagrange Multiplier-Frequency Based Substructuring (LM-FBS) with weakened interface [14].

The joint admittance can be isolated by rearranging eq. (6):

$$\mathbf{Y}_c^J = \mathbf{B}\mathbf{Y}_{22}^{A|B}\mathbf{B}^T(\mathbf{B}(\mathbf{Y}_{22}^{AJB} - \mathbf{Y}_{22}^{A|B})\mathbf{B}^T)^{-1}\mathbf{B}\mathbf{Y}_{22}^{A|B}\mathbf{B}^T - \mathbf{B}\mathbf{Y}_{22}^{A|B}\mathbf{B}^T \tag{7}$$

To maintain consistency with the definition of the quasi-static joint in section 2, the final step involves selecting a specific topology for the joint by defining the compatibility Boolean matrix as $\mathbf{B} = \begin{bmatrix}-\mathbf{I} & \mathbf{I}\end{bmatrix}$. This choice enforces a DoF to DoF connection between the two sides of the interface, effectively eliminating any cross-coupling term. As a result, this assumption allows to write the following explicit expression:

$$\mathbf{Y}_c^J = \left(\mathbf{Y}_{22}^A + \mathbf{Y}_{22}^B\right)\left(\mathbf{Y}_{2^A2^A}^{AJB} + \mathbf{Y}_{2^B2^B}^{AJB} - \mathbf{Y}_{2^A2^B}^{AJB} - \mathbf{Y}_{2^B2^A}^{AJB} - \mathbf{Y}_{22}^A - \mathbf{Y}_{22}^B\right)^{-1}\left(\mathbf{Y}_{22}^A + \mathbf{Y}_{22}^B\right) - \left(\mathbf{Y}_{22}^A + \mathbf{Y}_{22}^B\right) \tag{8}$$

An astute reader will notice the following:

- To match the topological assumptions implicitly made in the primal joint derivation, a specific choice of the compatibility function $\mathbf{B} = \begin{bmatrix}-\mathbf{I} & \mathbf{I}\end{bmatrix}$ is required. While more complex connections with cross-coupling spring-like terms are possible, non-Boolean shape functions for $\mathbf{B}$ can also be used to model more advanced mechanisms, such as directional constraints. Note that one can also devise the corresponding topology for the isolation of the joint component in primal form, offering an alternative to the quasi-static joint in eq. (3).
- Although both the primal impedance and dual admittance formulations integrate equivalent joint topologies, the result in eq. (8) appears to necessitate knowledge of the dynamics of the isolated components. This requirement obscures the advantageous features of the intended quasi-static principle, which favors a practical and efficient identification method, specifically, the measurement of the joint in its assembled configuration (so called 'in-situ'). As a consequence, the mathematical equivalence between the primal and dual derivations is not readily evident.

The next section will address the open questions raised in the second bullet point.

## On the equivalence between primal and dual quasi-static formulations

To demonstrate the equivalence of the quasi-static formulations in section 2.1 and section 2.2, the most straightforward approach is to begin with the dual quasi-static equations in eq. (4a) and eq. (5), and invert all the admittance terms. This leads to the following system:

$$\mathbf{Z}_{22}^{A|B}\mathbf{u}_2 = \mathbf{f}_2 - \mathbf{B}^T\boldsymbol{\lambda} \tag{9a}$$

$$\mathbf{Z}_c^J\mathbf{B}\boldsymbol{u}_2 = \mathbf{Z}_c^J\Delta\boldsymbol{u}_2^J = \boldsymbol{\lambda} \tag{9b}$$

By solving eq. (9) for $\boldsymbol{\lambda}$, an impedance-based formulation for the interface force-displacement relationship is derived:

$$\left[\mathbf{Z}_{22}^{A|B} + \mathbf{B}^T\mathbf{Z}_c^J\mathbf{B}\right]\mathbf{u}_2 = \mathbf{f}_2 \tag{10}$$

Analogous to section 2.2, the application of $\mathbf{B} = \begin{bmatrix}-\mathbf{I} & \mathbf{I}\end{bmatrix}$ leads to:

$$\left[\begin{bmatrix}\mathbf{Z}_{22}^A & \mathbf{0}\\ \mathbf{0} & \mathbf{Z}_{22}^B\end{bmatrix} + \begin{bmatrix}\mathbf{Z}_c^J & -\mathbf{Z}_c^J\\ -\mathbf{Z}_c^J & \mathbf{Z}_c^J\end{bmatrix}\right]\mathbf{u}_2 = \mathbf{f}_2 \tag{11}$$

This leads to the isolation of the joint and sub-components impedances, demonstrating the equivalence of the dual derivation with the primal (see eq. (2) and eq. (3)).

What remains unanswered is the formulation of the joint (and sub-components) admittances solely as a function of the assembled system admittance, as this aspect remains somewhat unclear in eq. (8). To the best of the author's knowledge, while a primal-based derivation exists in the literature, a dual-based derivation has not yet been published in full form. A concise summary of the primal derivation by [4] is provided in section 3.1, followed by the presentation of a dual-based expression in section 3.2.

### *Admittance-based primal derivation*

A concise summary of the derivation steps from [4] is as follows:

1. The primal quasi-static impedance-based formulation as in eq. (2) and eq. (3) is used as the starting point.
2. The block matrix inversion formula in eq. (34) is applied to obtain the assembled system dynamics in admittance notation $\mathbf{Y}_{22}^{AJB} = (\mathbf{Z}_{22}^{AJB})^{-1}$
3. A system of four equations (one per block element of the assembled system $\mathbf{Y}_{22}^{AJB}$) is solved to find the three unknowns $\mathbf{Y}_{22}^{A}, \mathbf{Y}_{22}^{B}, \mathbf{Y}_{c}^{J}$ as a function of the assembled admittances $\mathbf{Y}_{2_A2_A}^{AJB}$, $\mathbf{Y}_{2_A2_B}^{AJB}$, $\mathbf{Y}_{2_B2_A}^{AJB}$, $\mathbf{Y}_{2_B2_B}^{AJB}$.

After lengthly derivations (applying matrix inversion properties as described in appendix A, along with the system's symmetry properties and various equation manipulations), this results in the following solutions for the isolated subcomponents and the joint:

$$\mathbf{Y}_{22}^{A} = \mathbf{Y}_{2_A2_B}^{AJB}\left(\mathbf{Y}_{2_B2_B}^{AJB} - \mathbf{Y}_{2_A2_B}^{AJB}\right)^{-1}\left(\mathbf{Y}_{2_A2_A}^{AJB}(\mathbf{Y}_{2_B2_A}^{AJB})^{-1}\mathbf{Y}_{2_B2_B}^{AJB} - \mathbf{Y}_{2_A2_B}^{AJB}\right) \tag{12a}$$

$$\mathbf{Y}_{22}^{B} = \left(\mathbf{Y}_{2_A2_A}^{AJB}\left(\mathbf{Y}_{2_B2_A}^{AJB}\right)^{-1}\mathbf{Y}_{2_B2_B}^{AJB} - \mathbf{Y}_{2_A2_B}^{AJB}\right)\left(\mathbf{Y}_{2_A2_A}^{AJB} - \mathbf{Y}_{2_A2_B}^{AJB}\right)^{-1}\mathbf{Y}_{2_A2_B}^{AJB} \tag{12b}$$

$$\mathbf{Y}_{c}^{J} = (\mathbf{Z}_{c}^{J})^{-1} = \mathbf{Y}_{2_A2_A}^{AJB}\left(\mathbf{Y}_{2_B2_A}^{AJB}\right)^{-1}\mathbf{Y}_{2_B2_B}^{AJB} - \mathbf{Y}_{2_A2_B}^{AJB} \tag{12c}$$

It is evident that there is an explicit and sole dependence on the dynamics of the assembled system. This result will be taken as reference for the dual-based derivation in section 3.2.

Further, upon inspecting the result in eq. (12c), the following observation can be made:

$$\mathbf{Y}_{c}^{J} = (\mathbf{Z}_{c}^{J})^{-1} \approx (-\mathbf{Z}_{2_A2_B}^{AJB})^{-1} = -\mathbf{Y}_{2_A2_B}^{AJB,\text{cond}} = \mathbf{Y}_{2_A2_A}^{AJB}\left(\mathbf{Y}_{2_B2_A}^{AJB}\right)^{-1}\mathbf{Y}_{2_B2_B}^{AJB} - \mathbf{Y}_{2_A2_B}^{AJB} \tag{13}$$

The admittance form of the joint dynamics can thus be derived using the condensation rule (indicated by the superscript $\star^{\text{cond}}$), leveraging the topology of the primal formulation in eq. (2).

### *Admittance-based dual derivation*

Starting from the quasi-static dual formulation in eq. (6), assuming the topology given by $\mathbf{B} = \begin{bmatrix}-\mathbf{I} & \mathbf{I}\end{bmatrix}$:

$$\mathbf{Y}_{22}^{AJB} = \begin{bmatrix}\mathbf{Y}_{22}^{A} & \mathbf{0}\\ \mathbf{0} & \mathbf{Y}_{22}^{B}\end{bmatrix} - \begin{bmatrix}-\mathbf{Y}_{22}^{A}\\ \mathbf{Y}_{22}^{B}\end{bmatrix}\left(\mathbf{Y}_{22}^{A} + \mathbf{Y}_{22}^{B} + \mathbf{Y}_{c}^{J}\right)^{-1}\begin{bmatrix}-\mathbf{Y}_{22}^{A} & \mathbf{Y}_{22}^{B}\end{bmatrix} \tag{14}$$

From here on, the notation will be simplified using $\mathbf{Y}^{\text{tot}} = \mathbf{Y}_{22}^{A} + \mathbf{Y}_{22}^{B} + \mathbf{Y}_{c}^{J}$ and omitting the subscript $\star_{22}$ for $\mathbf{Y}_{22}^{A}$ and $\mathbf{Y}_{22}^{B}$.

Similarly to the primal-based steps described in section 3.1, a system of four equations can be derived from the block elements of eq. (14):

$$\mathbf{Y}_{2_A2_A}^{AJB} = \mathbf{Y}^{A} - \mathbf{Y}^{A}\left(\mathbf{Y}^{\text{tot}}\right)^{-1}\mathbf{Y}^{A} \tag{15a}$$

$$\mathbf{Y}_{2_A2_B}^{AJB} = \mathbf{Y}^{A}\left(\mathbf{Y}^{\text{tot}}\right)^{-1}\mathbf{Y}^{B} \tag{15b}$$

$$\mathbf{Y}_{2_B2_B}^{AJB} = \mathbf{Y}^{B} - \mathbf{Y}^{B}\left(\mathbf{Y}^{\text{tot}}\right)^{-1}\mathbf{Y}^{B} \tag{15c}$$

$$\mathbf{Y}_{2_B2_A}^{AJB} = \mathbf{Y}^{B}\left(\mathbf{Y}^{\text{tot}}\right)^{-1}\mathbf{Y}^{A} \tag{15d}$$

Note that, thanks to the symmetry of the admittances, it is true that $\mathbf{Y}_{2_A2_B}^{AJB} = \left(\mathbf{Y}_{2_B2_A}^{AJB}\right)^{T}$.

The stated goal is to find the three unknowns $\mathbf{Y}^{A}$, $\mathbf{Y}^{B}$, $\mathbf{Y}_{c}^{J}$ as a function of the admittances of the assembled system $\mathbf{Y}_{2_A2_A}^{AJB}$, $\mathbf{Y}_{2_A2_B}^{AJB}$, $\mathbf{Y}_{2_B2_A}^{AJB}$, $\mathbf{Y}_{2_B2_B}^{AJB}$. The derivation involves a combination of matrix/equation manipulations, matrix inversion operations (as detailed in appendix A), and the application of the system's symmetry properties.

The starting point is isolating $\mathbf{Y}^{\text{tot}}$ from eq. (15a):

$$\mathbf{Y}^{\text{tot}} = -\left(\mathbf{Y}^{A}\right)^{-1}\left(\mathbf{Y}_{2_A2_A}^{AJB} - \mathbf{Y}^{A}\right)\left(\mathbf{Y}^{A}\right)^{-1} \tag{16}$$

Then, eq. (16) is substituted into eq. (15b), eq. (15c) and eq. (15d) to remove the dependency on $\mathbf{Y}^{\text{tot}}$:

$$\mathbf{Y}^{AJB}_{2^A 2^B} = -\left(\mathbf{Y}^{AJB}_{2^A 2^A} - \mathbf{Y}^{A}\right)\left(\mathbf{Y}^{A}\right)^{-1}\mathbf{Y}^{B} \tag{17a}$$

$$\mathbf{Y}^{AJB}_{2^B 2^A} = -\mathbf{Y}^{B}\left(\mathbf{Y}^{A}\right)^{-1}\left(\mathbf{Y}^{AJB}_{2^A 2^A} - \mathbf{Y}^{A}\right) \tag{17b}$$

$$\mathbf{Y}^{AJB}_{2^B 2^A} = \mathbf{Y}^{B} + \mathbf{Y}^{B}\left(\mathbf{Y}^{A}\right)^{-1}\left(\mathbf{Y}^{AJB}_{2^A 2^A} - \mathbf{Y}^{A}\right)\left(\mathbf{Y}^{A}\right)^{-1}\mathbf{Y}^{B} \tag{17c}$$

The next step involves isolating $\mathbf{Y}^B$ from eq. (17a):

$$\mathbf{Y}^{B} = -\mathbf{Y}^{A}\left(\mathbf{Y}^{AJB}_{2^A 2^A} - \mathbf{Y}^{A}\right)^{-1}\mathbf{Y}^{AJB}_{2^A 2^B} \tag{18}$$

Further, by substituting eq. (17b) into eq. (17a) to eliminate $\mathbf{Y}^B$, the following is obtained:

$$\left(\mathbf{Y}^{AJB}_{2^A 2^B}\right)^{-1} = \left(\mathbf{Y}^{A}\right)^{-1}\left(\mathbf{Y}^{AJB}_{2^A 2^A} - \mathbf{Y}^{A}\right)\left(\mathbf{Y}^{AJB}_{2^B 2^A}\right)^{-1}\mathbf{Y}^{A}\left(\mathbf{Y}^{AJB}_{2^A 2^A} - \mathbf{Y}^{A}\right)^{-1} \tag{19}$$

Then, eq. (18) is substituted into eq. (17c):

$$\mathbf{Y}^{AJB}_{2^B 2^B} = \left(\left(\mathbf{Y}^{AJB}_{2^A 2^B}\right)^{-1}\left(\mathbf{Y}^{AJB}_{2^A 2^A}\left(\mathbf{Y}^{A}\right)^{-1} - \mathbf{I}\right)\right)^{-1}\left(\left(\mathbf{Y}^{A}\right)^{-1}\mathbf{Y}^{AJB}_{2^A 2^B} - \mathbf{I}\right) \tag{20}$$

This can be rearranged as:

$$\left(\mathbf{Y}^{AJB}_{2^A 2^B}\right)^{-1}\left(\mathbf{Y}^{AJB}_{2^A 2^A}\left(\mathbf{Y}^{A}\right)^{-1} - \mathbf{I}\right)\mathbf{Y}^{AJB}_{2^B 2^B} = \left(\mathbf{Y}^{A}\right)^{-1}\mathbf{Y}^{AJB}_{2^A 2^B} - \mathbf{I} \tag{21}$$

Next, eq. (19) is substituted into eq. (21):

$$\left(\mathbf{Y}^{A}\right)^{-1}\mathbf{Y}^{AJB}_{2^A 2^A}\left(\mathbf{Y}^{AJB}_{2^B 2^A}\right)^{-1}\mathbf{Y}^{AJB}_{2^B 2^B} - \left(\mathbf{Y}^{AJB}_{2^B 2^A}\right)^{-1}\mathbf{Y}^{AJB}_{2^B 2^B} = \left(\mathbf{Y}^{A}\right)^{-1}\mathbf{Y}^{AJB}_{2^A 2^B} - \mathbf{I} \tag{22}$$

Finally, $\mathbf{Y}^A$ can be isolated:

$$\mathbf{Y}^{A} = \left(\mathbf{Y}^{AJB}_{2^A 2^A}\left(\mathbf{Y}^{AJB}_{2^B 2^A}\right)^{-1}\mathbf{Y}^{AJB}_{2^B 2^B} - \mathbf{Y}^{AJB}_{2^A 2^B}\right)\left(\mathbf{Y}^{AJB}_{2^B 2^B} - \mathbf{Y}^{AJB}_{2^B 2^A}\right)^{-1}\mathbf{Y}^{AJB}_{2^B 2^A} \tag{23}$$

By leveraging the symmetry of the admittance and its inverse (impedance), eq. (23) can be rewritten as:

$$\mathbf{Y}^{A} = \mathbf{Y}^{AJB}_{2^A 2^B}\left(\mathbf{Y}^{AJB}_{2^B 2^B} - \mathbf{Y}^{AJB}_{2^A 2^B}\right)^{-1}\left(\mathbf{Y}^{AJB}_{2^A 2^A}\left(\mathbf{Y}^{AJB}_{2^B 2^A}\right)^{-1}\mathbf{Y}^{AJB}_{2^B 2^B} - \mathbf{Y}^{AJB}_{2^A 2^B}\right) \tag{24}$$

This result is equivalent to the primal derivation in eq. (12a).
In analogy to the above derivation, by mirroring the operations with respect to $\mathbf{Y}^B$, the equivalent of eq. (12b) is obtained:

$$\mathbf{Y}^{B} = \left(\mathbf{Y}^{AJB}_{2^A 2^A}\left(\mathbf{Y}^{AJB}_{2^B 2^A}\right)^{-1}\mathbf{Y}^{AJB}_{2^B 2^B} - \mathbf{Y}^{AJB}_{2^A 2^B}\right)\left(\mathbf{Y}^{AJB}_{2^A 2^A} - \mathbf{Y}^{AJB}_{2^A 2^B}\right)^{-1}\mathbf{Y}^{AJB}_{2^A 2^B} \tag{25}$$

Finally, manipulating eq. (15b) to isolate the joint admittance yields:

$$\mathbf{Y}^{J}_{c} = \mathbf{Y}^{B}\left(\mathbf{Y}^{AJB}_{2^A 2^B}\right)^{-1}\mathbf{Y}^{A} - \mathbf{Y}^{A} - \mathbf{Y}^{B} \tag{26}$$

where $\mathbf{Y}^A$ and $\mathbf{Y}^B$ can be expressed as functions of the assembled system (see eq. (24) and eq. (25)). Consequently, the isolated joint must also be a function of the assembled system dynamics alone. Given the complexity of deriving the simplified form of eq. (26) via substitution of eq. (24) and eq. (25), a different approach will be employed to complete the proof of equivalence with the primal expression in eq. (12c) (or eq. (13)).

A proof by verification is used, where the ansatz solution in eq. (12c) is substituted into eq. (26) to confirm that it satisfies the system.

First, $\mathbf{Y}^A$ from eq. (23) (equivalent to eq. (24) ) and $\mathbf{Y}^B$ from eq. (25) are rewritten using the ansatz for the joint admittance (eq. (12c)) and exploiting the symmetry of the cross-coupling terms of the assembled admittance:

$$\mathbf{Y}^{A} = \mathbf{Y}^{J}_{c}\left(\mathbf{Y}^{AJB}_{2^B 2^B} - \left(\mathbf{Y}^{AJB}_{2^A 2^B}\right)^{T}\right)^{-1}\left(\mathbf{Y}^{AJB}_{2^A 2^B}\right)^{T} \tag{27a}$$

$$\mathbf{Y}^{B} = \mathbf{Y}^{J}_{c}\left(\mathbf{Y}^{AJB}_{2^A 2^A} - \mathbf{Y}^{AJB}_{2^A 2^B}\right)^{-1}\mathbf{Y}^{AJB}_{2^A 2^B} \tag{27b}$$

Substituting eq. (27a) and eq. (27b) into eq. (26) results in:

$$\begin{aligned}\mathbf{Y}_c^J = \mathbf{Y}_c^J \left(\mathbf{Y}_{2_A2_A}^{AJB} - \mathbf{Y}_{2_A2_B}^{AJB}\right)^{-1} \mathbf{Y}_{2_A2_B}^{AJB} \left(\mathbf{Y}_{2_A2_B}^{AJB}\right)^{-1} \mathbf{Y}_c^J \left(\mathbf{Y}_{2_B2_B}^{AJB} - \left(\left(\mathbf{Y}_{2_A2_B}^{AJB}\right)^T\right)\right)^{-1} \left(\mathbf{Y}_{2_A2_B}^{AJB}\right)^T \\ - \mathbf{Y}_c^J \left(\mathbf{Y}_{2_B2_B}^{AJB} - \left(\mathbf{Y}_{2_A2_B}^{AJB}\right)^T\right)^{-1} \left(\mathbf{Y}_{2_A2_B}^{AJB}\right)^T - \mathbf{Y}_c^J \left(\mathbf{Y}_{2_A2_A}^{AJB} - \mathbf{Y}_{2_A2_B}^{AJB}\right)^{-1} \mathbf{Y}_{2_A2_B}^{AJB}\end{aligned} \tag{28}$$

After cancellation of redundant terms:

$$\begin{aligned}\mathbf{I} = \left(\mathbf{Y}_{2_A2_A}^{AJB} - \mathbf{Y}_{2_A2_B}^{AJB}\right)^{-1} \mathbf{Y}_c^J \left(\mathbf{Y}_{2_B2_B}^{AJB} - \left(\mathbf{Y}_{2_A2_B}^{AJB}\right)^T\right)^{-1} \left(\mathbf{Y}_{2_A2_B}^{AJB}\right)^T \\ - \left(\mathbf{Y}_{2_B2_B}^{AJB} - \left(\mathbf{Y}_{2_A2_B}^{AJB}\right)^T\right)^{-1} \left(\mathbf{Y}_{2_A2_B}^{AJB}\right)^T - \left(\mathbf{Y}_{2_A2_A}^{AJB} - \mathbf{Y}_{2_A2_B}^{AJB}\right)^{-1} \mathbf{Y}_{2_A2_B}^{AJB}\end{aligned} \tag{29}$$

By further elimination:

$$\begin{aligned}\left(\mathbf{Y}_{2_A2_A}^{AJB} - \mathbf{Y}_{2_A2_B}^{AJB}\right) = \mathbf{Y}_c^J \left(\mathbf{Y}_{2_B2_B}^{AJB} - \left(\mathbf{Y}_{2_A2_B}^{AJB}\right)^T\right)^{-1} \left(\mathbf{Y}_{2_A2_B}^{AJB}\right)^T \\ - \left(\mathbf{Y}_{2_A2_A}^{AJB} - \mathbf{Y}_{2_A2_B}^{AJB}\right) \left(\mathbf{Y}_{2_B2_B}^{AJB} - \left(\mathbf{Y}_{2_A2_B}^{AJB}\right)^T\right)^{-1} \left(\mathbf{Y}_{2_A2_B}^{AJB}\right)^T - \mathbf{Y}_{2_A2_B}^{AJB}\end{aligned} \tag{30}$$

Further simplification results in:

$$\mathbf{Y}_{2_A2_A}^{AJB} \left(\left(\mathbf{Y}_{2_A2_B}^{AJB}\right)^T\right)^{-1} \left(\mathbf{Y}_{2_B2_B}^{AJB} - \left(\mathbf{Y}_{2_A2_B}^{AJB}\right)^T\right) = \mathbf{Y}_c^J - \left(\mathbf{Y}_{2_A2_A}^{AJB} - \mathbf{Y}_{2_A2_B}^{AJB}\right) \tag{31}$$

This leads to the final simplified form:

$$\mathbf{Y}_c^J = \mathbf{Y}_{2_A2_A}^{AJB} \left(\mathbf{Y}_{2_B2_A}^{AJB}\right)^{-1} \mathbf{Y}_{2_B2_B}^{AJB} - \mathbf{Y}_{2_A2_B}^{AJB} \tag{32}$$

By substituting the ansatz solution, this is reduced to the trivial identity:

$$\mathbf{Y}_c^J = \mathbf{Y}_c^J \tag{33}$$

This concludes the verification.

As a remark, the results in eq. (24), eq. (25) and eq. (32) are not unique. Equivalent expressions can be obtained by reordering or rearranging the equations, applying matrix inversion properties, and leveraging the symmetry inherent in the matrices. This has particular significance from an experimental perspective for two key reasons:

- Measurement errors may propagate through the inversion operations. As a result, the choice of the mathematical expression used for identification could impact the accuracy of the outcome.
- While the primary goal of the derivation in section 3.2 is to obtain an 'in-situ' expression for the joint admittance, alternative formulations for joint isolation, based on the dynamics of the assembled system and its subcomponents, can be readily derived by inverting the equations in eq. (15). These alternative expressions may be advantageous in certain scenarios where access to component locations is limited.

## Conclusion

This paper presents a concise review of popular substructuring quasi-static joint identification methods, focusing on the primal impedance-based and dual admittance-based formulations. The assumptions underlying both approaches are clearly outlined, with attention to their significance and the specific stages of the derivations where they are introduced. The analysis explores the equivalence of the two formulations and revisits the dual admittance-based derivation to highlight key aspects of isolating joint and subcomponents 'in-situ' within the admittance domain.

Ongoing work focuses on generalizing the quasi-static identification framework to incorporate non-interface measurements, aiming to further enhance the methodology.

## Some Matrix Inversion Properties

Some relevant matrix inversion properties used for the derivations in section 3.1 and section 3.2 are provided in the following:

$$\mathbf{S} := \begin{bmatrix} \mathbf{A} & \mathbf{B} \\ \mathbf{C} & \mathbf{D} \end{bmatrix}, \quad \mathbf{S}^{-1} = \begin{bmatrix} \mathbf{A}^{-1} + \mathbf{A}^{-1}\mathbf{B}(\mathbf{D} - \mathbf{C}\mathbf{A}^{-1}\mathbf{B})^{-1}\mathbf{C}\mathbf{A}^{-1} & -\mathbf{A}^{-1}\mathbf{B}(\mathbf{D} - \mathbf{C}\mathbf{A}^{-1}\mathbf{B})^{-1} \\ -(\mathbf{D} - \mathbf{C}\mathbf{A}^{-1}\mathbf{B})^{-1}\mathbf{C}\mathbf{A}^{-1} & (\mathbf{D} - \mathbf{C}\mathbf{A}^{-1}\mathbf{B})^{-1} \end{bmatrix} \tag{34}$$

$$\left(\mathbf{A}^{-1} + \mathbf{B}^{-1}\right)^{-1} = \mathbf{A}\left(\mathbf{A} + \mathbf{B}\right)^{-1}\mathbf{B} \tag{35}$$

$$(\mathbf{ABC})^{-1} = \mathbf{C}^{-1}\mathbf{B}^{-1}\mathbf{A}^{-1} \tag{36}$$

$$(\mathbf{AB})^{-1} = \mathbf{B}^{-1}\mathbf{A}^{-1} \tag{37}$$

## References

[1] Dennis Klerk, Daniel Rixen, and Sven Voormeeren. "General Framework for Dynamic Substructuring: History, Review, and Classification of Techniques". In: *Aiaa Journal - AIAA J* 46 (May 2008), pp. 1169–1181. DOI: 10.2514/1.33274.

[2] Matthew S. Allen et al. *Substructuring in Engineering Dynamics*. Vol. 594. 2018. ISBN: 978-3-030-25531-2.

[3] H. Y. Hwang. "Identification techniques of structure connection parameters using frequency response functions". In: *Journal of Sound and Vibration* 212.3 (1998), pp. 469–479. ISSN: 0022460X. DOI: 10.1006/jsvi.1997.1433.

[4] Jiantie Zhen, Teik C. Lim, and Guangqing Lu. "Determination of system vibratory response characteristics applying a spectral-based inverse sub-structuring approach. Part I: analytical formulation". In: *International Journal of Vehicle Noise and Vibration* 1.1-2 (2004), pp. 1–30. ISSN: 1479148X. DOI: 10.1504/ijvnv.2004.004066.

[5] Zhi Wei Wang and Jun Wang. "Inverse substructure method of three-substructures coupled system and its application in product-transport-system". In: *JVC/Journal of Vibration and Control* 17.6 (2011), pp. 943–952. ISSN: 10775463. DOI: 10.1177/1077546310376083.

[6] Laurent Keersmaekers et al. "Decoupling of mechanical systems based on in-situ frequency response functions: The link-preserving, decoupling method". eng. In: *Mechanical Systems and Signal Processing* 58 (2015), pp. 340–354. ISSN: 10961216, 08883270. DOI: 10.1016/j.ymssp.2014.11.016.

[7] J. W.R. Meggitt and A. T. Moorhouse. "In-situ sub-structure decoupling of resiliently coupled assemblies". eng. In: *Mechanical Systems and Signal Processing* 117 (2019), pp. 723–737. ISSN: 10961216, 08883270. DOI: 10.1016/j.ymssp.2018.07.045.

[8] M. Haeussler, S. W.B. Klaassen, and D. J. Rixen. "Experimental twelve degree of freedom rubber isolator models for use in substructuring assemblies". In: *Journal of Sound and Vibration* 474 (2020). ISSN: 10958568. DOI: 10.1016/j.jsv.2020.115253.

[9] J. S. Tsai and Y. F. Chou. "The identification of dynamic characteristics of a single bolt joint". In: *Journal of Sound and Vibration* 125.3 (1988), pp. 487–502. ISSN: 10958568. DOI: 10.1016/0022-460X(88)90256-8.

[10] Y. Ren and C. F. Beards. "Identification of joint properties of a structure using FRF data". In: *Journal of Sound and Vibration* 186.4 (1995), pp. 567–587. ISSN: 10958568. DOI: 10.1006/jsvi.1995.0469.

[11] Tachung Yang, Shuo Hao Fan, and Chorng Shyan Lin. "Joint stiffness identification using FRF measurements". In: *Computers and Structures* 81.28-29 (2003), pp. 2549–2556. ISSN: 00457949. DOI: 10.1016/S0045-7949(03)00328-6.

[12] Damjan Čelič and Miha Boltežar. "Identification of the dynamic properties of joints using frequency-response functions". In: *Journal of Sound and Vibration* 317.1-2 (2008), pp. 158–174. ISSN: 0022460X. DOI: 10.1016/j.jsv.2008.03.009.

[13] Taner Kalaycioglu and H. Nevzat Özgüven. "New FRF based methods for substructure decoupling". In: *Conference Proceedings of the Society for Experimental Mechanics Series*. Vol. 4. Springer New York LLC, 2016, pp. 463–472. ISBN: 9783319297620. DOI: 10.1007/978-3-319-29763-7_46.

[14] Ahmed El Mahmoudi, Daniel J. Rixen, and Christian H. Meyer. "Comparison of Different Approaches to Include Connection Elements into Frequency-Based Substructuring". In: *Experimental Techniques* 44.4 (2020), pp. 425–433. ISSN: 17471567. DOI: 10.1007/s40799-020-00360-1.

# Chapter 14
# On the Use of Frequency-Dependent Modal Basis for Interface Modeling in Frequency-Based Substructuring

**Domen Ocepek, Miha Pogačar, Jure Korbar, and Gregor Čepon**

**Abstract** In experimental dynamic substructuring, coupling of substructures sharing a line- or surface-like interface proves to be a challenge due to the difficulties in interface modeling. Modeling a high number of degrees of freedom at the common interface can be too stringent when imposing compatibility and equilibrium conditions, thereby causing redundancy and ill-conditioning. To mitigate the effects of over-determination and experimental errors several techniques have been developed, proposing different reduction spaces to weaken the interface conditions. In the context of component-mode synthesis (also known as modal substructuring), modal constraints for fixture and subsystem is a commonly used approach the use of which was also suggested for frequency-based substructuring. A flexible fixture is introduced to the coupling workflow and the reduction basis is defined as a truncated set of physical mode shapes of the fixture. Both the selection and the number of modes considered in the reduction step can significantly affect the quality of the substructuring prediction. This work investigates the benefits of applying frequency-dependent reduction spaces to weaken the interface conditions in the context of frequency-based substructuring. The approach is investigated using an experimental case study on the Dynamic Substructuring Round Robin Benchmark.

**Keywords** Dynamic Substructuring · Frequency-based Substructuring · Transmission Simulator · Modal Constrain for Fixture and Subsystems · Round Robin Benchmark

## Introduction

In the context of experimental frequency-based substructuring (FBS) [1], the main challenge for a successful substructure-coupling implementation remains the modeling of the common interface between the substructures. Coupling of the substructures requires compatibility of the displacements along the interface between two substructures and the equilibrium of interface forces to be satisfied. Practical difficulties relate to the measurement of an appropriate number of degrees of freedom (DoFs) at which interface conditions are imposed. Coupling too many DoFs can results in redundancy and consequently bad conditioning of the interface coupling equations. Ill-conditioning combined with inevitably present measurement errors (random or systematic in nature) can lead to a high error amplification. Limiting the number of interface measurements and imposing weak compatibility avoids those problems, but can significantly deteriorate the accuracy of the coupled dynamics.

A solution can be applied by projecting the measured dynamics into a representative subspace [2]. This mitigates the effects of measurement errors since the interface conditions are now imposed in the reduced space and are thus related to as weak. Redundant and insignificant dynamic information that is not included in the reduced subspaces is left uncoupled. This is beneficial since this part is commonly strongly affected by measurement errors, is badly controlled or badly observed.

In the context of component-mode synthesis (CMS), modal constraints for fixture and subsystem (MCFS) [3] is a commonly used approach [4] the use of which was also suggested for FBS in [5]. A flexible fixture is introduced to the coupling workflow and the reduction basis is defined as a truncated set of physical mode shapes of the fixture. A method developed with the aim to handle flexible interface behavior in the frequency domain is the singular vector transformation (SVT) [6], where force and displacement reduction spaces are extracted from the measured response models by the means of singular value decomposition.

Domen Ocepek · Miha Pogačar·Jure Korbar· Gregor Čepon, Faculty of Mechanical Engineering, University of Ljubljana, Aškerčeva 6, 1000 Ljubljana, Slovenia
e-mail: domen.ocepek@fs.uni-lj.si; miha.pogacar@fs.uni-lj.si; jure.korbar@fs.uni-lj.si; gregor.cepon@fs.uni-lj.si



Walter D'Ambrogio, et al. (eds.), *Dynamic Substructuring & Transfer Path Analysis, Vol. 4*, Conference Proceedings of the Society for Experimental Mechanics Series, https://doi.org/10.13052/9788743801498-14

This work can be seen as the extension of [7], where authors investigated MCFS and SVT to couple experimental response models of the substructures sharing a continuous interface. In [7], authors suggested the benefits of applying frequency-dependent reduction spaces to weaken the interface conditions in the context of frequency-based substructuring and particular MCFS, which is elaborated within this contribution.

## Theoretical Background and Notation

Consider two substructures A and B assembled at the interface DoFs $(\star)_2^{\mathrm{A}}$ and $(\star)_2^{\mathrm{B}}$, as depicted in Fig. 1.

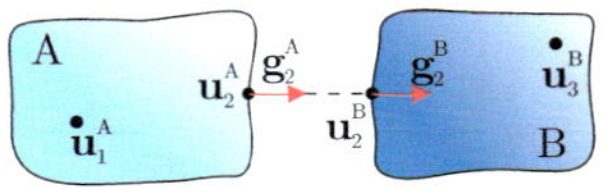


**Fig. 1** Substructures A and B to be coupled at the common interface.

With admittances of the individual subsystems known ($\mathbf{Y}^{\mathrm{A}}$ and $\mathbf{Y}^{\mathrm{B}}$) and partitioned in internal ($(\star)_1^{\mathrm{A}}$ and $(\star)_3^{\mathrm{B}}$) and interface DoFs, the governing equation of motion for the uncoupled system can be written as:

$$\mathbf{u} = \mathbf{Y}^{\mathrm{A|B}}\left(\mathbf{f} + \mathbf{g}\right). \tag{1}$$

The vector $\mathbf{u}$ represents the displacements to the external force vector $\mathbf{f}$, and $\mathbf{g}$ is the vector of interface forces between the substructures that exist only at the interface DoFs, keeping the substructures together. $\mathbf{Y}^{\mathrm{A|B}}$ is a block-diagonal matrix of subsystem's admittances. All subsystems considered are assembled through the proper application of interface conditions[1]. The compatibility of the displacements at the common boundary is recast in the general formulation:

$$\mathbf{B}\,\mathbf{T}_{\mathrm{u}}\,\mathbf{u} = \mathbf{0} \quad \text{where} \quad \mathbf{B} = \begin{bmatrix}\mathbf{0} & -\mathbf{I} & \mathbf{I} & \mathbf{0}\end{bmatrix}. \tag{2}$$

The equilibrium condition is imposed by replacing the interface forces using a set of Lagrange multiplier vectors $\boldsymbol{\lambda}$:

$$\mathbf{g} = -\mathbf{T}_{\mathrm{f}}^{\mathrm{H}}\,\mathbf{B}^{\mathrm{T}}\boldsymbol{\lambda}. \tag{3}$$

By eliminating the Lagrange multiplier vector from the set of Eqs. (1 - 3) we obtain:

$$\mathbf{u} = \underbrace{\left[\mathbf{I} - \mathbf{Y}^{\mathrm{A|B}}\,\mathbf{T}_{\mathrm{f}}^{\mathrm{H}}\,\mathbf{B}^{\mathrm{T}}\left(\mathbf{B}\,\mathbf{T}_{\mathrm{u}}\,\mathbf{Y}^{\mathrm{A|B}}\,\mathbf{T}_{\mathrm{f}}^{\mathrm{H}}\,\mathbf{B}^{\mathrm{T}}\right)^{-1}\mathbf{B}\,\mathbf{T}_{\mathrm{u}}\right]\mathbf{Y}^{\mathrm{A|B}}}_{\mathbf{Y}^{\mathrm{AB}}}\,\mathbf{f}. \tag{4}$$

where $\mathbf{Y}^{\mathrm{AB}}$ is the admittance of the assembled system, $\mathbf{T}_{\mathrm{u}}$ is the transformation matrix to project interface displacements to the reduced domain, and $\mathbf{T}_{\mathrm{f}}$ is the transformation matrix for the interface forces which fulfill a weak compatibility in a space of lower dimension. Note that different Boolean matrices can be used to apply the compatibility $\mathbf{B}_{\mathrm{u}}$ and equilibrium $\mathbf{B}_{\mathrm{f}}$ conditions.

A flexible fixture (named transmission simulator or TS) can be introduced to the coupling workflow and attached to substructure B (Fig. 2). The combined (sub)structure is denoted as BTS in the following which is coupled to the substructure A from which the TS structure is then decoupled to obtain the response model for AB. The procedure is schematically depicted in Fig. 2.

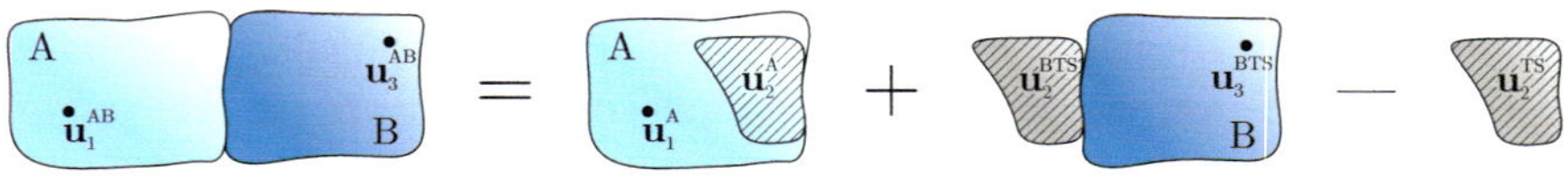


**Fig. 2** Coupling application with a TS substructure. Common interface for all substructures is presented as hatched.

[1] Weak compatibility is considered here. For more information, an interested reader is refereed to [7]

## Case Study

The coupling approach with the MCFS and SVT method adopted the workflow presented in Fig. 2. The coupling prediction in [7] where the number of mode shapes and singular modes retained in the reduction step varied from 8 to 10 are presented in Fig. 3. A reasonably higher number of DoFs retained in the reduction improves prediction in a higher frequency range and vice versa.

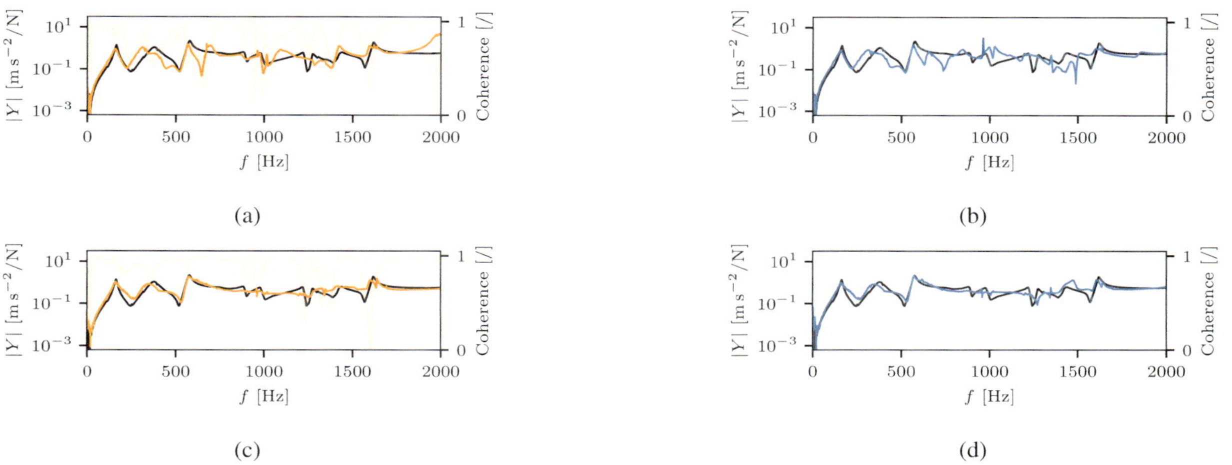


**Fig. 3** Amplitude and coherence of a FRF of $\mathbf{Y}^{AB}$ using MCFS and SVT approach with various number of mode shapes/singular modes retained in the reduction bases; a) first 8 mode shapes, b) first 8 singular modes, c) first 10 mode shapes, d) first 10 singular modes. (— reference, — TS approach, — SVT approach, — coherence on TS approach, — coherence on SVT approach)

The authors argue that the use of frequency-dependent reduction bases for MCFS and SVT, involving a varying number of mode shapes and singular modes respectively, would likely result in more consistent response prediction. The aim of this approach is to define reduction bases that adjust for the dominant/representative physical modes at each frequency line, hence provide a representative interface description yet avoid redundancy in terms of the number of interface DoFs.

## References

1. de Klerk, D., Rixen, D.J., and Voormeeren, S. "General framework for dynamic substructuring: history, review and classification of techniques". *AIAA journal*, 46(5):1169–1181 (2008)
2. Allen, M.S., Rixen, D., Van der Seijs, M., Tiso, P., Abrahamsson, T., and Mayes, R.L. *Substructuring in Engineering Dynamics: Emerging Numerical and Experimental Techniques*, volume 594. Springer (2019)
3. Allen, M.S., Mayes, R.L., and Bergman, E.J. - "Experimental modal substructuring to couple and uncouple substructures with flexible fixtures and multi-point connections". *Journal of Sound and Vibration*, 329(23):4891–4906 (2010)
4. Roettgen, D., Seeger, B., Tai, W.C., Baek, S., Dossogne, T., Allen, M., Kuether, R., Brake, M.R., and Mayes, R. "A comparison of reduced order modeling techniques used in dynamic substructuring". In *Dynamics of Coupled Structures, Volume 4: Proceedings of the 34th IMAC, A Conference and Exposition on Structural Dynamics 2016*, pages 511–528. Springer (2016)
5. Mayes, R.L. "Tutorial on experimental dynamic substructuring using the transmission simulator method". In *Topics in Experimental Dynamics Substructuring and Wind Turbine Dynamics, Volume 2: Proceedings of the 30th IMAC, A Conference on Structural Dynamics, 2012*, pages 1–9. Springer (2012)
6. Trainotti, F., Bregar, T., Klaassen, S., and Rixen, D. "Experimental decoupling of substructures by singular vector transformation". *Mechanical Systems and Signal Processing*, 163:108092 (2022)
7. Ocepek, D., Trainotti, F., Čepon, G., Rixen, D.J., and Boltežar, M. "On the experimental coupling with continuous interfaces using frequency based substructuring". *Mechanical Systems and Signal Processing*, 217:111517 (2024)

# Chapter 15
# Assessing the Importance of Contact Joints Relative to Other Sources of Uncertainty in Dynamic Substructuring

**Samuel Choi, Manuel Vega, Tyler Alvis, and Teresa Portone**

**Abstract** Dynamic substructuring enables the analysis of complex mechanical systems by partitioning them into system subcomponents. These subcomponents can be either modeled as numerical or experimental pieces. There are many sources of uncertainty in the dynamic substructuring process, including those arising from numerical modeling choices. Contact joint modeling choices specifically may be influential, especially for high-frequency dynamic quantities of interest. This paper presents a sensitivity analysis to assess the importance of contact joint modeling relative to other sources of uncertainty in the dynamic substructuring process. The study focuses on SEM's frame-wing benchmark structure, with the aim of providing suggestions to improve the fidelity of dynamic substructuring simulations.

**Keywords** Dynamic substructuring · Contact joint modeling · Uncertainty Quantification · SEM Round Robin · Contact

## Introduction

Modal analysis provides critical insight into the dynamic behavior of engineered systems. However, numerous uncertainties can influence the outcomes of this analysis in practice. Traditionally, modal analysis involves testing the complete system as a unified entity, meaning that any modifications necessitate a comprehensive rebuild and retesting of the entire system. In this context, numerical modal analysis serves as a complementary approach, utilizing computational methods such as finite element analysis (FEA) to determine the dynamic characteristics of mechanical systems [1]. This technique involves discretizing a structure into smaller elements, enabling the calculation of natural frequencies, mode shapes, and damping ratios.

Dynamic substructuring offers an alternative method when experimental analysis of the entire structure is impractical, particularly in cases where the structure is too large or complex. In some instances, it may not be feasible to experimentally test the entire structure, leading to scenarios where part of the structure is modeled experimentally while other parts are modeled numerically [2, 3, 4]. Dynamic substructuring enables the integration of subcomponents to accurately predict the dynamics of the assembled system, thereby reducing both testing time and associated costs.

Despite its advantages, dynamic substructuring is accompanied by various uncertainties. These uncertainties can arise from multiple sources, including measurement errors when using experimental data for a subcomponent, input uncertainties in finite element (FE) models for numerical data, and ambiguous choices in the substructuring process—such as where to apply constraints, how many constraints to implement, and how many modes to include from each subcomponent. Understanding the influence of these uncertainties on our quantities of interest is essential for assessing the confidence we can place in the results. Additionally, identifying the key factors driving uncertainty is crucial for focusing efforts on their reduction.

Samuel Choi · Manuel Vega
Verification, Validation, and Uncertainty Quantification Processes Department, Sandia National Laboratories, Albuquerque, NM 87123
e-mail: samchoi@sandia.gov; mavega@sandia.gov

Tyler Alvis
Analytical Structural Dynamics Department, Sandia National Laboratories, Albuquerque, NM 87123
e-mail: tylalvi@sandia.gov

Teresa Portone
Optimization and Uncertainty Quantification Department, Sandia National Laboratories, Albuquerque, NM 87123
e-mail: tporton@sandia.gov



Walter D'Ambrogio, et al. (eds.), *Dynamic Substructuring & Transfer Path Analysis, Vol. 4*, Conference Proceedings of the Society for Experimental Mechanics Series, https://doi.org/10.13052/9788743801498-15

Real-world contact surfaces are rarely flat, and when a surface is fastened by multiple bolts, uneven pressure distribution occurs due to variations in surface roughness and torquing specifications. These variations can significantly affect the modal characteristics of the system. In [5] these joints were modeled as spring contacts tuned to experimental data. In this paper, we will discuss the results that compare the importance of contact joint modeling relative to other sources of uncertainty within the dynamic substructuring framework.

Last year, we investigated the impact of FE model input uncertainties on dynamic substructuring quantities of interest in the context of the Dynamic Substructuring SEM Round Robin problem [6]. However, we did not explore the effects of substructuring uncertainties, such as the number of modes to include from each subcomponent, nor did we examine the implications of joint model choices. Joint modeling presents particular challenges due to the complex interactions at the interfaces, which can significantly influence the overall dynamic response of the system. This year, we expand our investigation to incorporate these additional sources of uncertainty, aiming to provide a more comprehensive understanding of their effects on dynamic substructuring outcomes.

## Background

The objective of dynamic substructuring is to analyze a multi-component structure by breaking it down into smaller, more manageable subcomponents or substructures. One method to do this is to use a transmission simulator [7, 8]. Let's assume we have two subcomponents, A and B. If A and B share a common fixture, they can be connected via this fixture. This common fixture, known as a transmission simulator (TS), facilitates the analysis of the integration between subcomponents A and B. The TS acts as a bridge, enabling the study of combined behavior of A and B. This concept is illustrated in Figure 1.

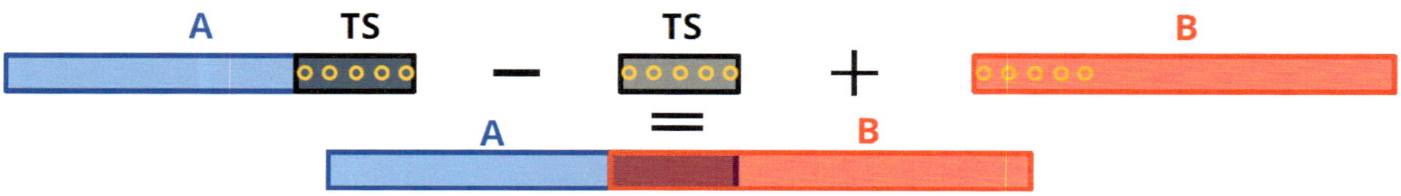


**Fig. 1** Visualization of Transmission Simulator (TS)[5]

To get the modal characteristics, we have to solve a modal system of equations. The general form of the uncoupled modal equations of motion for a subcomponent is:

$$\mathbf{M}_i\ddot{\mathbf{q}}_i(t) + \mathbf{C}_i\dot{\mathbf{q}}_i(t) + \mathbf{K}_i\mathbf{q}_i(t) = \mathbf{F}_i(t) \tag{1}$$

where:

- $\mathbf{M}_i$ is the mass matrix,
- $\mathbf{C}_i$ is the damping matrix,
- $\mathbf{K}_i$ is the stiffness matrix,
- $\mathbf{q}_i(t)$ is the modal displacement vector,
- $\mathbf{F}_i(t)$ is the external force vector.

Each subcomponent is characterized by a set of uncoupled modal equations, each defined by its own modal characteristics: linear natural frequency ($\omega$), damping ratio ($\zeta$), and mode shape ($\phi$). By combining the three equations and moving the mass terms, Equation (1) can be written as Equation (2) in terms of subcomponents A, B and TS. Note that the negative TS terms correlate to the removal of the TS subcomponent in Figure 1.

$$\begin{aligned}
&\begin{bmatrix} \mathbf{I}_A & 0 & 0 \\ 0 & \mathbf{I}_B & 0 \\ 0 & 0 & -\mathbf{I}_{TS} \end{bmatrix} \begin{bmatrix} \ddot{\mathbf{q}}_A(t) \\ \ddot{\mathbf{q}}_B(t) \\ \ddot{\mathbf{q}}_{TS}(t) \end{bmatrix} + \begin{bmatrix} [\ddots 2\zeta_A\omega_A \ddots] & 0 & 0 \\ 0 & [\ddots 2\zeta_B\omega_B \ddots] & 0 \\ 0 & 0 & -[\ddots 2\zeta_{TS}\omega_{TS} \ddots] \end{bmatrix} \begin{bmatrix} \dot{\mathbf{q}}_A(t) \\ \dot{\mathbf{q}}_B(t) \\ \dot{\mathbf{q}}_{TS}(t) \end{bmatrix} \\
&+ \begin{bmatrix} [\ddots \omega_A^2 \ddots] & 0 & 0 \\ 0 & [\ddots \omega_B^2 \ddots] & 0 \\ 0 & 0 & -[\ddots \omega_{TS}^2 \ddots] \end{bmatrix} \begin{bmatrix} \mathbf{q}_A(t) \\ \mathbf{q}_B(t) \\ \mathbf{q}_{TS}(t) \end{bmatrix} = \begin{bmatrix} \phi_A^{\top}\mathbf{F}_A(t) \\ \phi_B^{\top}\mathbf{F}_B(t) \\ \phi_{TS}^{\top}\mathbf{F}_{TS}(t) \end{bmatrix}
\end{aligned} \tag{2}$$

To couple the subcomponent, the TS method enforces constraints on the displacements at a subset of degrees of freedom (DOFs) in subcomponent $A$ and $B$, ensuring they match those of the TS. These constraints are transformed into modal space using a modal transformation and softened using the pseudo-inverse of the TS shapes [6]. The coordinate transformation $\mathbf{q} = \mathbf{L}\eta$ is used to enforce these constraints, where $\mathbf{L}$ resides in the null space of the modal constraint matrix. A more in-depth description of the math is mentioned in [5]. The aim of this work is to understand the influence of uncertainties in the dynamic substructuring process.

Uncertainty quantification is the process of identifying, characterizing, and quantifying uncertainty in order to reduce uncertainties in mathematical models. Representing real world phenomena in exact terms is nearly impossible. Uncertainties in models arise from two sources: aleatory uncertainties and epistemic uncertainties. Aleatory uncertainties are due to inherent variability in known variables. In the manufacturing of aluminum, the processes of manufacturing and post-treating metals do not produce an exact Young's modulus or density. The reported values are averages derived from samples taken throughout the entire process. Within the same block, due to grain structure, the values exhibit a range, let alone the variance from bar to bar. This variability increases when the composition and post-treatment processes are unknown. Aleatory uncertainties are most often modeled using a probability distribution. In this case, as in the previous study, since only the range is known, a uniform random variable is assumed over this range. On the other hand, epistemic uncertainties arise from the lack of knowledge. An example in dynamic substructuring is interface modeling uncertainties that arise from the lack of knowledge such as the effects of friction or gaps.

In this study, we employed a forward propagation method to analyze the influence of uncertainties on the system under investigation. By evaluating random samples of uncertain inputs, we determined their corresponding outputs, capturing the range of possible outcomes and associated uncertainties. To achieve this, we utilized Latin Hypercube Sampling (LHS) to generate the input samples. FE analysis was conducted using Sierra/SD [9], while dynamic substructuring was performed using the Python library Sdynpy [10]. Uncertainty propagation was carried out using the open-source uncertainty quantification toolkit Dakota [11], with a total of 6,400 samples taken.

Correlation coefficients measures the linear correlation between the input the output. These coefficients range from -1 to 1 where 1 represents a perfect positive linear relationship and -1 represents a perfect negative linear relationship. It is important to note that the Pearson correlation coefficient does not represent the slope of the best-fit line; it only measures the strength and direction of the linear relationship between the variables. Additionally, coefficients may not detect a nonmonotonic or strongly nonlinear dependence of the output on an input. The benefit of using Pearson correlation coefficients over Sobol' indices, for instance, is that they do not require a large sample size.

The Pearson correlation coefficient $r$ between two variables $X$ and $Y$ is defined as follows:

$$r = \frac{\mathrm{Cov}(X, Y)}{\sigma_X \sigma_Y} \tag{3}$$

where:

- $\mathrm{Cov}(X, Y)$ is the covariance between $X$ and $Y$,
- $\sigma_X$ is the standard deviation of $X$,
- $\sigma_Y$ is the standard deviation of $Y$.

## Methodology

The frame-wing example problem represents a second iteration of modeling the Dynamic Substructuring SEM Round Robin problem [6, 5]. The models and the location of the 4 bolts are displayed in Figure 2. During the initial study, the dynamic substructuring problem modeled the highlighted connection using springs. In this iteration, it is modeled as a contact surface. The initial configuration (i.e., Frame + Thin Wing) featured a frame with a thin wing bolted via four bolts. The thin wing, which served as a transmission simulator, is replaced with a thicker wing.

For the material properties, including Young's modulus and density, known ranges were sourced from Granta MI [12]. Due to the unknown specific composition of the aluminum, a range encompassing a wide range of compositions was used. The ranges are listed in Table 1.

The contact areas were determined by the radii at the four bolt locations, with each bolt having an independent contact radius. The contact constraints were defined by the size of the bolt head on the lower range and the frame on the upper range. Tied contact was employed to join the two surfaces together for the duration of the simulation. This method constrains the

---

Number of modes were found using the random variable (RV) and Equation (4).

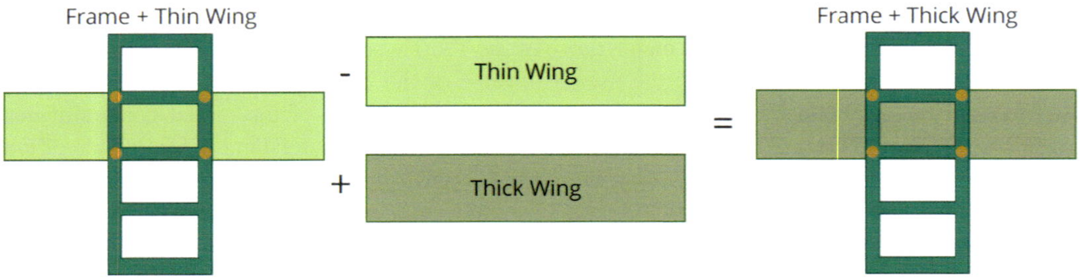


**Fig. 2** Visualization of the Model used for Dynamic Substructuring [5]

**Table 1** Uncertainty types and their corresponding uncertain input distributions.

| Uncertainty Type | Uncertain Input | Distribution |
|---|---|---|
| **Material properties** | Young's modulus $E_{\text{al}}, E_{\text{al\_wing\_thick}}, E_{\text{al\_wing\_thin}}$ [GPa] | $\text{U}[66.6, 74.5]$ |
| | Density $\nu_{\text{al}}, \nu_{\text{al\_wing\_thick}}, \nu_{\text{al\_wing\_thin}}$ [kg/m³] | $\text{U}[2694.3, 2824.6]$ |
| **Geometric properties** | Contact radius $R_1, R_2, R_3, R_4$ [mm] | $\text{U}[4.19, 5.46]$ |
| **Substructuring** | Number of modes $RV_{\text{ThinFrame}}, RV_{\text{Thick}}, RV_{\text{Thin}}$ [-] | $\text{U}[0, 1]$[1] |

nodes on one surface to match the displacement of the corresponding nodes on the opposing surface. Figure 3(a) shows the maximum contact area, with the red section indicating the region where contact occurs. Conversely, Figure 3(b) illustrates the minimum contact area, also highlighted in red. The green sections indicate partial contact, while the blue sections indicate no contact. The irregular shapes of the other sections are due to meshing artifacts and should be disregarded.

(a) Maximum contact radius

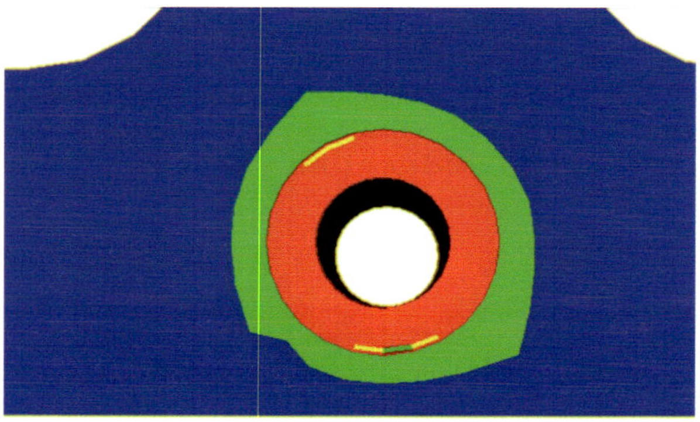

(b) Minimum contact radius

**Fig. 3** Minimum and maximum contact area

To determine the appropriate number of modes for each configuration, we considered several methodologies, including manual selection and the use of a loss function. However, given the complexity and the reliance on subject matter expert (SME) judgment typically required for mode selection, we opted for sampling and systematic evaluation of a wide range of possible modes. For each number of mode variable, a uniform random variable between 0 and 1 is sampled. This uniform random variable is used to select a point within predefined bounds. The actual number of modes selected for substructuring is then set based on these bounds and the sampled uniform random variable. For the frame plus thin wing configuration, the bounds are defined to ensure that the number of modes includes at least one mode at 1600 Hz in the shape set, with an upper limit of 50.

The Number of Modes (Nmodes) value is calculated using the formula:

$$\text{Nmodes} = U \times (\text{upper bound} - \text{lower bound}) + \text{lower bound} \tag{4}$$

where U is the sampled uniform random variable between 0 and 1. For the thick wing and thin wing configurations, the lower bound is set to 7 (accounting for 6 rigid modes plus one flexible mode), and the upper bound is the number of modes selected for the frame plus thin wing configuration. Nmodes value is then rounded to the nearest integer.

Before conducting a detailed analysis, we first investigated the effects of uncertainties in the FE model. There are a few key differences between the FE model and the experimental model: the FE model utilized simplified bolts with the threads removed and without nuts, whereas the experimental model featured an aluminum frame with steel threaded inserts and standard bolts. This analysis was performed using Sierra/SD FE models [9]. A Modal Assurance Criterion (MAC) was calculated between the mode shapes obtained from each of the subcomponent FEM models and the experimental mode shapes. The resulting MAC values are displayed in Figure 4. Darker squares represent higher MAC scores, meaning there is a stronger similarity between the numerical and experimental mode shapes. This indicates a higher degree of similarity in the modal shapes, suggesting that the modeled FE models closely match the experimental results. The high scores across the diagonal for both the thin wing and thick wing configurations indicate a similarity in all mode shapes. However, in the frame plus thin wing configuration, the MAC score drops along the diagonal as the frequency increases, particularly around the frequency of 1400 Hz. The first 44 bending modes were taken into consideration.

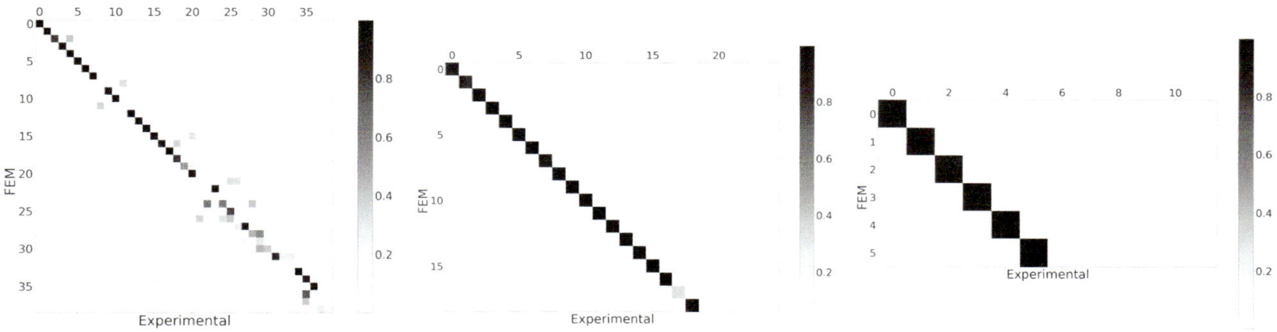


(a) MAC: Frame + Thin Wing (b) MAC: Thin Wing (c) MAC: Thick Wing

**Fig. 4** Modal Assurance Criterion for subcomponents of dynamic substructuring

To gain deeper insights into the input parameters, a Pearson correlation analysis was performed to assess the relationship between the input parameters (i.e., model parameters and substructuring parameters) and the natural frequencies across each of the considered flexible modes. Figure 5 illustrates the correlation coefficients between the input parameters and the non-rigid natural frequencies, with dark red indicating a perfect positive correlation and dark blue representing a perfect negative

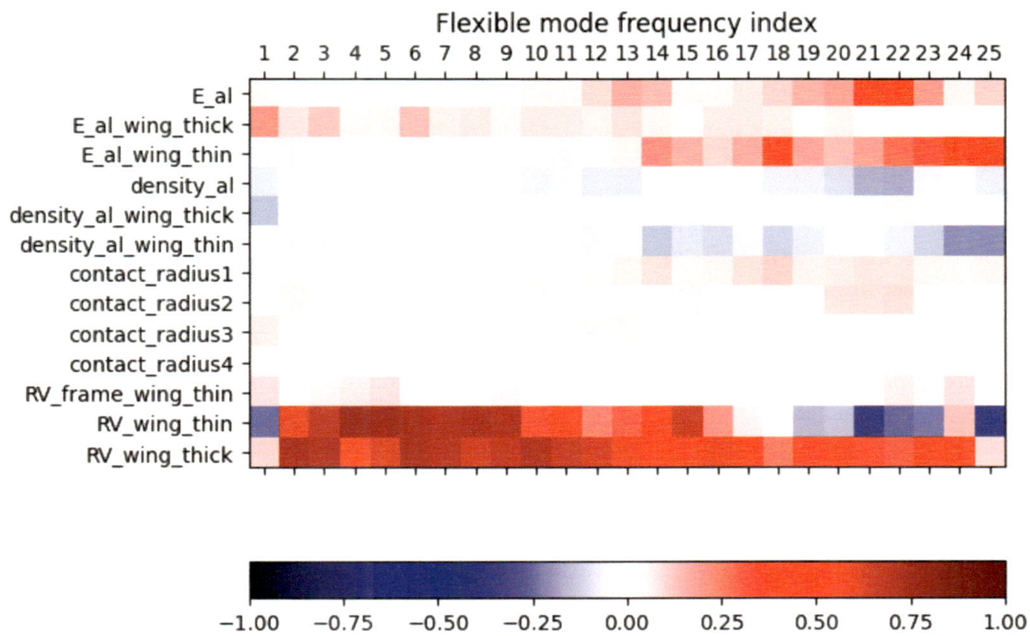


**Fig. 5** Correlation coefficients across flexible modes

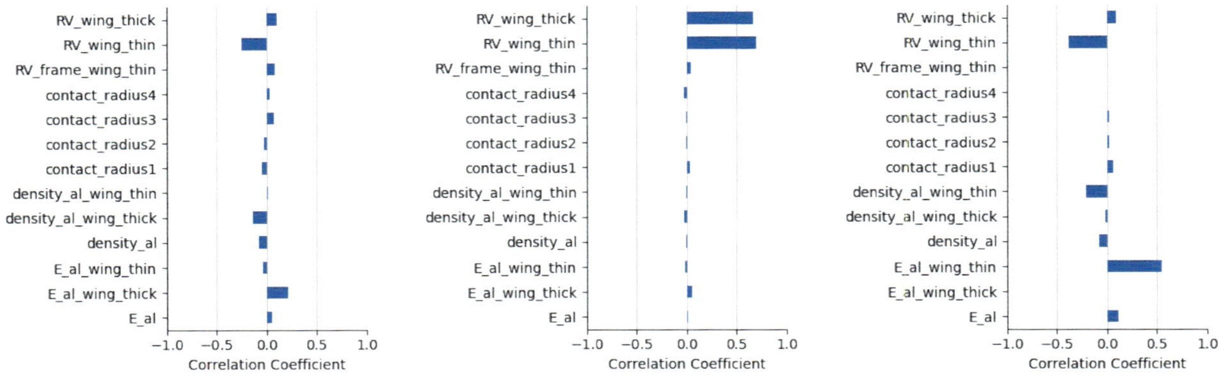


(a) Flexible mode 1 (b) Flexible mode 9 (c) Flexible mode 25

**Fig. 6** Correlation coefficient for individual flexible modes

correlation. In most flexible modes, a significant positive correlation was observed between the random variables selected for the substructuring of the removed thin wing and the added thick wing. This indicates that variations in the substructuring parameters are linked to changes in the natural frequencies, underscoring the importance of these variables in the system's dynamic behavior. Moreover, at higher frequencies, the correlation between the number of modes selected from the thin wing and the natural frequencies turns negative. Additionally, Young's modulus in the thin wing shows a stronger positive correlation with the higher natural frequencies.

Visualizing the correlation coefficients for flexible modes 1, 9, and 25 individually highlights representative patterns, as shown in Figure 6(a). The correlation coefficients for the first flexible mode (or first natural frequency) do not exhibit high values, indicating a lack of strong linear correlation with the input variables. In contrast, Figure 6(b) demonstrates a strong correlation between the natural frequency and the number of modes chosen when removing the thin wing and adding the thick wing, suggesting that substructuring parameters play a significant role in this natural frequency. Furthermore, Figure 6(c) illustrates that for the 25th natural frequency, there is a strong negative correlation between the natural frequency and the Young's modulus of the thin wing, highlighting the influence of material properties on the dynamic behavior of the system at this frequency.

## Conclusion

In the previous study [5], the number of modes selected for substructuring was not included. We predicted that this would be a major source of uncertainty, and this was verified during this study. Based on this study, the other variables did not have a significant effect compared to the number of modes selected. The correlation coefficients between the number of modes and other variables were found to be close to zero, indicating minimal association under the current conditions. This may change at higher frequencies as the influence of the number of modes diminishes. It is also possible that the high correlation coefficient of mode selection could obscure the correlations between other variables and the natural frequencies.

We could validate our selection of modes because we can compare the substructured model against a full model of a system with the updated component (e.g., after adding component B and removing component A). However, in scenarios where full validated models are unavailable, validating the substructured model becomes challenging. Therefore, we need to find a method to select the modes more intelligently. This will help prevent the dominant effect of uncertainty in mode selection from overshadowing other uncertainties. Historically, modes have been manually picked or sampled from a distribution, but these approaches may not be optimal or feasible.

Although contact radius don't seem to have a high correlation for this model, models that incorporate a greater number of joints or more complex geometries could potentially exhibit different behaviors. The increased complexity may introduce additional interactions and dependencies among the variables, which could lead to a stronger correlation between contact uncertainties and natural frequencies. Therefore, further exploration into models with varying configurations and joint arrangements is warranted. This could enhance our understanding of how contact conditions influence dynamic performance and may reveal critical insights that were not apparent in this study.

Dynamic substructuring and its associated uncertainties remain the subject of ongoing work. We plan on conducting more samples and comparing correlations using other methods, such as Sobol indices. Future work involves increasing the contact radius and studying the effects this has on the modal frequency. In the current study, the contacts were circular and of uniform shape; this was to simplify initial calculations. However, in practice, contact is rarely nicely rounded and constant. Many times, contact occurs in other locations along the frame, not just near the bolts. Additionally, future research will focus on determining the effects of these uncertainties on the frequency response functions, which will be the major focus of our future work.

## Acknowledgments

This article has been authored by an employee of National Technology & Engineering Solutions of Sandia, LLC under Contract No. DE-NA0003525 with the U.S. Department of Energy (DOE). The employee owns all right, title and interest in and to the article and is solely responsible for its contents. The United States Government retains and the publisher, by accepting the article for publication, acknowledges that the United States Government retains a non-exclusive, paid-up, irrevocable, worldwide license to publish or reproduce the published form of this article or allow others to do so, for United States Government purposes. The DOE will provide public access to these results of federally sponsored research in accordance with the DOE Public Access Plan https://www.energy.gov/downloads/doe-public-access-plan.

## References

1. Black, J.K., Callis, S.J., Feizy, A., Johnson, C.L., Lieven, N.A., and Vega, M.A. "A simplified finite element joint model updated with experimental modal features". In *Proceedings of the Society for Experimental Mechanics Annual Conference and Exposition*, pages 107–123. Society for Experimental Mechanics, Springer (2023)
2. De Klerk, D., Rixen, D.J., and Voormeeren, S. "General framework for dynamic substructuring: history, review and classification of techniques". *AIAA journal*, 46(5):1169–1181 (2008)
3. Vega, M.A., Schellenberg, A.H., Caudana, H., and Mosqueda, G. "Implementation of real-time hybrid shake table testing using the ucsd large high-performance outdoor shake table". *International Journal of Lifecycle Performance Engineering*, 4(1-3):80–102 (2020)
4. Allen, M.S., Gindlin, H.M., and Mayes, R.L. "Experimental modal substructuring to estimate fixed-base modes from tests on a flexible fixture". *Journal of Sound and Vibration*, 330(18-19):4413–4428 (2011)
5. Portone, T., Roettgen, D., Neal, K., and Debusschere, B. "A preliminary quantification of uncertainties in dynamic substructuring and the frame-wing problem". In *Proceedings of the IMAC Conference*. Society for Experimental Mechanics (2024)
6. Roettgen, D.R., Lopp, G.K., and Linderholt, A. "Technical division benchmark structure for dynamic substructuring.". Technical report, Sandia National Lab.(SNL-NM), Albuquerque, NM (United States) (2020)
7. Mayes, R.L. and Arviso, M. "Design studies for the transmission simulator method of experimental dynamic substructuring". In *international seminar on modal analysis (ISMA2010), Lueven, Belgium* (2010)
8. Mayes, R., Rixen, D., Griffith, D., De Klerk, D., Chauhan, S., Voormeeren, S., and Allen, M. "Topics in experimental dynamics substructuring and wind turbine dynamics, volume 2: Proceedings of the 30th imac, a conference on structural dynamics". In *Proceedings of the 30th IMAC, A Conference on Structural Dynamics*, volume 2, pages 21–31. Springer (2012)
9. Crane, N., Day, D., Dohrmann, C., Stevens, B., Lindsay, P., Plews, J., Vo, J., Bunting, G., Walsh, T., and Joshi, S. "Sierra/sd-user's manual-5.10". Technical report, Sandia National Laboratories (2022)
10. Rohe, D.P. *SDynPy* (2021) https://github.com/sandialabs/sdynpy.
11. Adams, B. et al. *Dakota 6.18.0 documentation* (2023) Technical Report SAND2023-03367O, Sandia National Laboratories, Albuquerque, NM.
12. Design, G. "Granta mi" (2023) Software. https://www.grantadesign.com/.

# Chapter 16
# High-frequency Dynamic Characterization of Rubber Mounts through an Enhanced Virtual Point Transformation

**Bart Forrier, Fabio Bianciardi, and Karl Janssens**

**Abstract** Rubber mounts are critical components in the optimization of vehicle designs towards noise, vibration and harshness performance. In virtual prototyping applications, they are often represented by experimentally identified dynamic stiffness curves. The latter are extracted, partly via inverse substructuring or decoupling, from frequency response function measurements. Dedicated fixtures are hereby instrumental: excitation and response measurements are distributed over the fixture. Afterward, the so-called virtual point transformation condenses them into six degrees of freedom at one central point on each side of the mount, under the assumption that the fixtures behave rigidly. As the frequency range of interest reaches up to or beyond the first natural frequencies of the fixtures, though, the dynamics of the fixtures pollute the identified mount stiffness curves. Therefore, this work aims to increase the accuracy of rubber mount characterizations toward high frequencies through an enhanced virtual point transformation. This enhanced transformation considers fixture deformation, though it finally preserves the condensation into six degrees of freedom per side. The latter feature is critical for smoothly integrating the experimental mount model into various virtual prototype assemblies. Through a numerical validation, this work shows, firstly, that finite element model based natural mode shapes of the fixture form a well-suited basis for capturing its deformation. A subsequent experimental validation deals with the characterization of an actual automotive rubber mount, accounting for one flexible mode of the fixture assembly. The results demonstrate that the proposed enhancement indeed extends the useful identification bandwidth, to beyond the first natural frequency introduced by the fixture assembly.

**Keywords** Mount stiffness · High-Frequency · Experimental · Dynamic substructuring

## Introduction

Noise, vibration, and harshness (NVH) performance is an important differentiator in today's automotive vehicle market. Whilst customer expectations increase continuously, engineering efforts and costs must kept as low as possible. In relation to NVH engineering, the field of virtual acoustic prototyping has emerged to tackle those conflicting requirements [1]. Nowadays, virtual prototype assembly (VPA) technology supports NVH engineers with predictive NVH performance analyses [2]. This technology is based on component-based transfer path analysis (C-TPA) methods [3]. Dynamic substructuring (DS) is hereby employed to determine the structural-vibration characteristics of an assembly from those of its constituent components [4, 5].

The advent of hybrid and battery electric vehicles (BEV) has brought additional NVH challenges, in large part originating from the enlarged frequency bandwidth over which an electric motor excites both acoustic and structural noise transfer paths [2]. Accordingly, post-factum or predictive TPA are now required to be accurate up to several kHz, also for what regards the structure-borne paths. Because these paths often include rubber mounts, the latter's vibration characteristics must be known accurately over an ever larger frequency bandwidth – exceeding those attainable using standard practice [6].

In a VPA framework, inter-component connections are represented as point connections with a maximum of 6 degrees of freedom (DOF), Therefore, a complete mount characterization relates applied forces and moments to vibration responses using 6 degrees of freedom (DOF) on each side, Several methods exist to determine such twelve DOF mount models experimentally [7]. They rely on the so-called virtual point transformation (VPT) to project a set of measured non-collocated frequency response functions (FRF) onto the six DOF associated with one connection point [8]. In practice, a fixture (often cross-shaped and made out of steel or aluminium) is attached to each side of the mount. The role of the fixture is to

Bart Forrier · Fabio Bianciardi · Karl Janssens
RTD Department of Test and Smart Vehicle Technology, Siemens Digital Industries Software NV, Leuven, Belgium
e-mail: bart.forrier@siemens.com; fabio.bianciardi@siemens.com; karl.janssens@siemens.com



Walter D'Ambrogio, et al. (eds.), *Dynamic Substructuring & Transfer Path Analysis, Vol. 4*, Conference Proceedings of the Society for Experimental Mechanics Series, https://doi.org/10.13052/9788743801498-16

facilitate excitation and response measurements, ensuring proper excitation and observation of all six DOF [7]. The addition of such fixtures does not change the obtained dynamic transfer stiffness of the mount [9]. The full dynamic stiffness matrix of only the mount can also be obtained, via the so-called round-trip identity method [9, 10], inverse substructuring [10, 7] or structural decoupling through frequency based substructuring (FBS)[7].

The aforementioned methods yield accurate dynamic characterizations at frequencies up to about 1 kHz. As the frequency range of interest reaches into the multi-kHz range, though, the fixtures start to behave as flexible components [7]. This violates the assumption of rigidity which is at the heart of the VPT [8]. Making a fixture more rigid, though, would yield additional drawbacks related to the addition of mass. Firstly, it attenuates the vibration response at higher frequencies, which leads to noisy FRF measurements [9]. Secondly, obtaining the point-wise dynamic mount stiffness values requires one to subtract the effect of the crosses from the measured assembly stiffness. The latter decoupling problem is ill-conditioned for a combination of higher frequencies and/or larger mass values, leading to amplification of phase measurement errors [7].

In the context of FBS, extensions to VPT have been proposed in order to relax on the requirement of local rigidity around the connection points [11, 12]. In essence, this is done by extending the original set of 6 rigid interface displacement modes with additional interface deformation modes (IDM). Depending on the application case, a limited set of pre-defined IDM may be cherry-picked from a library of general deformation modes [11]. Alternatively, the expected strain field around the interface may motivate one to assume local beam- or plate-like behavior. In such case, the corresponding IDM can be obtained by fitting polynomial beam or plate deformation models onto the measured data [12]. These extensions have targeted FBS applications where matching additional flexible DOF are introduced to both components to be coupled.

The addition of extra DOF is not feasible when using VPA for design optimization, however. As different component models may originate from different teams or vendors, the inclusion of the same amount of flexible DOF on either side would likely still result in incompatible interface deformation representations. Moreover, the entire library of variants or replacement components should be made compatible with the extra DOF introduced in one adjacent component, which is not at all feasible. Keeping in mind the VPA application, this work therefore proposes to complement the extension of the VPT with flexible DOF, with a subsequent reduction to maintain six DOF per connection point. We propose to extend the VPT with normal modes obtained from a finite element model (FEM) of the fixtures. The FEM based shapes can be accurate yet obtained with relative ease, thanks to the simple geometry and monolithic nature of typical fixtures. The subsequent reduction to six DOF, though not advisable in the general context of DS, will prove to be well-suited for the purpose of mount characterization.

The remainder of this paper is organized as follows. The following section reviews current practice of mount characterization via the VPT. Subsequently, an enhanced VPT is proposed, accounting for fixture flexibility while maintaining six DOF per connection point. Next, this approach is validated numerically and experimentally. Finally, the main conclusions are summarized.

## Dynamic mount characterization through the virtual point transformation

In the context of this work, mount characterization refers to determining the dynamic mount stiffness matrix $\mathbf{Z} \in \mathbb{R}^{12\times 12}$ for a pre-determined frequency range of interest [6]. This matrix is defined per frequency $f$, or pulsation $\omega = 2\pi f$, by equation 1. It describes a linear steady-state relation between the generalized force vectors $m_1, m_2 \in \mathbb{R}^{6\times 1}$ acting on both connection points and the corresponding displacement vectors $q_1, q_2 \in \mathbb{R}^{6\times 1}$. Both interfaces to adjacent components are thus represented in a lumped manner, such that a complete description requires 3 translational and 3 rotational DOF per connection point.

$$\begin{bmatrix} m_1 \\ m_2 \end{bmatrix} = \underbrace{\begin{bmatrix} \mathbf{Z_{11}} & \mathbf{Z_{12}} \\ \mathbf{Z_{21}} & \mathbf{Z_{22}} \end{bmatrix}}_{\mathbf{Z}(\omega)} \begin{bmatrix} q_1 \\ q_2 \end{bmatrix} \tag{1}$$

Besides the frequency of excitation, the mount stiffness also depends on the static preload, the temperature, and the vibration amplitude ( see e.g. [7]). In general, equation 1 thus defines a linearized stiffness matrix, obtained e.g. under the static pre-load corresponding to the assembly design, for a grid of different pre-load and temperature values, etcetera.

The main diagonal blocks describe the relation between an excitation and a response on the same side (with the other side held fixed), whereas the off-diagonal blocks describe the *transfer dynamics* from one side to another. Under the assumption of conservation of force, i.e. $m_1 = -m_2$, a direct consequence is that $\mathbf{Z_{11}} = -\mathbf{Z_{21}}$ and $\mathbf{Z_{22}} = -\mathbf{Z_{12}}$. This assumption is valid below the first internal resonance of the mount. In that frequency range, it simplifies the mount characterization because only the transfer stiffness dynamics must be identified [10]. Moreover, the transfer stiffnesses ($\mathbf{Z_{21}}$ and $\mathbf{Z_{12}}$) are not affected by the eventual addition of mass, or even more complex dynamics, on either side of the mount [9, 10].

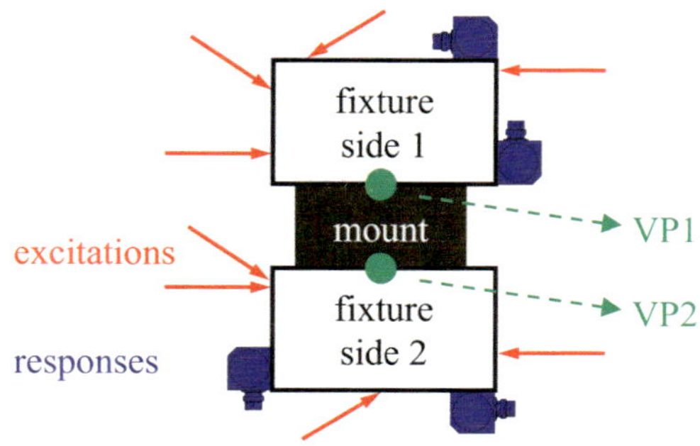


**Fig. 1** Schematic of a measurement setup, with two virtual points and their associated excitation and response DOF.

Experimental methods exist to identify parts of $\mathbf{Z}$ in a direct manner, e.g. fixing $q_2 = 0$ and measuring parts of $q_1, m_2$ on off-the-shelf available hydraulic testing machines (see e.g. [6, 7]). Two notable diffulties arise, though. The first is the requirement to fix all DOF on one side. The second is the fact that exciting and measuring rotational DOF is problematic in practice. The first difficulty is relaxed by noting that in any assembly, such as the one of Figure 1, the characteristics of the mount itself can be determined indirectly from measured FRF data, via sub-structure decoupling or inverse substructuring techniques [10, 7]. The second difficulty is circumvented by exciting and measuring in various points and directions, cfr. Figure 1, such that all 12 DOF of the model (equation 1) are excited and measured. For each side of the mount and for both references and responses, the measured data must then be projected onto a 6-dimensional subspace corresponding to the 6 DOF of a virtual point. A widely used projection is the so-called virtual point transformation [8]. Because one VPT, linked to one virtual point, is defined per side of the mount, we use the generic subscript $k \in \{1, 2\}$ to distinguish between both sides. It is noteworthy that when not excited, and unlike the general depiction of Figure 1, both virtual points usually coincide by definition.

The original VPT assumes that all $n_u > 6$ measurement DOF behave as one rigid body, and likewise for all $n_f > 6$ excitation DOF. Under this assumption, all physically measured data on side $k$ can be thought of as being extrapolated from six (virtual) response DOF and six (virtual) excitation DOF, cfr. equation 2.

$$u_k = \mathbf{R}_{u,k} q_k, \qquad m_k = \mathbf{R}_{f,k}^T f_k, \qquad k \in \{1, 2\} \tag{2}$$

In the matrices $\mathbf{R}_u \in \mathbb{R}^{n_u \times 6}$ and $\mathbf{R}_f \in \mathbb{R}^{n_f \times 6}$, each column corresponds to a rigid-body mode. These matrices are constructed from geometric information, namely the positions and orientations of the evaluated DOF with respect to the virtual point. Their pseudo-inverses, denoted with $\cdot^\dagger$, form the projection matrices $\mathbf{T}_u$ and $\mathbf{T}_f^T$. Those projections are applied respectively on the responses ($u$) and references (forces $f$), as in equation 3.

$$q_k = \mathbf{R}_{u,k}^\dagger u_k \triangleq \mathbf{T}_{u,k} u_k, \qquad f_k = \left(\mathbf{R}_{f,k}^T\right)^\dagger m_k \triangleq \mathbf{T}_{f,k}^T m_k \tag{3}$$

By applying the VPT to the FRF data as in equation 4, one obtains the projected FRF matrix $\mathbf{H}_{qm} \in \mathbb{R}^{12 \times 12}$. By inverting the latter, one may readily obtain the coupled assembly's stiffness matrix $\mathbf{Z}^\mathbf{C}$ [10]. Assuming acceleration-level responses, this is made explicit in equation 5. The off-diagonal blocks in $\mathbf{Z}^\mathbf{C}$ already describe the transfer dynamics of the mount itself, i.e. $\mathbf{Z}_{12}^\mathbf{C} = \mathbf{Z}_{12}$ and $\mathbf{Z}_{21}^\mathbf{C} = \mathbf{Z}_{21}$. One may derive $\mathbf{Z}_{11}$ and $\mathbf{Z}_{22}$ from $\mathbf{Z}^\mathbf{C}$ via substructure decoupling or inverse substructuring [10].

$$\mathbf{H}_{qm} = \begin{bmatrix} \mathbf{H}_{11} & \mathbf{H}_{12} \\ \mathbf{H}_{21} & \mathbf{H}_{22} \end{bmatrix} = \mathbf{T}_u \mathbf{H}_{uf} \mathbf{T}_f^T = \begin{bmatrix} \mathbf{T}_{u,1} & \mathbf{0} \\ \mathbf{0} & \mathbf{T}_{u,2} \end{bmatrix} \begin{bmatrix} \mathbf{H}_{u_1 f_1} & \mathbf{H}_{u_1 f_2} \\ \mathbf{H}_{u_2 f_1} & \mathbf{H}_{u_2 f_2} \end{bmatrix} \begin{bmatrix} \mathbf{T}_{f,1} & \mathbf{0} \\ \mathbf{0} & \mathbf{T}_{f,2} \end{bmatrix}^T \tag{4}$$

$$\mathbf{Z}^\mathbf{C}(\omega) = \begin{bmatrix} \mathbf{Z}_{11}^\mathbf{C} & \mathbf{Z}_{12}^\mathbf{C} \\ \mathbf{Z}_{21}^\mathbf{C} & \mathbf{Z}_{22}^\mathbf{C} \end{bmatrix} = \begin{bmatrix} \mathbf{Z}_{11}^\mathbf{C} & \mathbf{Z}_{12} \\ \mathbf{Z}_{21} & \mathbf{Z}_{22}^\mathbf{C} \end{bmatrix} = -\omega^2 \mathbf{H}_{qm}^{-1} = -\omega^2 \begin{bmatrix} \mathbf{H}_{11} & \mathbf{H}_{12} \\ \mathbf{H}_{21} & \mathbf{H}_{22} \end{bmatrix}^{-1} \tag{5}$$

Any error resulting from the VPT (equation 4) is detrimental to the accuracy of the dynamic stiffness characteristics obtained through equation 5. This limits the maximum frequency of identification, because of the assumption that the (instrumented and excited) fixtures behave rigidly. Typically, one verifies this assumption using the so-called *sensor consistency* metric $\rho_u^2$ and *impact consistency* metric $\rho_f^2$ [8]. In equation 6, $H_{f_h}$ denotes a column of $\mathbf{H}$ that corresponds to a distant excitation DOF $f_h$ and $H_{u_i}$ denotes one row, corresponding to the $i$-th response DOF. In equation 7, the superscript $\cdot^H$

denotes the Hermitian transpose.

$$\rho_u^2(f_h) \triangleq \text{MAC}\left(\mathbf{R}_u\mathbf{T}_u H_{f_h}, H_{f_h}\right), \qquad \rho_f^2(u_i) \triangleq \text{MAC}\left(\mathbf{R}_f\mathbf{T}_f H_{u_i}, H_{u_i}\right) \tag{6}$$

$$\text{MAC}(a,b) \triangleq \frac{\left(a^H b\right)\left(b^H a\right)}{\left(a^H a\right)\left(b^H b\right)}, \qquad a, b \in \mathbb{R}^{n\times 1} \tag{7}$$

In this work, we define the average sensor consistency of a VPT to be the sensor consistency (as function of frequency) averaged over all excitations linked to it. *Vice versa*, the average impact consistency of a VPT is the average over all sensors linked to it.

## Enhanced Virtual Point Transformation to Account for Fixture Flexibility

To extend the validity of the VPT towards frequencies near the first eigenfrequency of a fixture, one must add flexible modes to $\mathbf{R}_u$ and $\mathbf{R}_f$. In the context of DS, an ad-hoc selection from a library of flexible modes has been proposed to this end [11]. This works well if one can accurately predict the deformation in the vicinity of the virtual point from engineering insight. Alternatively, a polynomial fit onto the measured flexible deformation may be applied. In general, this leads to an overly large amount of extra DOF $\eta$, being the participation factors to the flexible modes. The allowed strain fields are therefore constrained to those of e.g. an idealized beam or thin plate [12]. Again, one must choose which constraint to apply based on engineering insight. In this case, one cannot add any amount of extra DOF $\eta$, because e.g. assuming plate-like behavior yields two bending modes and one torsion mode, and hence 3 additional DOF.

Besides the choice of deformation mode shapes, the introduction of flexible modes to the VPT causes another major difficulty in the context of virtual prototyping. Namely, to enforce compatibility between two components requires that the mode shapes on both components are matched. In other words, it is not sufficient to decide on a common number of flexible modes. This requirement is not straightforward to fulfill, because different component models may originate from different teams or vendors. For rubber mount models, however, a subsequent projection onto 6 DOF per side may be applied, as explained in the following.

Firstly, we propose that the extension to the VPT be based on eigenmodes of the fixtures. This is an attractive choice because (i) the fixtures used for mount identification are usually easy to model, (ii) the first few eigenmodes of the fixture are not expected to be overly affected by the presence of a rubber mount, and they are insensitive to e.g. a wrongly assumed material density or Young's modulus, and (iii) the problem of selecting which modes to include is trivial; one should just add one or multiple eigenmodes in order of increasing eigenfrequency. Secondly, we propose that the addition of extra (flexible) DOF merely be used to improve the consistency between measured (physical) and generalized (lumped, virtual) responses or excitations. In other words, we aim to consider the effect of the fixtures' deformation on the 6 DOF on each side of the mount model, without adding flexible DOF to the mount model. This approach is justified as long as the local interface between fixture and mount does not deform significantly in the first few eigenmodes of the fixtures, or as long as any such local deformation on one side of the mount does not cause energy to transfer through the mount.

In the following, the FE-based eigenmodes are assumed to be partitioned as in equation 8, where the subscripts indicate the required selection of DOF. One may note the requirement to evaluate the mode shape vectors in the DOF associated to the virtual point. This partition $\phi_q\,(=\phi_m) \in \mathbb{R}^{6\times n_\eta}$ is generally not available from test data, for the same reasons that motivate the use of the VPT. One may straightforwardly extract them from a FEM, though.

$$\phi_k^T \triangleq \begin{bmatrix}\phi_{u,k}^T & \phi_{f,k}^T & \phi_{q,k}^T\end{bmatrix}, \qquad k \in \{1,2\} \tag{8}$$

Through concatenating $\mathbf{R}_u$ and $\mathbf{R}_f$ with the corresponding partitions of the $n_\eta$ flexible modes, as in equation 9, the extended VPT is defined as in equation 10. Its application will require the extended projection matrices $\mathbf{T}^*$, obtained via equation 11. In these equations, the modal participation factors $\rho$ and $\eta$ correspond to the rigid and flexible modes, respectively, and $g_\rho$ and $g_\eta$ denote the associated generalized forces.

$$\mathbf{R}_{u,k}^* \triangleq \begin{bmatrix}\mathbf{R}_{u,k} & \phi_{u,k}\end{bmatrix} \qquad \mathbf{R}_{f,k}^* \triangleq \begin{bmatrix}\mathbf{R}_{f,k} & \phi_{f,k}\end{bmatrix} \tag{9}$$

$$u_k = \mathbf{R}_{u,k}^*\begin{bmatrix}\rho_k\\ \eta_k\end{bmatrix}, \qquad \begin{bmatrix}g_{\rho,k}\\ g_{\eta,k}\end{bmatrix} = \left(\mathbf{R}_{f,k}^*\right)^T f_k \tag{10}$$

$$\begin{bmatrix}\rho_k\\ \eta_k\end{bmatrix} = \left(\mathbf{R}_{u,k}^*\right)^\dagger u_k \triangleq \mathbf{T}_{u,k}^* u_k, \qquad f_k = \left(\left(\mathbf{R}_{u,k}^*\right)^T\right)^\dagger \begin{bmatrix}g_{\rho,k}\\ g_{\eta,k}\end{bmatrix} \triangleq \left(\mathbf{T}_{f,k}^*\right)^T \begin{bmatrix}g_{\rho,k}\\ g_{\eta,k}\end{bmatrix} \tag{11}$$

A mere substitution of the projection matrices $\mathbf{T}$ with $\mathbf{T}^*$ in equation 4 would not yield the desired FRF matrix $\mathbf{H}_{qm} \in \mathbb{R}^{12\times 12}$. Instead, the projected FRF matrix would contain $12 + n_\eta$ rows and columns, due to the additional flexible DOF. Aiming to preserve only 6 DOF per virtual point, the flexible contribution to the response vector $q \in \mathbb{R}^{6\times 1}$ is evaluated at the virtual point and added to the rigid contribution $\rho$, cfr. equation 12. Regarding the excitation, equation 12 states that the excitation vector $m \in \mathbb{R}^{6\times 1}$, while acting on the virtual point, excites both rigid and flexible modes of the fixture. In relation to the measured FRF matrix, this addition boils down to a subsequent projection that maps the contribution of the (measured and excited) flexible modes onto the preserved 6 DOF of each virtual point, as in equation 13.

$$q_k = \rho_k + \phi_{q,k}\eta_k, \qquad \begin{bmatrix} g_{\rho,k} \\ g_{\eta,k} \end{bmatrix} = \begin{bmatrix} m_k \\ \phi_{q,k}^T m_k \end{bmatrix} \tag{12}$$

$$\mathbf{H}^*_{qm} = \begin{bmatrix} \mathbf{I} & \phi_{q,1} & \mathbf{0} & \mathbf{0} \\ \mathbf{0} & \mathbf{0} & \mathbf{I} & \phi_{q,2} \end{bmatrix} \mathbf{T}^*_u \mathbf{H}_{uf} \left(\mathbf{T}^*_f\right)^T \begin{bmatrix} \mathbf{I} & \phi_{q,1} & \mathbf{0} & \mathbf{0} \\ \mathbf{0} & \mathbf{0} & \mathbf{I} & \phi_{q,2} \end{bmatrix}^T \tag{13}$$

The here obtained FRF matrix $\mathbf{H}^*_{qm}$ describes how the excitations $m_k$ and (displacement-level) responses $q_k$, *as introduced in equation 1*, relate to one another in the test assembly. It is thus fully equivalent to $H_{qm}$ as obtained through equation 4. Equation 5 thus remains valid, i.e. one must merely substitute $\mathbf{H}_{qm}$ with $\mathbf{H}^*_{qm}$ to obtain the transfer dynamics of the mount itself. In conclusion, the proposed variant to the VPT is not an extended VPT, in the sense that the projected dynamics are captured using 6 DOF on each side of the interface, like is the case for the original VPT. It is enhanced, because in contrast to the original VPT, the flexibility of the fixtures is taken into account.

Finally, one may remark – from the block-diagonal structure of all projection matrices in equation 13 – that the amount of flexible DOF $n_\eta$ may differ from side to side. One may thus apply the here proposed approach on side 1 and the original VPT on side 2. To maintain the generally recommended over-determination factor of 2, one should add two response channels and two excitations per flexible DOF considered. Similarly to the original VPT, the instrumentation and excitation layout can be checked based on the condition numbers of the matrices $\mathbf{R}^*$.

The following section presents a numerical validation, highlighting the benefits of extending the reduction basis with flexible eigenmodes. The experimental validation of the subsequent section demonstrates the added value of the enhanced VPT, which includes the second projection onto 6 DOF, for dynamic mount stiffness identification.

## Numerical Validation of the Enhanced VPT

Figure 2 illustrates the numerical validation setup. It represents a 30 mm thick plate, modeled using linear hexahedral elements of size 5 mm. This 3D model was used previously for validating a polynomial fitting approach, assuming an idealized thin-plate strain field. The here presented validation includes a comparison against that polynomial fitting approach [12]. We consider only the projection of $8 \times 3 = 24$ response DOF ($u_1$) onto the 6 DOF ($q_1$) of 1 virtual point, for a single excitation DOF $f_h$, cfr. equation 14. For this single-sided projection, equation 13 simplifies into 14. This simplification does not affect the validation, because dealing with a perfectly reciprocal model, one may interchange the notions of response DOF and reference DOF.

$$\mathbf{H}^*_{q_1 f_h} = \begin{bmatrix} \mathbf{I} & \phi_{q,1} \end{bmatrix} \mathbf{T}^*_{u,1} \mathbf{H}_{u_1 f_h} = \begin{bmatrix} \mathbf{I} & \phi_{q,1} \end{bmatrix} \begin{bmatrix} \mathbf{R}_{u,1} & \phi_{u,1} \end{bmatrix}^\dagger \mathbf{H}_{u_1 f_h} = \begin{bmatrix} \mathbf{I} & \phi_{q,1} \end{bmatrix} \begin{bmatrix} \mathbf{H}_{\rho_1 f_h} \\ \mathbf{H}_{\eta_1 f_h} \end{bmatrix} \tag{14}$$

The focus of this numerical validation is on the choice of extending $\mathbf{R}_u$ with flexible eigenmodes $\phi_u$. Figure 3 shows that with only a single extra DOF, being the participation factor to one flexible eigenmode, a significantly higher sensor consistency is obtained than with three extra DOF in the case of the polynomial plate model fit. It must be noted that in this case, the single extra DOF is chosen to correspond to the second bending mode. This choice is made because, as one can see in Figure 2, the selected responses are located such that the first bending mode introduces little deformation. Even so, the benefits of using flexible eigenmodes are clear: with an equal amount of flexible DOF ($n_\eta = 3$), the reduction basis extended with eigenmodes clearly outperforms the one based on the polynomial plate model. Using $m$ perfectly identified mode shapes, the consistency is virtually perfect up to the $m$-th eigenfrequency. Noting that in Figure 3, eigenfrequencies $f_1$ to $f_{12}$ are marked by vertical lines, this is illustrated also for $m = 12$. In case of closely spaced modes (around 900 Hz), this holds approximately.

Another significant benefit in practical applications, is the fact that one can freely select the amount of relevant DOF to consider. In Figure 3, for example, it is clear that by accounting for only two bending modes ($n_\eta = 2$), one can already capture very accurately the local flexibility observed in the region of indicator points up to about 1500 Hz. In this case,

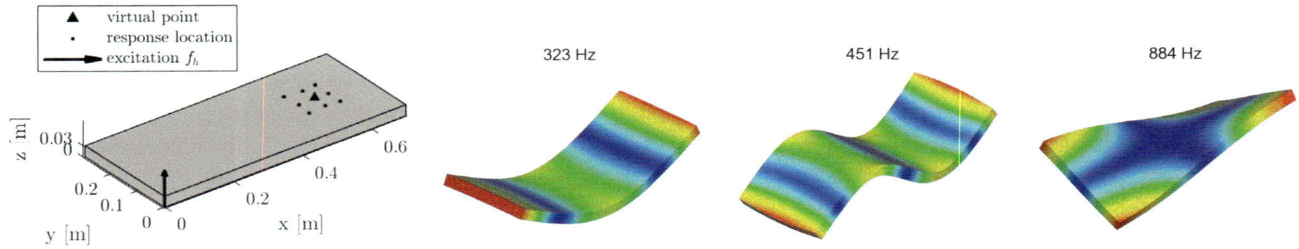


**Fig. 2** Numerical validation setup, and first three flexible eigenmodes.

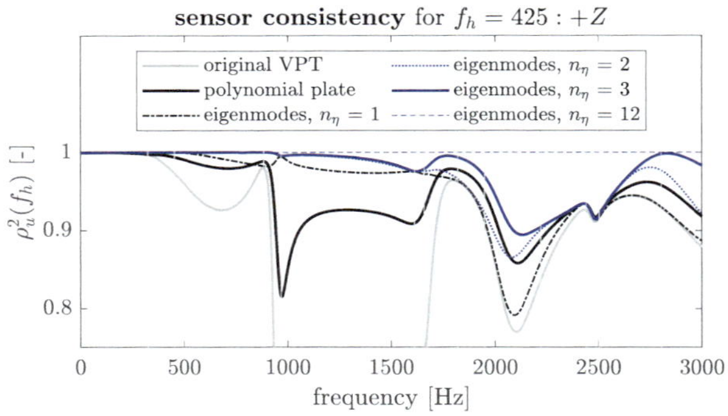


**Fig. 3** Sensor consistency obtained with different reductions onto $6 + n_\eta$ response DOF.

adding the third extra DOF would either add to the required instrumentation effort, or reduce the level of over-determination. One such disadvantage would be unavoidable in case of the polynomial plate model.

Using FE based modeshapes also has advantages over a generic library of shapes [11]. Firstly, the here proposed FE based approach is targeted towards the specific fixtures by definition. The corresponding shapes may thus be expected to yield better consistencies than the generic shapes, or allow to obtain similar consistencies with fewer extra DOF. Secondly, an FE based approach allows to synthesize consistency curves like those of Figure 3 prior to the test campaign. These can guide one to select required mode shapes, as well as the appropriate amount, locations and orientations of sensors (and excitations).

## Experimental Validation: High-Frequency Dynamic Mount Stiffness Identification

Figure 4 shows the experimental validation setup. Its intent is to characterize an automotive rubber mount, of which the inner core (side 1) is connected to a vehicle body and the outer cylinder (side 2) to a powertrain subframe. Figure 4a also shows the two cross-shaped fixtures on side 1 and one square-shaped fixture on side 2. The expected result is the 12-by-12 matrix of mount stiffness curves of equation 1, with both virtual points (denoted as VP1 and VP2) located at the center of the mount.

As visible in Figure 4b, side 1 is instrumented with two tri-axial accelerometers on each cross, hence 12 response DOF are measured on this side. An impact hammer has been used for the excitation; 9 DOF were excited on side 1. On side 2, 11 DOF are excited, and 12 response DOF are measured using four tri-axial accelerometers. This amounts to 432 FRF, acquired from 0 Hz to 3.2 kHz with a 1 Hz resolution. Based on their input-output coherences, the useful bandwidth reaches up to 2.5 kHz. Within this bandwidth, measured operational deflection shapes on side 1 reveal a torsion mode at about 1.77 kHz and two bending modes at 2.19 kHz and 2.47 kHz, respectively. As will be shown below, these flexibilities cause the average sensor and impact consistencies of the original VPT to fall below 0.8 at 1.3 kHz. On side 2, the average sensor and impact consistencies of the original VPT fall below 0.8 and 0.9, respectively, at the frequency of 2 kHz.

Following the above, the enhanced VPT is applied on side 1 of the mount, while the original VPT is applied on side 2. As such, the frequency range of mount stiffness identification can potentially be extended from 1.3 kHz to 2 kHz. Figure 4c shows the therefore constructed FEM of side 1. Although the case at hand is an assembly, the FEM is obtained straightforwardly. It consists of linear tetrahedral elements, does not include the threaded rod, washers and nuts holding the real assembly together, and it was not updated in any way. From it, modal displacements are extracted in the sensor and impact

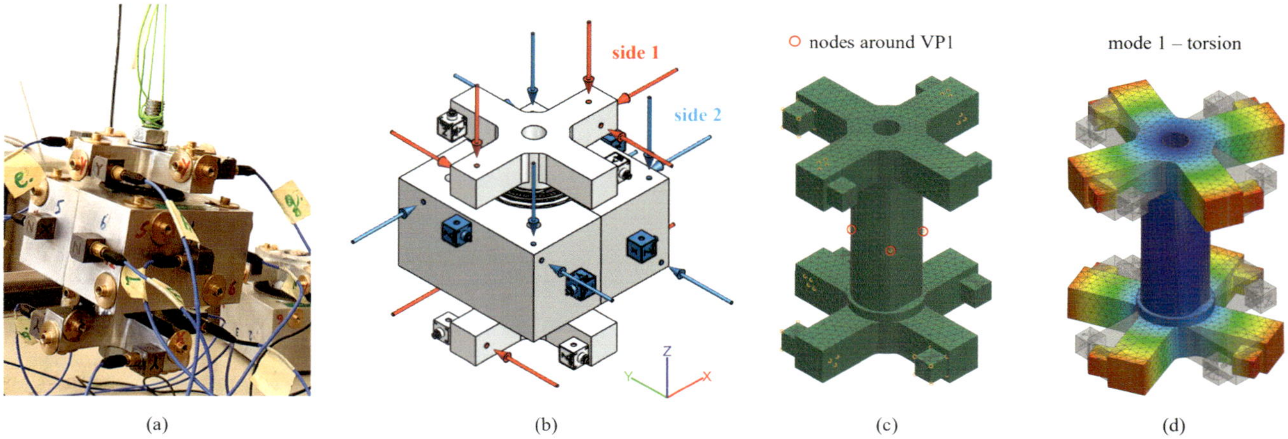


**Fig. 4** Experimental setup (a) with geometry definition (b), and FEM of side 1 (c) with flexible eigenmode prediction (d).

locations, cfr. equation 9, and modal displacements and rotations in VP1 (equation 13) are obtained from 4 nodes nearby the center of the mount. Figure 4d shows the first predicted flexible (torsion) eigenmode. The subsequent two eigenmodes are bending modes.

Figure 5 shows the effect of extending the VPT, as in equation 11, with $n_\eta \in \{1, 2, 3\}$ flexible eigenmodes on the averaged sensor and impact consistencies. In these graphs, the vertical lines indicate the approximate eigenfrequencies as obtained from analyzing the operational deflection shapes. Between 1.3 kHz and 2 kHz, a significant improvement is observed even when considering only one flexible (torsion) eigenmode. As expected, the consistency further increases when introducing more than one flexible DOF. For the given dataset, however, doing so would lead to very low levels of over-determination for the inversions required in equation 11[1]. Because of this, combined with the lower data quality above 2 kHz, only the torsion mode is considered in the remainder of this preliminary experimental validation.

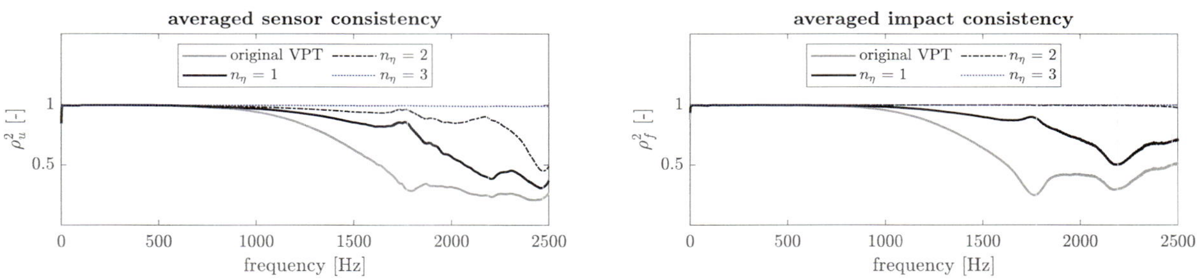


**Fig. 5** Sensor consistency (left) and impact consistency (right) obtained on side 1, with different reductions onto 6+$n_\eta$ DOF.

Figure 6 shows a selection of FRF obtained from, respectively, the original VPT (equation 4) and the enhanced VPT (equation 13). The enhanced VPT does not suffer from the spurious effect that the first torsion mode has on the results of the original VPT. This is indeed an improvement, because exciting the side 1 assembly in the considered DOF of VP1 should not excite the first torsion mode. As expected, the effect of bending behavior still affects the driving point FRF, notably in the '+X' direction. Figure 7 shows the corresponding selection of transfer FRF. Unlike the FRF of Figure 6, these transfer FRF have remained invariant to the enhancement of the VPT. This may indicate that excitation of the considered torsion mode does not lead to energy transfer through the mount, as such validating the approach of projecting $6 + n_\eta$ DOF onto 6 DOF.

Finally, Figure 8 compares the enhanced VPT to the original VPT in terms of mount transfer stiffness curves. The latter are derived from the 12-by-12 matrix of projected FRF as per equation 5. Above 1.75 kHz onwards, these results are still affected by the presence of bending. However, the enhanced VPT avoids the significant excursions from the expected (smooth and flat) mount stiffness curves up to this frequency, whereas the original VPT leads to problematic deviations already from 1.3 kHz onwards. Thus, by accounting for the first torsion mode, the proposed approach has extended the

---

The perfect impact consistency for $n_\eta = 3$ is even a trivial result, because the projection onto $6 + 3 = 9$ excitation DOF is not a reduction in this case.

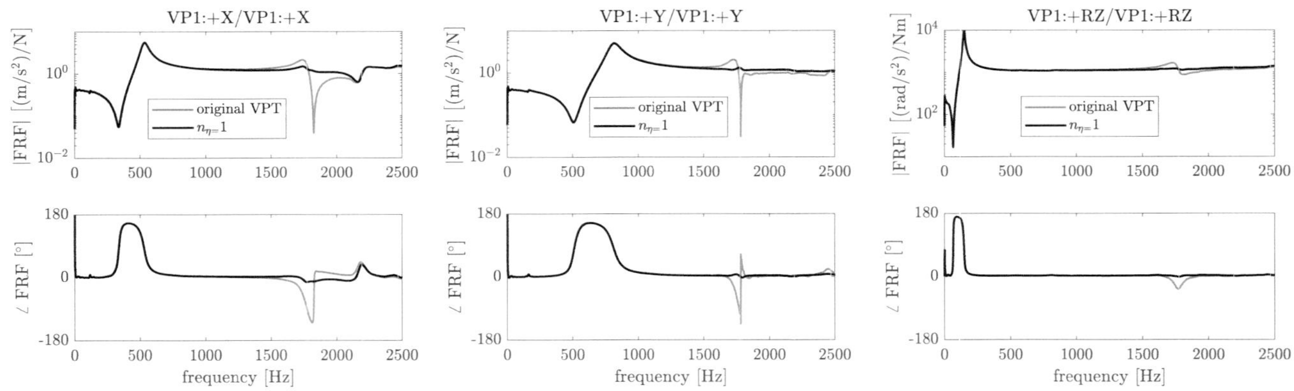


**Fig. 6** Direct FRF resulting from either the original VPT, or the proposed approach using $n_\eta = 1$ flexible torsion mode.

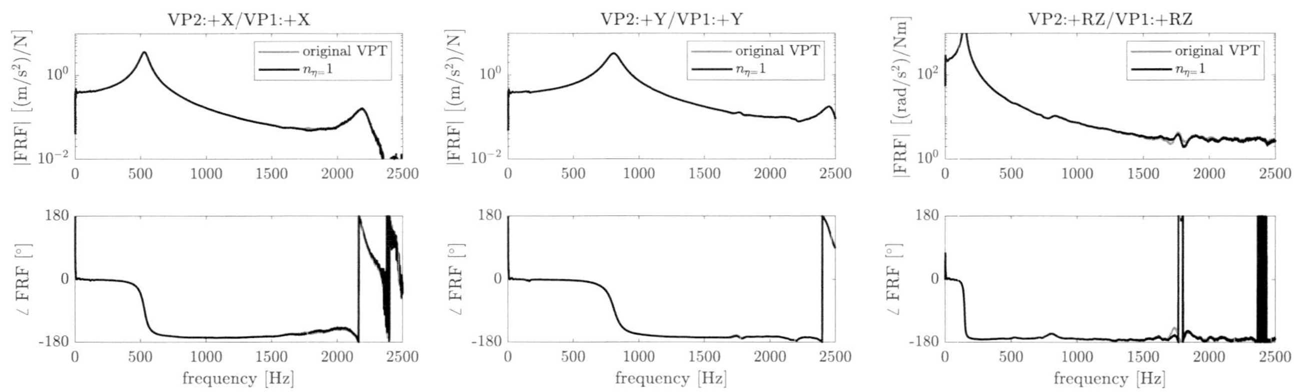


**Fig. 7** Transfer FRF resulting from either the original VPT, or the proposed approach using $n_\eta = 1$ flexible torsion mode.

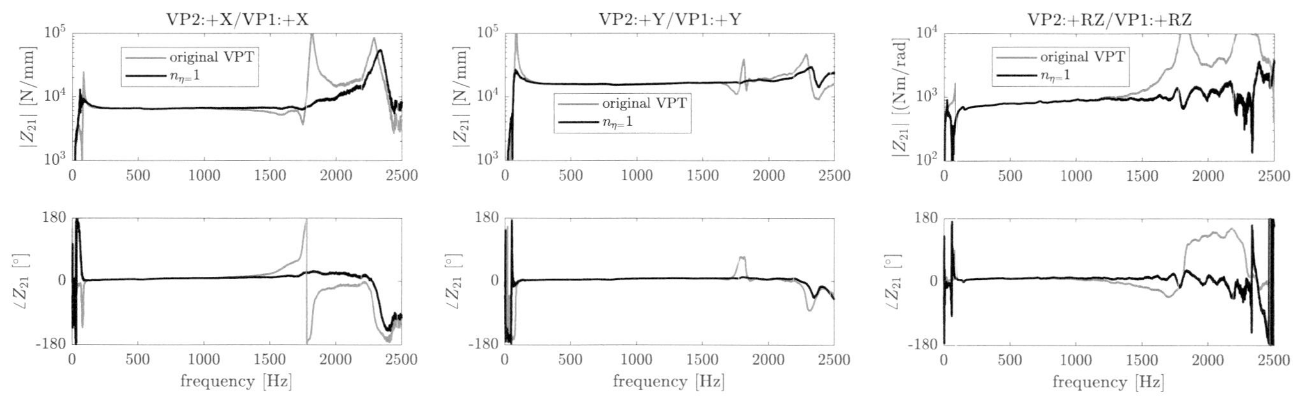


**Fig. 8** Transfer stiffness curves that result from either the original VPT, or the proposed approach with $n_\eta = 1$.

useful bandwidth for mount characterization by about 450 Hz (34 %). This result is promising, considering that the FEM based nature of the proposed enhancement is well suited for optimizing instrumentation and/or fixture designs employed in future test campaigns.

## Conclusion

This work has targeted the indirect identification of mount stiffness characteristics from FRF measurements over an extended frequency bandwidth. To achieve this, an enhancement to the virtual point transformation (VPT) has been proposed and validated. It accounts for flexibility of the fixtures, which violates the original VPT's primary assumption of rigidity. The proposed enhancement is based on two novelties, namely (i) the use of finite element model based eigenmodes for enriching the basis of the VPT's reduction space, and (ii) an additional reduction onto the six DOF associated to the virtual point. The latter reduction is critical in order to obtain a mount model which can be readily applied in virtual prototype assemblies. A numerical validation has highlighted the value of extending the reduction basis with eigenmodes, in terms of the consistency improvement per additional DOF to identify. A preliminary experimental validation of the proposed approach has shown that, by accounting for the first torsion mode of the fixture assembly at hand, the enhanced VPT is valid over a +30 % larger bandwidth than the original VPT. As such, the proposed approach has allowed to extend the validity of the identified 12-by-12 stiffness matrix, to beyond the first eigenfrequency introduced by the fixture assembly. Future work will consider the identification of multiple flexible DOF already in the test setup definition, aiming to demonstrate the full potential of the here proposed enhanced VPT.

**Acknowledgments** This research was partially supported by VLAIO (Flanders Innovation & Entrepreneurship Agency) within the O&O project ADRENALINE (Advanced modular approaches for cabin noise analysis in EVs), VLAIO O&O project no. 2022.1027.

## References

1. Moorhouse, A., Elliott, A., and Meggitt, J. "Virtual acoustic prototyping - a story of four decades". In *Proceedings of Inter-noise 2023*, Glasgow, Scotland (2022).
2. Bianciardi, F., Salamone, N., Colangeli, C., Ortega Almirón, J., Verrecas, B., Corbeels, P., and Janssens, K. "NVH prediction and real-time auralization for virtually assembled vehicles: An overview of the technology". In *Proc. SAE Noise and Vibration Conference and Exhibition*, MI, United States (2023).
3. van der Seijs, M., de Klerk, D., and Rixen, D. "General framework for transfer path analysis: History, theory and classification of techniques". *Mechanical Systems and Signal Processing*, 68-69:217–244 (2016).
4. Jetmundsen, B., Bielawa, R.L., and Flannelly, W.G. "Generalized frequency domain substructure synthesis". *Journal of the American Helicopter Society*, 33(1):55–64 (1988).
5. de Klerk, D., Rixen, D.J., and Voormeeren, S.N. "General framework for dynamic substructuring: History, review and classification of techniques". *AIAA Journal*, 46(5):1169–1181 (2008).
6. International Organization for Standardization. *DIN EN ISO 10846-1:2008 Acoustics and vibration — Laboratory measurement of vibroacoustic transfer properties of resilient elements – Part 1: Principles and guidelines*. ISO (2008).
7. Haeussler, M., Klaassen, S., and Rixen, D. "Experimental twelve degree of freedom rubber isolator models for use in substructuring assemblies". *Journal of Sound and Vibration*, 474:115253 (2020).
8. van der Seijs, M.V., van den Bosch, D.D., Rixen, D.J., and de Klerk, D. "An improved methodology for the virtual point transformation of measured frequency response functions in dynamic substructuring". In *4th ECCOMAS thematic conference on computational methods in structural dynamics and earthquake engineering*, volume 4 (2013).
9. Meggitt, J.W.R., Elliott, A.S., Moorhouse, A.T., and Lai, H.K. "In situ determination of dynamic stiffness for resilient elements". *Proceedings of the institution of mechanical engineers, Part C: Journal of mechanical engineering science*, 230(6):986–993 (2016).
10. Meggitt, J. and Moorhouse, A. "In-situ sub-structure decoupling of resiliently coupled assemblies". *Mechanical Systems and Signal Processing*, 117:723–737 (2019).
11. Pasma, E., van der Seijs, M.V., Klaassen, S., and van der Kooij, M. "Frequency based substructuring with the virtual point transformation, flexible interface modes and a transmission simulator". In *Dynamics of Coupled Structures: Proceedings of the 36th IMAC, A Conference and Exposition on Structural Dynamics*, volume 4, pages 205–213. Springer (2018).
12. Ortega Almirón, J., Bianciardi, F., and Desmet, W. "Flexible interface models for force/displacement field reconstruction applications". *Journal of Sound and Vibration*, 534:117001 (2022)

For Product Safety Concerns and Information please contact our
EU representative GPSR@taylorandfrancis.com Taylor & Francis
Verlag GmbH, Kaufingerstraße 24, 80331 München, Germany